T0314083

Source Separation in Physical-Chemical Sensing

Source Separation in Physical-Chemical Sensing

Edited by

Christian Jutten
Université Grenoble Alpes, Grenoble-INP, CNRS, GIPSA-lab
France

Leonardo Tomazeli Duarte
University of Campinas
Limeira, Brazil

Saïd Moussaoui
Nantes Université, Ecole Centrale Nantes, CNRS, LS2N
Nantes, France

This edition first published 2024.

Registered Offices
John Wiley & Sons, Inc., 111 River Street, Hoboken, NJ 07030, USA
John Wiley & Sons Ltd, The Atrium, Southern Gate, Chichester, West Sussex, PO19 8SQ, UK

For details of our global editorial offices, customer services, and more information about Wiley products visit us at www.wiley.com.

Wiley also publishes its books in a variety of electronic formats and by print-on-demand. Some content that appears in standard print versions of this book may not be available in other formats.

Library of Congress Cataloging-in-Publication Data applied for:

Hardback ISBN: 9781119137221

Cover Design: Wiley
Cover Image: © MR.Cole_Photographer/Getty Images
Set in 9.5/12.5pt STIXTwoText by Straive, Chennai, India
Printed and bound by CPI Group (UK) Ltd, Croydon, CR0 4YY

C9781119137221_081023

Contents

About the Editors

Christian Jutten received PhD (1981) and Doctor ès Sciences (1987) degrees from Grenoble Institute of Technology, France. He was Associate Professor (1982–1989), Professor (1989–2019) at University Grenoble Alpes, where he is now Emeritus Professor since September 2019. Since 1980s, his research interests are in machine learning and source separation, including theory and applications (biomedical engineering, hyperspectral imaging, chemical sensing, speech). He is author/co-author of four books, 125+ papers in international journals, and 250+ publications in international conferences.

Jutten was a visiting professor at EPFL, RIKEN labs, and University of Campinas. He served as director or deputy director of the signal/image processing laboratory in Grenoble (1993 to 2010), as scientific advisor for signal/image processing at the French Ministry of Research (1996–1998), and at CNRS (2003–2006 and 2012–2019).

Jutten was organizer or program chair of many international conferences, including the first Independent Component Analysis conference in 1999 (ICA'99) and IEEE MLSP 2009. He was the technical program co-chair of ICASSP 2020. He was a member of the IEEE MLSP and SPTM Technical Committees. He was associate editor for *Signal Processing and IEEE Trans. on Circuits and Systems*, and guest co-editor for *IEEE Signal Processing Magazine* (2014) and the *Proceedings of the IEEE* (2015). From 2021 to 2023, he was editor-in-chief of *IEEE Signal Processing Magazine*.

Jutten received many awards, e.g. best paper awards of EURASIP (1992) and IEEE GRSS (2012), Medal Blondel (1997) from the French Electrical Engineering society, and one Grand Prix of the French Académie des Sciences (2016). He was elevated as IEEE fellow (2008), EURASIP fellow (2013), and as a Senior Member of Institut Universitaire de France for

10 years since 2008. He was the recipient of a 2012 ERC Advanced Grant for the project Challenges in Extraction and Separation of Sources (CHESS).

Leonardo Tomazeli Duarte received the BS and MSc degrees in electrical engineering from the University of Campinas (UNICAMP), Brazil, in 2004 and 2006, respectively, and the PhD degree from the Grenoble Institute of Technology (Grenoble INP, Université Grenoble Alpes), France, in 2009. Since 2011, he has been with the School of Applied Sciences (FCA) at UNICAMP, Limeira, Brazil, where he is currently an associate professor. He is a Senior Member of the IEEE. In 2016, he was a Visiting Professor at the École de Génie Industriel (GI-Grenoble INP, France). Since 2015, he has been recipient of the National Council for Scientific and Technological Development (CNPq, Brazil) productivity research grant. Since 2023, he is one of the principal investigators within the Brazilian Institute of Data Science (BI0S), one of the Brazilian Applied Research Centers on Artificial Intelligence. In 2017, he was the recipient of UNICAMP "Zeferino Vaz" Academic Recognition Award (for research and teaching performance at UNICAMP). In 2022, he was elected Affiliated Member (up to 40 years old) of the Brazilian Academy of Sciences (ABC). His research interests center around the broad area of data science and lie primarily in the fields of signal processing, decision aiding and machine learning, and also in the interplays between these fields.

Saïd Moussaoui received the MEng degree in electrical engineering from Ecole Nationale Polytechnique, Algiers, Algeria, in 2001; the MSc degree in Control, Signals and Communication from the University Henri Poincaré, Nancy, France, in 2004; and the PhD degree in Signal and Image Processing from Université Henri Poincaré, Nancy, France, in 2005. Since 2006, he has been with the Department of Automatics and Robotics at Ecole Centrale Nantes where he is currently a full Professor. He is a member of the group of Signals, Images and Sounds (SIMS) of the Laboratory of Digital Sciences of Nantes (LS2N, CNRS UMR 6004). His research interests are related to the field of signal and image processing, including methodological aspects of statistical inference, numerical optimization, and the application in various real-life contexts such as chemical data analysis, remote sensing, and biological imaging.

List of Contributors

José M. Bioucas-Dias
Instituto de Telecomunicacoes
Instituto Superior Técnico
Lisboa
Portugal

David Brie
CRAN
Lorraine University
Nancy
France

Rasmus Bro
Department of Food Science
University of Copenhagen
Frederiksberg
Denmark

Emilie Chouzenoux
OPIS, Inria Saclay
University Paris-Saclay
Gif-sur-Yvette
France

Jérémy Cohen
CNRS, CREATIS
Lyon
France

Pierre Comon
CNRS
GIPSA-lab
Université Grenoble Alpes
France

Nicolas Dobigeon
IRIT
INP-ENSEEIHT
IRIT University of Toulouse
Toulouse
France

Leonardo Tomazeli Duarte
School of Applied Sciences (FCA)
University of Campinas
Limeira
Brazil

Nicolas Gillis
Department of Mathematics and
Operational Research
University of Mons
Mons
Belgium

Christian Jutten
GIPSA-lab
Univ. Grenoble Alpes
CNRS, Institut Univ.
de France
Grenoble
France

Wing-Kin Ma
Department of Electronic
Engineering
The Chinese University of
Hong Kong
Hong Kong SAR
China

Saïd Moussaoui
LS2N, Nantes Université
Ecole Centrale Nantes
Nantes
France

Jean-Christophe Pesquet
CVN, CentraleSupélec
University Paris Saclay
France

Foreword
In Memoriam: José M. Bioucas-Dias (1960–2020), a Humble Giant

José Manuel Bioucas-Dias (1960–2020) was a Professor at Instituto Superior Técnico (IST), the engineering school of the University of Lisbon, and a senior researcher at Instituto de Telecomunicações. After obtaining his PhD degree in Electrical and Computer Engineering, in 1995, from IST, he dedicated his research life to the area of signal and image processing, in problems related to the reconstruction, restoration, and analysis of images. In particular, he focused deeply on processing and analysis of remote sensing observations, a subject in which he was widely considered a world-leading authority. José Bioucas-Dias contributed to highly relevant and influential scientific and technical advances in synthetic aperture radar, interferometric radar, and hyperspectral imaging.

Figure 1 features a word cloud constructed from the titles of his numerous publications available from IEEE Xplore. Clearly, "hyperspectral" appears as a central topic of interest and he was indeed a key member of the hyperspectral imaging research community, be it from the geoscience and remote sensing point of view or from the signal and image processing point of view. The direct and indirect implications and applications of his contributions are numerous, namely in the processing and analysis of satellite images, whose impact on modern society is enormous.

For his contributions to image processing and analysis in remote sensing, José Bioucas-Dias was elevated to Fellow of the IEEE (Institute of Electrical and Electronics Engineers) in 2016. In 2017, he received the first *David Landgrebe Award*, (Figure 2a) from the Geoscience and Remote Sensing Society (GRSS) of the IEEE, with the citation "for outstanding contributions in the field of remote sensing image analysis." According to the GRSS, "the David Landgrebe award is a career award, given for extraordinary

Figure 1 Word cloud based on the titles of all publications of José Bioucas-Dias, extracted from IEEE Xplore.

(a) (b)

Figure 2 (a) José Bioucas-Dias receiving the David Landgrebe Award. (b) José Bioucas-Dias, general chair of the 3rd Workshop on Hyperspectral Image and Signal Processing, held in 2011 in Lisbon, Portugal.

contributions in the field of remote observation image analysis." In 2018 he was named Distinguished Lecturer by the IEEE GRSS.

Throughout his career, José Bioucas-Dias published more than 100 articles in the most prestigious journals in his areas of work, as well as more than

200 articles in all the best signal and image processing and remote sensing conferences, where he received several "best paper awards." The high international impact of his work is evident in the important distinctions mentioned in the previous paragraph, as well as in the fact that he was included in the 2015, 2019, 2020, and 2021 editions of the prestigious "Highly Cited Researchers" list (by Clarivate Analytics), which is reserved to the world elite of researchers with the greatest impact in their fields.

His recognition and prestige in the international scientific community are also evident in the fact that he was invited to join the editorial boards of the best journals in the area: the *IEEE Transactions on Circuits and Systems*, the *IEEE Transactions on Image Processing* (in which he was a senior editor), and the *IEEE Transactions on Geoscience and Remote Sensing*. He was also a guest editor for several special issues in these and other journals. He was the General Co-Chair of the third IEEE GRSS Workshop on Hyperspectral Image and Signal Processing, Evolution in Remote sensing ((WHISPERS'2011, (Figure 2b)) and has been a member of program/technical committees of several international conferences. Even more revealing of his recognition is the high number (over 60) of invited presentations he gave at institutions and prestigious conferences around the world, from which around 20 plenary and keynote presentations stand out. He also held several visiting professor positions, at the invitation of several European universities: Tampere University of Technology (Finland, in 2008 and 2012), Université Grenoble Alpes (France, 2014), Université de Toulouse (France, 2013 and 2015), Università di Pavia (Italy, 2016).

In addition to being an excellent researcher, José Bioucas-Dias was also a devoted teacher and educator. He supervised and co-supervised 20 PhD theses and more than 10 post-doctoral internships for researchers from several countries around the world. At IST, he created the first PhD course in inverse problems in image processing and, during his career, he taught many other courses (in the areas of signal processing, telecommunications, information and communication theory, remote sensing, to mention only the most significant), always with great dedication and high pedagogical and scientific quality. He was also engaged in various university management tasks, having joined the Scientific Council of IST.

José Bioucas-Dias was one of the brightest and most influential scientists in the field of remote sensing image processing and analysis. His contributions were not only fundamental, but also of very significant practical applicability. His great scientific stature was matched by an exceptional modesty, courteous affability, and selfless readiness to help others. He was an exceptionally dedicated mentor to so many and a fantastic human being, totally driven by his passion for science, his creativity, and his rigor. He was always available to engage in scientific discussions, sit down with a student to dig into a problem until it was solved fair and square, sharing his joy and optimism, never giving up. On the international scientific scene, he was a giant; on his daily interactions, his humility was a striking feature. This rare blend made him a truly inspirational human being, the perfect role model, colleague, friend.

With his untimely death, we lost an outstanding researcher, a great educator, and a truly good man. We are truly honored to dedicate this book to him, highlighting his long-lasting legacy to our community.

Mário A. T. Figueiredo
Instituto de Telecomunicações
Instituto Superior Técnico
Universidade de Lisboa, Portugal

Jocelyn Chanussot
Grenoble INP
University of Grenoble Alpes, CNRS
GIPSA-Lab, France

Preface

Source separation is a very ubiquitous concept in signal processing, whose main objective, essential in signal processing, is to recover an information of interest. When the acquired signal is a mixture of sources, e.g., sounds from different speakers recorded by a microphone, or different ions measured by an ion sensitive sensor, due to the poor selectivity of sensors, recovering the different sources (speakers or ions, in the previous examples) is a tricky question. The main approach for solving the problem is to use several sensors instead of a unique one. Indeed, if the sensors are at different locations for sounds, or with different sensitivities for ions, each sensor provides a different instance of the mixture of the sources and, under mild conditions, a sufficient amount of information will thus be available for recovering each source.

With technological progresses, sensors are more and more numerous, common, and low cost, and many physical quantities are measured by arrays of sensors, so that source separation can be very useful in various domains. As a few examples, sensors have been used (i) in biomedical signal processing, for noninvasive fetal electrocardiogram extraction using a few chest electrodes, for extracting and locating brain sources from electroencephalogram scalp electrodes or magnetic resonance imaging, (ii) in satellite and airborne remote sensing for estimating the composition of ground surface, of planets, or of the universe from hyperspectral images, (iii) in chemical engineering for measuring the composition of electrolytes based on record with ion-sensitive sensors or optical spectrometers for the inspection of molecular composition of materials.

Separating different sources from a set of observations (sensors) is still an ill-posed signal processing problem that cannot be handled appropriately without additional hypotheses on the sought sources. If mutual statistical independence of sources can be an efficient property, which leads to very powerful classes of algorithms, namely independent component analysis, it cannot be used in all domains. Especially, for signals coming from physical/chemical sensors, the mutual independence property may be totally irrelevant. Conversely, non-negativity is a very common property for such signals: it is typical when measuring spectra, counting ions, or assessing molecular concentrations and abundances. In addition to non-negativity, sum-to-one and sparsity are also relevant properties. These properties can be used for designing very powerful algorithms able to solve the source separation problem.

This book proposes a tour of approaches and algorithms of source separation in physical/chemical sensing applications. The first chapter presents a comprehensive view of the source separation problem and points out the main approaches for its resolution. Since all the source separation algorithms are based on the optimization of a cost function constituted of a data fitting term and few regularization terms promoting the desired source properties or imposing some physical constraints, the second chapter is focused on basic principles and recent advances in mathematical optimization theory and algorithms. The four other chapters address different approaches for source separation: non-negative matrix factorization (NMF), Bayesian estimation approaches, geometrical formulation based approaches, and tensor factorization methods. These four chapters present both a methodological view of each approach and examples of applications for illustrating its ubiquitous interest in various physical/chemical sensing applications. The authors of the chapters of this book are worldwide recognized scientists, with a strong expertise and many valuable contributions in the field of signal processing and particularly in developing source separation methods based on various approaches. They also have a long-time experience in dealing with multidisciplinary signal processing applications in chemical sensing, hyperspectral imaging, and physical applications.

Life is short and hard. During the writing of this book, our friend José Bioucas-Dias passed away, swept away by illness. All the co-authors of this book dedicate this work to José, a great scientist, and colleague or friend. Our grateful thanks to Mário A. T. Figueiredo and Jocelyn Chanussot for writing the foreword of this book as a tribute to José.

Grenoble

Christian Jutten
Leonardo Tomazeli Duarte
Saïd Moussaoui

Notation

Signals, Vectors and Matrices

$\boldsymbol{x}$	column vector of components x_p, $1 \leq p \leq P$
$\mathrm{Diag}\{\boldsymbol{a}\}$	diagonal matrix whose entries are those of vector $\boldsymbol{a}$
$\boldsymbol{G}, \boldsymbol{W}, \boldsymbol{Q}$	global, whitening, and separating unitary matrices
$\boldsymbol{A}$	matrix with components A_{ij}
$\boldsymbol{A}, \boldsymbol{B}$	mixing and separation matrices
$\mathcal{A}, \mathcal{B}$	mixing and separating (nonlinear) operators
R	number of sources
P	number of sensors
T	number of observed samples
s, x, y	sources, observations, separator outputs
$\mathrm{diag}\{\boldsymbol{A}\}$	vector whose components are the diagonal of matrix $\boldsymbol{A}$

Operators on Signals, Vectors, Matrices, Tensors, and Functions

$\boldsymbol{Q}^*$	complex conjugation of matrix Q
$\star$	convolution
$\boldsymbol{Q}^H$	conjugate transposition of matrix Q
$\bullet_j$	contraction over index j
$\det \boldsymbol{A}$	determinant of matrix $\boldsymbol{A}$
$\check{s}(\nu), \check{g}$	Fourier transform of signal $s(t)$ and of g
$\boxdot$	Hadamard (entry-wise) product between arrays
$\square$	infimum-convolution
$\odot$	Khatri-Rao (column-wise Kronecker) product between matrices
$\boxtimes$	Kronecker product between matrices
$\mathrm{krank}\{\boldsymbol{A}\}$	Kruskal's rank of matrix $\boldsymbol{A}$
$\boldsymbol{Q}^\dagger$	pseudo-inverse of matrix Q
$\mathrm{rank}\{\boldsymbol{A}\}$	rank of matrix $\boldsymbol{A}$
$\otimes$	tensor product
$\mathrm{trace}\{\boldsymbol{A}\}$	trace of matrix $\boldsymbol{A}$
$\boldsymbol{Q}^\mathsf{T}$	transposition of matrix Q

Random Variables and Vectors, Statistical Operators

Υ	contrast function
$\hat{\boldsymbol{s}}$	estimate of quantity $\boldsymbol{s}$
$\text{cum}\{x_1, \ldots, x_P\}$	joint cumulant of variables $\{x_1, \ldots, x_P\}$
ψ	joint score function
$K\{\boldsymbol{x};\boldsymbol{y}\}$ or $K(f(\boldsymbol{x}); f(\boldsymbol{y}))$	Kullback divergence between $f(\boldsymbol{x})$ and $f(\boldsymbol{y})$
$\mathcal{L}$	likelihood
$\text{cum}_R\{y\}$	marginal cumulant of order R of variable y
φ_i	marginal score function of source s_i
$\mathbb{E}[\boldsymbol{x}]$, $\mathbb{E}\{\boldsymbol{x}\}$	mathematical expectation of $\boldsymbol{x}$
$I\{\boldsymbol{y}\}$ or $I(\boldsymbol{y})$	mutual information of $\boldsymbol{y}$
$f_X(x)$ or $f(x)$	probability density function of random variable X
$H\{\boldsymbol{x}\}$ or $H\{p_{\boldsymbol{x}}\}$	Shannon entropy of $\boldsymbol{x}$

Sets

$\mathbb{C}$	complex field
$\text{dom}\, g$	domain of function g
$\text{epi}\, g$	epigraph of function g
ι_C	indicator function of set C
P_C	projector on set C
$\mathbb{R}$	real field
$\mathcal{S}, \mathcal{G}$	sets

Operators for Optimization

g^*	conjugate of function g
$\|\cdot\|_F$	Frobenius norm
$\text{prox}_g^{\boldsymbol{A}}(\boldsymbol{x})$	proximity operator of function g within the metric induced by $\boldsymbol{A}$ computed at $\boldsymbol{x}$
∇g	gradient of g
$\nabla^2 g$	Hessian of g
$\underset{C}{\text{argmin}}\ g$	minimum argument of g over set C
$\underset{C}{\text{argmax}}\ g$	maximum argument of g over set C
${}^{\gamma}g$	Moreau envelope of g of parameter γ
$\inf_C g$	infimum of g over set C
$\square$	infimum-convolution
$\vert\vert\vert\cdot\vert\vert\vert$	spectral norm
$\partial g(\boldsymbol{u})$	subdifferential set of g at $\boldsymbol{u}$
$\sup_C g$	supremum of g over set C

1

Overview of Source Separation

Christian Jutten[1], Leonardo Tomazeli Duarte[2], and Saïd Moussaoui[3]

[1] GIPSA-lab, Univ. Grenoble Alpes, CNRS, Institut Univ., de France, Grenoble, France
[2] School of Applied Sciences (FCA), University of Campinas, Limeira, Brazil
[3] LS2N, Nantes Université, Ecole Centrale Nantes, Nantes, France

1.1 Introduction

The purpose of this chapter is to give a general overview of the source separation problem and its underlying hypotheses, and of methods and algorithms for solving the problem, with an emphasis on the context of data processing in physical/chemical sensing applications.

1.1.1 Brief Introduction to Source Separation

Source separation is a very general problem in signal processing, and, more generally, in sensing. In fact, a basic task in signal processing consists in separating useful information (called signal) from non-useful one (called noise) in noisy measurements. Measurements (also called observations) are frequently obtained through sensors, sensitive to some physical or chemical properties of the object which is analyzed. However, the sensor selectivity is limited so that its output depends on various phenomena (called sources).

As a first example, the signal captured by a microphone is the superimposition of signals emitted by all the acoustic sources in the neighborhood. Similarly, the signal measured by a scalp electrode in electroencephalography (EEG) is assumed to be the superimposition of the synchronous electrical activity of neural assemblies located in various areas in the brain. In remote sensing by hyperspectral imaging, due to low spatial resolution, the measured reflectance spectrum in each pixel is an aggregation of the reflectance spectra of all physical materials present in the ground surface

Source Separation in Physical-Chemical Sensing, First Edition.
Edited by Christian Jutten, Leonardo Tomazeli Duarte, and Saïd Moussaoui.

related to this pixel. Finally, in chemical sensing, ion-sensitive electrodes have been designed for measuring activity of specific ions; however, due to a limited selectivity, the sensor output depends on the activity of the main ion and on the activities of interfering ions which can be present in the solution.

Source separation problems are considered in a blind (unsupervised) framework, i.e., by assuming that only sensor measurements (called mixtures) are available, but neither the source signals nor the mixing process (the superimposition in the above examples) are known. The main concept for solving blind source separation (BSS) problems is based on *diversity*, whose simplest implementation is to use a large number of sensors, thus providing spatial diversity. Solving source separation requires first to model the observations, i.e., how the signals received by the sensors are related to the sources, and then to add some (weak) priors and hypotheses on these sources in order to ensure their *separability*, and therefore the separation problem becomes well-posed.

The problem of source separation has been formulated in the middle of 1980s by Jutten and Hérault [1] for modeling motion decoding in vertebrates. Then, theoretical foundations have been developed mainly in the signal processing community by different researchers like Comon [2], Cardoso and Souloumiac [3, 4], Pham and Garat [5], Pham and Cardoso [6], Delfosse and Loubaton [7], and in parallel in the Neural Networks and Machine Learning communities by researchers like Bell and Sejnowski in USA [8], Hyvärinen in Finland [9], Cichocki, Amari, and their team in Japan [10].

The interest for methods of source separation is due to its strong theoretical foundations [11] (Chapters 2–14) and to its very wide application domains [11] (Chapter 16). In 2022, with the keywords *source separation*, Google recalled more than 584 million entries!

From the application point of view, due to expansion of both low-cost sensors and powerful computers that are able to process very fast huge data sets, the problem of source separation appears in several domains, like communications [11] (Chapters 15 and 17), audio and music processing [11, 12] (Chapter 19), biomedical engineering [11] (Chapter 18), and remote sensing and hyperspectral imaging [13]. First applications of source separation appeared in the middle of 1990s and were focused on biomedical problems: non-invasive fetal electrocardiogram (ECG) separation [14] in 1994 and ECG processing [15] in 1996. Source separation has its success stories, too. As an example, applying the spectral matching independent component analysis algorithm [16] on the data provided by the Plank space mission, Cardoso was able to extract wonderful images of the Cosmic

Microwaves Background, i.e., a very early image of our universe, which is a very important material for cosmologists.

In chemical engineering, it seems that source separation has been applied first for nuclear magnetic resonance spectroscopy [17] in 1998, and a larger number of works have been published in middle of 2000s as detailed by Monakhova *et al.* in the review [18]. In this context, even if Monte Carlo statistical methods have been proposed for mixture analysis [19], it must be noted that source separation is strongly related to positive matrix factorization [20] and to algebraic methods of tensor factorization, popular in Chemometrics and quoted PARAFAC [21].

1.1.2 Chapter's Organization

This chapter is a general introduction to the main concepts related to source separation. It is organized as follows. Section 1.2 presents the mathematical problem of source separation and a few basic solution principles. Section 1.3 focuses on source separation methods based on mutual statistical independence. Section 1.4 gives some examples of the various applications in physics and chemistry that can be formulated in the source separation framework. Section 1.5 is an overview of approaches that can be used for solving problems of Section 1.4. Section 1.6 details the organization of the book.

1.2 The Problem of Source Separation

1.2.1 Mathematical Description

Let us first consider a unique sensor and denote $x(t), t = 1, \dots T$ its output, at sample t. Due to the poor selectivity of the sensor, one can model $x(t)$ as a function of R unknown sources, denoted $s_r(t)$, $r = 1, \dots, R$:

$$x(t) = \mathcal{A}(s_1(t), \dots, s_R(t)), \tag{1.1}$$

where $\mathcal{A}$ models the mixing operator, unknown too.

In the simplest case, i.e. if the mapping $\mathcal{A}$ is assumed linear, one could write:

$$x(t) = A_1 s_1(t) + \dots + A_R s_R(t) = \sum_{r=1}^{R} A_r s_r(t). \tag{1.2}$$

Although this model is simple, we are faced with two problems: (i) the mixing coefficients A_r are unknown, as well as the sources $s_r(t)$ (ii) even if the mixing coefficients A_r were known, the separation of the contributions coming from the various sources remains an ill-posed problem.

The first problem can be solved through modeling or identification methods. Modeling requires to write propagation equations of signals from sources to sensors, and implicitly to know the locations of sources and sensors. Then, identification can be achieved but it requires a training set of many samples $(s_1(t), \dots, s_R(t); x(t))$, $t = 1, \dots, T$, i.e., T pairs of inputs/output, for performing supervised parameter estimation.

The second problem could be solved if we have additional information concerning the different sources. For instance, in the above example of acoustic sources recorded by a microphone, if we know that the useful source and non-useful ones are characterized by different frequency bands, we can separate them by simple spectral filtering. Subtraction methods [22] can be also used, provided that one has a reference of the background noise (i.e. non-useful sources).

As a conclusion, solving the source separation problem with a unique sensor is impossible without very strong priors.

The basic concept of methods for solving source separation problems is *diversity*. A simple way to enhance diversity is to use a few sensors, say P, instead of only one. In that case, at each sample t, instead of having a scalar observation, we have a P-dimensional observation vector $\boldsymbol{x}(t)$:

$$\boldsymbol{x}(t) = \mathcal{A}(s_1(t), \dots, s_R(t)), \forall t = 1, \dots, T, \tag{1.3}$$

where $\mathcal{A}$ denotes the multidimensional mapping between the R sources and the P sensors. Denoting $\boldsymbol{s}(t)$ the R-size vector of sources, one can also write:

$$\boldsymbol{x}(t) = \mathcal{A}(\boldsymbol{s}(t)) \quad \forall t = 1, \dots, T. \tag{1.4}$$

For instance, in the above examples, *spatial diversity* is obtained by using several microphones or few EEG electrodes located at different places, or several image pixels, or several ion-sensitive sensors, specific to different ions. But another diversity is required, *the sample diversity* (time diversity in the above examples), i.e., the shapes of functions $s_r(t)$ must be actually different. This sample diversity can be ensured by assuming that sources (considered as random variables) are mutually independent, or have different spectra or different variance time courses. Note that the discrete character of sources can be used instead [23].

The problem of source separation is often said *blind* in the signal processing community: in this context, *blind* refers to the fact that we only have T observations $\boldsymbol{x}(t)$, without a precise knowledge either on the sources or on the mapping $\mathcal{A}$. In other words, *blind* must be understood as *unsupervised* in machine learning. In such a case, identification of the mapping is not directly possible since we do not have input/output pairs for mapping estimation, and an accurate physical modeling is not possible since one has no idea about source locations.

However, weak information concerning the mapping and the sources are sometimes given and can be used for enhancing source separation. Here *weak* means that information is mainly qualitative, and it would be more realistic to qualify the methods as semi-blind (or semi-supervised).

1.2.2 Different Types of Mixing Models

Even if an accurate modeling of the mixing process is impossible to achieve, a simple study of the relationship between sources and sensors (through physical modeling of the propagation from sources to sensors) can specify the generic observation model (1.4) and consequently simplify the resolution of the source separation problem.

1.2.2.1 Linear Mixtures

If the mappings can be assumed as linear, the general Eq. (1.4) simplifies, and can be formulated and solved using linear algebra tools. Especially, the multidimensional mapping $\mathcal{A}$ can be represented by a $P \times R$ *mixing matrix*. In the case of linear mixtures, one can distinguish two situations, according to the signal propagation from sources to sensors: linear mixing and convolutive mixing.

If the propagation speed is very fast (instantaneous), the mapping between source r and sensor p is a simple scalar entry A_{pr}. At each sample t, the observation on sensor p is then a weighted sum of the sources $s_r(t)$, $r = 1, \dots, R$:

$$x_p(t) = \sum_r A_{pr} s_r(t), \quad \forall t = 1, \dots, T. \tag{1.5}$$

For the set of sensors, one can model the observations in compact form as:

$$\boldsymbol{x}(t) = \boldsymbol{A}\boldsymbol{s}(t), \quad \forall t = 1, \dots, T. \tag{1.6}$$

For instance, linear instantaneous mixing models nicely fit the signal measured on scalp electrodes in EEG, since the propagation of electrical activity can be neglected following the quasi-static approximation of Maxwell's equations.

Conversely, if the propagation delay cannot be neglected, the mapping between source r and sensor p has to be modeled as a linear filter with impulse response $A_{pr}(t)$. The measurement on sensor p is then the convolution product:

$$x_p(t) = \sum_{r=1}^{R} [A_{pr} \star s_r](t), \quad \forall t = 1, \dots, T, \tag{1.7}$$

where ⋆ denotes the convolution operator. Globally, for the set of sensors,

$$\boldsymbol{x}(t) = [\boldsymbol{A} \star \boldsymbol{s}](t), \quad \forall t = 1, \ldots, T. \tag{1.8}$$

Due to the discrete nature of sampled data, the above model is usually written in z-domain or in the frequency domain to handle the convolution operator. Actually, linear convolutive mixing models are required for modeling mixtures of acoustic sources. A survey on convolutive source separation models and methods in the context of speech and audio processing is given in [24]. The case of pure delays is addressed in [25]. In the framework of chemical sensing, it could be used for taking into account the diffusion of ions in a solution, or of gas molecules in air [26], from sources to sensors.

1.2.2.2 Nonlinear Mixtures

In some applications, the linear approximation is not sufficient to give an accurate description of the mixtures. The observations must be modeled as a nonlinear mappings of the sources, eventually taking into account propagation. As we will see in more detail in Section 1.4 and also in Chapter 4 (Section 4.4), nonlinear mappings must be considered for modeling observations provided by ion-sensitive sensor arrays [27, 28] where the p-th sensor output, at sample t, can be written as:

$$x_p(t) = e_p + d_p \log\left(s_p(t) + \sum_{r, r \neq p} A_{pr} s_r(t)^{z_p/z_r} \right), \tag{1.9}$$

where d_p and e_p are physical constants and z_p denotes the valence of the p-th ion. The selectivity coefficients, A_{pr}, are unknown mixing parameters that model the interference between the ions of index r and the electrode dedicated to ion p.

Nonlinear models are also usual in hyperspectral imaging [13, 29] when considering multiple reflections in 3D scenes, which can be approximated (if restricting to double reflections) by bilinear models. The light reflectance $x_p(\lambda)$ in a given pixel p at wavelength λ is expressed as a nonlinear mixture of the reflectance spectra (denoted $s_r(\lambda)$, $r = 1, \ldots R$) of the R materials present in the surface area covered by this pixel:

$$x_p(\lambda) = \sum_{r=1}^{R} A_{pr} s_r(\lambda) + \sum_{r=1}^{R-1} \sum_{\ell=r+1}^{R} B_{pr\ell} s_r(\lambda) s_\ell(\lambda). \tag{1.10}$$

where coefficient $B_{pr\ell}$ represents the amount of interaction between different components.

1.2.2.3 Overdetermined, Determined, or Underdetermined Models

Beyond the nature of the mixing models, one can consider three different situations, depending on the number of sources and number of sensors.

If the number of sensors is equal to the number of sources, i.e. $P = R$, and the mixing mapping $\mathcal{A}$ is invertible, the problem is said to be determined as: estimation of $\mathcal{A}$ (or of its inverse) leads directly to source separation, at least in the noiseless case.

If the number of sensors is larger than the number of sources, i.e. $P > R$, and the mixing mapping $\mathcal{A}$ is still invertible, the problem is overdetermined: a first step of dimension reduction and signal subspace estimation [3] is necessary for leading to a determined problem. In the determined or overdetermined cases, source separation methods can be based on the estimation of either the mixing mapping or the *unmixing* mapping (i.e. $\mathcal{B}$ which is an inverse of $\mathcal{A}$, up to acceptable undeterminacies). Then, the source separation is achieved without extra effort. If noise is present, a matched filter should be used.

Conversely, if the number of sensors is smaller than the number of sources, i.e. $P < R$, the problem is underdetermined: then the mixing mapping $\mathcal{A}$ cannot be inverted ($\mathcal{B}$ does not exist!). Consequently, we can only estimate $\mathcal{A}$. Moreover, after knowing $\mathcal{A}$, source separation is a problem, not at all trivial, which requires extra priors for being solved. For underdetermined mixtures, estimation of mapping and estimation of sources are two different and tricky problems that require additional hypotheses on signals [30]. Practically, sparsity is a nice prior, realistic in many applications, which can lead to unique solution for underdetermined linear instantaneous mixtures. Other approaches, in particular deterministic, permit to recover the sources by exploiting extraneous diversities [31].

Actually, the main idea of source separation methods is not to directly estimate the sources, but just to exploit some weak properties of the sources and of the mixing model, inspired by the diversity concept, for estimating the mixing mapping $\mathcal{A}$, or its inverse if it exists. For this purpose, one can use indirect or direct approaches:

- to either estimate the inverse of $\mathcal{A}$ (indirect approach), so that the estimated sources satisfy a desired property,
- or to estimate the mixing $\mathcal{A}$ (direct approach) optimizing a cost function (e.g. maximum likelihood) in which the desired property of the sources is explicitly used.

One can also estimate both mixing $\mathcal{A}$ and sources if at least three diversities are available (see Section 1.2.4 and Chapter 6).

Of course, whatever the adopted approach (direct or indirect), the choice of a suitable mixing model and of actual priors on source signals is essential for the success of source separation. Finally, as it will be detailed in Sections 1.2.3 and 1.3, the counterpart of weak priors is the typical ambiguities of the solution, which is usually not unique.

1.2.2.4 Noisy Mixtures

In fact, due to model simplification and measurement errors, it is mandatory to take into account a noise term in the mixing models. Usually, the noise is modeled by an additive term in Eqs. (1.6), (1.8), and (1.4). Zero mean Gaussian independent and identically distributed noise is generally assumed, even if the relevance of such hypothesis can be questioned, especially when considering physical/chemical measurements which are essentially non-negative. In the case of applications involving photon counting and statistics, noise models based on Poisson distribution are taken instead.

Practically, even for the simple linear instantaneous model, source separation from noisy mixtures becomes more tricky. In fact, a noisy observation can be written as:

$$\boldsymbol{x}(t) = \boldsymbol{A}\boldsymbol{s}(t) + \boldsymbol{e}(t), \tag{1.11}$$

where $\boldsymbol{e}(t)$ denotes the noise due to measurement errors.

In practice, the estimation of $\boldsymbol{A}$ (or of its inverse) is altered in the presence of the noise $\boldsymbol{n}(t)$. However, even if we are able to perfectly estimate $\boldsymbol{A}$ and then deduce its inverse $\boldsymbol{A}^{-1}$, the source estimation becomes:

$$\hat{\boldsymbol{s}}(t) = \boldsymbol{A}^{-1}\boldsymbol{x}(t) = \boldsymbol{s}(t) + \boldsymbol{A}^{-1}\boldsymbol{e}(t). \tag{1.12}$$

Clearly, the inversion problem is still ill-posed [32] and the last right-side term points out possible noise amplification, due to ill-conditioning of the mixing matrix, so that the perfect source separation can lead to a poor signal noise ratio (SNR). This is the reason why a regularization using a matched filter is preferred [11] (Chapter 1).

1.2.3 From Source Separation to Matrix Factorization

Let us focus on the section on the particular case of linear instantaneous mixtures, i.e., the case where the observations can be modeled according to Eq. (1.6):

$$\boldsymbol{x}(t) = \boldsymbol{A}\boldsymbol{s}(t), \quad \forall t = 1, \dots, T.$$

In fact, by merging all the T samples of the observations, one can denote $\boldsymbol{X}$ the 2-D observation array of size $P \times T$ and $\boldsymbol{S}$ the 2-D source array of size

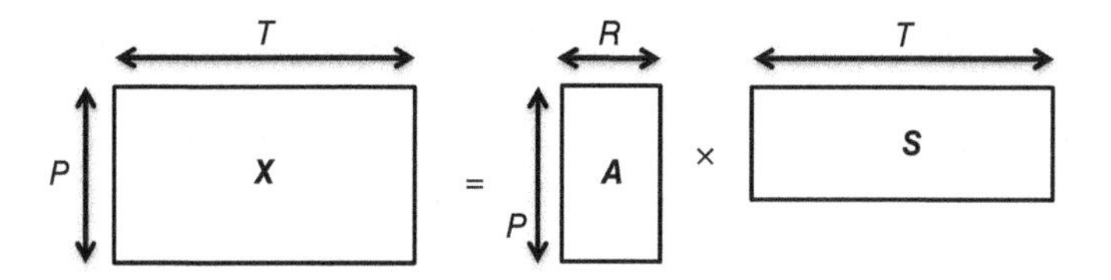

Figure 1.1 Factorization of a 2-D array as a product of two matrices.

$R \times T$, so that the above equation can be written as:

$$\boldsymbol{X} = \boldsymbol{AS}. \tag{1.13}$$

This factorization is illustrated in Figure 1.1

Remember that, in the *blind* source separation problem, we only observe the data $\boldsymbol{X}$, and the mixing matrix and the sources are unknown. It means that solving source separation problem is equivalent to factorizing the matrix $\boldsymbol{X}$, of size $P \times T$ as a product of two matrices, $\boldsymbol{A}$ and $\boldsymbol{S}$, of size $P \times R$ and $R \times T$, respectively.

1.2.3.1 Factorization Ambiguity

Two problems are related to this factorization. First, since we do not know the sources, probably we do not know the number, R, of sources: choosing the size of $\boldsymbol{A}$ and $\boldsymbol{S}$ is then a first issue. The second problem is much more annoying: it is clear that the above factorization problem is ill-posed , due to its non-uniqueness. In fact, for any invertible $R \times R$ matrix, $\boldsymbol{C}$:

$$\boldsymbol{X} = \boldsymbol{ACC}^{-1}\boldsymbol{S} = \boldsymbol{A}'\boldsymbol{S}', \tag{1.14}$$

the factorization remains unchanged. It means that there are an infinite number of possible factorizations of $\boldsymbol{X}$, up to any invertible matrix of size $R \times R$.

It is thus mandatory to add priors on sources for achieving uniqueness in the factorization, or at least restricting to an admissible set of solutions. The first idea which has been used in this purpose is to assume that the sources are mutually statistically independent: it leads to independent component analysis (ICA) methods, which will be detailed in Section 1.3. With ICA, the uniqueness is not fully achieved, but the matrix $\boldsymbol{C}$ is reduced to the product of a permutation matrix by a diagonal scaling matrix:

$$\boldsymbol{C} = \boldsymbol{PD}, \tag{1.15}$$

i.e., the sources are mainly recovered up to a scale in an arbitrary order.

Other priors have been investigated. Non-negativity of sources and/or of mixing coefficients is a suitable prior for many data (especially in image processing or with spectral measurement data), which can restrict the set of solutions: exploiting this prior leads to a large class of non-negative matrix factorization (NMF) methods.

Source (or mixture) sparsity is also an efficient prior. A signal $s(t), t = 1, \ldots, T$, is sparse if most of its samples are equal (or very close) to zero. Sparse component analysis (SCA) [33] has been intensively investigated and is especially very efficient for underdetermined mixtures, i.e., when the number of sensors is (much) less than the number of sources.

1.2.3.2 Data Representation

Finally, it is useful to note a nice property of the linear instantaneous model. Let us denote $\mathcal{T}$ a transform which preserves linearity. Then, using $\mathcal{T}$, Eq. (1.6) becomes:

$$\mathcal{T}[\mathbf{x}(t)] = \mathcal{T}[\mathbf{A}\mathbf{s}(t)] = \mathbf{A}\mathcal{T}[\mathbf{s}(t)], \forall t = 1, \ldots, T. \tag{1.16}$$

This means that, in the transformed space, the same linear relation (with the same mixing matrix $\mathbf{A}$) remains true between the transformed observations and the transformed sources. Consequently, the factorization problem can be considered in the initial space or in the transformed space. The preservation of linearity especially holds with many transforms, e.g., Fourier Transform, Discrete Cosine Transform, and Wavelet Transforms.

As an example of the interest of solving the problem in another space, consider again the sparsity prior. If the signals in the initial space are not sparse, we can enhance their sparsity by applying Discrete Cosine and Wavelet Transforms. Then, in the transformed space, sparsity is enhanced and the factorization problem can be solved based on sparsity. Then, using the inverse transform, one can come back to the initial space.

1.2.3.3 Factorization Algorithms

Practically, there are many algorithms for factorizing the data $\mathbf{X}$ as the product of two matrices. For instance, assuming source independence, one can compute a contrast function $\Upsilon(\mathbf{y}) \in \mathbb{R}$, which is maximal if the components of $\mathbf{y}(t) = \mathbf{B}\mathbf{x}(t)$ are mutually independent. The factorization can then be driven by the maximization of the contrast function with respect to the separating matrix $\mathbf{B}$. More details on contrast functions, and examples of algorithms, will be given in Section 1.3.2, or can be found in [11].

For other priors, like non-negativity or sparsity, factorization algorithms are generally based on constrained optimization methods. For example, for non-negativity constraints on $\mathbf{A}$ and $\mathbf{S}$, the factorization can be achieved by solving:

$$\min_{\mathbf{A},\mathbf{S}} \|\mathbf{X} - \mathbf{A}\mathbf{S}\|_F^2 \text{ subject to } \mathbf{A} \geqslant 0, \mathbf{S} \geqslant 0, \tag{1.17}$$

where $||\boldsymbol{M}||_F$ denotes the Frobenius norm of the matrix $\boldsymbol{M}$ and $\boldsymbol{A} \geq 0$ means that all the entries of the matrix $\boldsymbol{A}$ are non-negative. The choice of the Frobenius norm could be discussed: it is optimal for an additive Gaussian distributed noise $\boldsymbol{e}(t)$ in Eq. (1.11), but this assumption is not very compatible with the non-negativity of $\boldsymbol{A}$ and $\boldsymbol{S}$. This will be discussed in Chapters 3 and 5.

Sparsity of $s_r(t)$ means that the number of nonzero samples of $s_r(t)$ is very small, i.e. $\mathrm{Card}\{s_r(t) \neq 0, t = 1, \ldots, T\} \ll T$, where Card denotes the cardinal of a set. Conversely, it also means, that at each sample t, the probability that more than one source is nonzero is very weak. Sparsity can then be measured by counting the number of nonzero entries, either in each column or in each row of the matrix $\boldsymbol{S}$. Denoting briefly this measure with the pseudo-norm $\mathcal{L}_0$, the pseudo-norm $\mathcal{L}_0$ of the t-th column of $\boldsymbol{S}$ is denoted $||\boldsymbol{s}_t||_0$. Then, the factorization can be obtained by solving the optimization problem:

$$\min_{\boldsymbol{A},\boldsymbol{S}} ||\boldsymbol{X} - \boldsymbol{A}\boldsymbol{S}||_F^2 \text{ subject to } \sum_t ||\boldsymbol{s}_t||_0 < K. \tag{1.18}$$

Typically, the two optimization problems (1.17) and (1.18) are non-convex, due to the minimization with respect to the two terms in the product $\boldsymbol{AS}$. Although simple alternating minimization algorithms can be used, it is desirable to investigate more powerful methods for optimization with constraints compatible with physical or chemical data, which are often non-negative, sometimes sparse (e.g., in mass spectrometry). Chapter 2 will focus on advanced methods for optimization subject to constraints.

1.2.4 Enhanced Diversity: Tensor Formulation and Factorization

Coming back to the linear model of Eq. (1.6):

$$\boldsymbol{x}(t) = \boldsymbol{A}\boldsymbol{s}(t), \quad \forall t = 1, \ldots, T.$$

Merging all the T observation samples, one can denote $\boldsymbol{X}$ the 2-D observation array of size $R \times T$, and merging all the T samples of the r-th source in the vector $\boldsymbol{s}_r = (s_r(1), \ldots, s_r(T))^T$, the previous equation can be written as:

$$\boldsymbol{X} = \sum_{r=1}^{R} \boldsymbol{a}_r \boldsymbol{s}_r^T, \tag{1.19}$$

where $\boldsymbol{a}_r$ denotes the r-th column vector of the matrix $\boldsymbol{A}$. This equation shows that the observations can be explained through R latent variables (sources) which appear in each product through the time course, $\boldsymbol{s}_r$, and

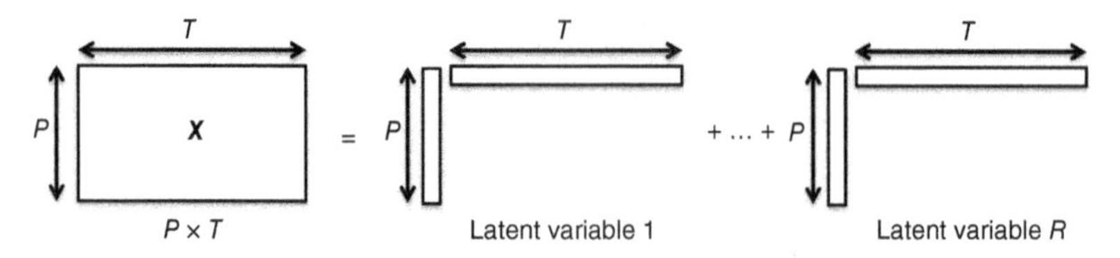

Figure 1.2 Decomposition of a 2-D array as a sum of R products of rank-1.

the weighting function on each sensor, $\boldsymbol{a}_r$. This formulation, illustrated in Figure 1.2, explicitly points out that $\boldsymbol{X}$ can be decomposed as a sum of R rank-1 terms. This representation shows that each term, i.e., each latent variable, is characterized by its time shape, and its scale shape, i.e., the weights of the time shape on each sensor: this explicitly points out time and spatial diversities.

Using tensorial product, one can write:

$$\boldsymbol{a}_r \boldsymbol{s}_r^T = \boldsymbol{a}_r \otimes \boldsymbol{s}_r, \tag{1.20}$$

so that Eq. (1.19) becomes:

$$\boldsymbol{X} = \sum_{r=1}^{R} \boldsymbol{a}_r \otimes \boldsymbol{s}_r. \tag{1.21}$$

Up to now, we considered observations related to only two diversities (time and space, in the above examples), i.e., as measures of values returned by a function of two variables, $f(u, v)$, for different discrete values, u_i, v_j, of these two variables:

$$X_{ij} = f(u_i, v_j), i = 1 \dots, P,\ j = 1, \dots, T. \tag{1.22}$$

The decomposition $\boldsymbol{X} = \boldsymbol{AS} = \sum_{r=1}^{R} \boldsymbol{a}_r \otimes \boldsymbol{s}_r$ assumes that:

- the two variables are separable so that $X_{ij} = f(u_i, v_j) = f_u(u_i) f_v(v_j)$,
- the 2-D array $\boldsymbol{X}$ can be decomposed as a sum of R rank-1 terms.

Now, let us assume that the observations are dependent on three variables, i.e. $\mathcal{X} = f(u, v, w)$. This means that the measurements are obtained through three diversities. Measuring $\mathcal{X}$ for all discrete values (u_i, v_j, w_k), $i = 1 \dots, I,\ j = 1, \dots, J; k = 1, \dots, K$ leads to a 3D-array, with a general term:

$$\mathcal{X}_{ijk} = f(u_i, v_j, w_k), \forall i = 1 \dots, I,\ j = 1, \dots, J; k = 1, \dots, K. \tag{1.23}$$

Assuming again the variables are separable, i.e. $\mathcal{X}_{ijk} = f(u_i, v_j, w_k) = f_u(u_i) f_v(v_j) f_w(w_k)$, we can decompose the 3-D array of observations as a sum of R terms (latent factors), each one being a tensorial product of three rank-1 vectors:

$$\mathcal{T}(u, v, w) = \sum_{r=1}^{R} \boldsymbol{u}_r \otimes \boldsymbol{v}_r \otimes \boldsymbol{w}_r, \tag{1.24}$$

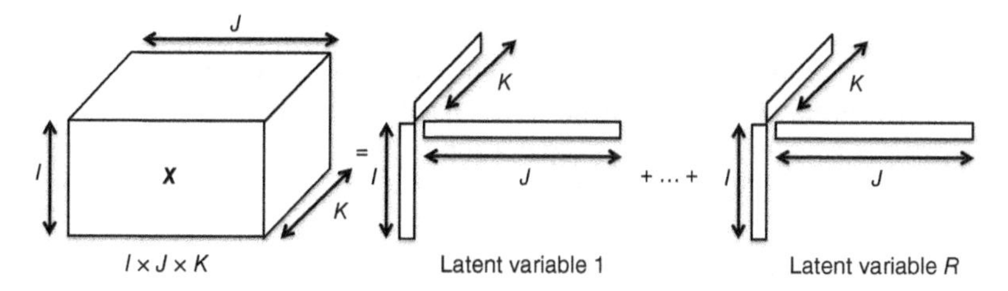

Figure 1.3 Factorization of a three-way array as a sum of R product of three rank-1 terms.

where $\boldsymbol{u}_r = (u_r(1), \ldots, u_r(I))^T$, $\boldsymbol{v}_r = (v_r(1), \ldots, v_r(J))^T$ and $\boldsymbol{w}_r = (w_r(1), \ldots, w_r(K))^T$. Each latent variable is clearly associated with its "shape" along the three diversities. Figure 1.3 illustrates this decomposition.

As an example of such three-way array of data associated with three-diversity measurements, one can consider fluorescence spectroscopy, where, at low concentration c_k of the sample k, the fluorescence intensity [34] can be written as a linear approximation of the Beer–Lambert law:

$$I(\lambda_f, \lambda_e, k) = I_0\gamma(\lambda_f)\epsilon(\lambda_e)c_k, \tag{1.25}$$

where ϵ denotes the absorption spectrum, γ the fluorescence emission spectrum, λ_f the fluorescence emission wavelength, and λ_e the excitation wavelength. Clearly, the fluorescence intensity is dependent on these three variables. Then, measuring the fluorescence intensity for a series of values of these three variables (emission spectrum, excitation spectrum, concentration) leads to a three-way array of data, $\mathcal{I}$, of size $I \times J \times K$ with a general term:

$$\mathcal{I}_{ijk} = I_0\gamma(\lambda_f(i))\epsilon(\lambda_e(j))c_r(k), \tag{1.26}$$

In the presence of R fluorescent solutes with low concentrations, additivity holds and the fluorescence intensity of the mixture can be written as a sum of R terms:

$$I(\lambda_f, \lambda_e, k) = I_0 \sum_{r=1}^{R} \gamma_r(\lambda_f)\epsilon_r(\lambda_e)c_r(k). \tag{1.27}$$

Following this model, the three-way array of data $\mathcal{I}_{ijk}$ can be decomposed as:

$$\mathcal{I} = \sum_{r=1}^{R} \boldsymbol{\gamma}_r \otimes \boldsymbol{\epsilon}_r \otimes \boldsymbol{c}_r. \tag{1.28}$$

This expression explains the data as a sum of R latent variables (see Figure 1.3), each one being characterized by its emission spectrum, $\boldsymbol{\gamma}_r$, its absorption spectrum, $\boldsymbol{\epsilon}_r$, and its concentration, $\boldsymbol{c}_r$. The above decomposition is an extension of source separation for three-way arrays (generalization to higher dimension array is straightforward), and PARAFAC [21] is an usual way for computing this factorization. Chapter 6 will explain in more detail theoretical interest of tensor decomposition and related algorithms, and will provide few application examples.

1.2.5 From Supervised to Blind Solutions

As explained in Section 1.2, the source separation problem is never actually blind. First, it is necessary to have a model of the mixing process, generally to know the nature (linear instantaneous or convolutive, nonlinear) of the model. However, this assumption is not sufficient for ensuring that source separation becomes well-posed. Additional priors, generally on sources, but sometimes on mixtures too, must be added. These priors are usually weak priors, in the sense that there are qualitative properties. Of course, this is the interest of all these methods, which are consequently very flexible. Even if we only observe sensor outputs, additional priors are necessary for getting solutions, and consequently source separation, although unsupervised, is not fully blind.

A simple idea was to assume that the unknown sources are coming from different devices, without relationships between them. This first prior used for solving the source separation problem was mutual statistical independence of the unknown sources. Its main advantage is that this assumption is true in many cases, and only considers sample distribution, without taking care of the sample order. Its main drawbacks are it requires high-order statistics and is unable to separate sources with Gaussian distributions.

In fact, following Darmois's results [11, 35] concerning factor analysis (see more details in Section 1.3), recovering independent factor $\boldsymbol{s}$ by only observing the sum $\boldsymbol{x} = \boldsymbol{A}\boldsymbol{s}$, is impossible if the random variables (components of $\boldsymbol{s}$) are independent and identically distributed (iid) and Gaussian. In the framework of source separation, this result leads to two possible approaches [36]:

- consider the random variables are iid but not Gaussian,
- consider the random variables are possibly Gaussian but not iid.

This first way leads to ICA, where one only considers the distribution of samples, without taking care of the sample order. In fact, ICA can work for non iid samples, but it does not use information between samples. ICA requires higher (than two) statistics, and thus cannot separate Gaussian distributed signals, since Gaussian distribution is fully characterized by its first- and second-order statistics.

The second way is to take into account relationships between samples. If sample index is time, a signal whose samples are not iid distributed, can be either colored (successive samples are not *independent*, i.e., the correlation function $\mathbb{E}\{x(t)x(t-\tau)\} \neq 0$, for $\tau \neq 0$), or non-stationary (successive samples are not *identically distributed*). In that case, as we will explain in more detail in Section 1.3, source separation is possible using only second-order statistics (variance–covariance matrices), and it works for separating Gaussian distributed sources, too. Unfortunately, independence assumption is not always satisfied especially with chemical and physical data. For instance, let us consider the spectra of light reflected by two different endmembers: although the shapes of the spectra are different, they are usually not independent. Moreover, considering interacting ions in a solution, their concentrations cannot be considered as independent.

In physical/chemical data, a very usual assumption is the non-negativity. For instance, concentrations of ions in a solution are non-negative; spectra of light reflected by a material are non-negative; mass of particles is non-negative. Moreover, in the above examples, mixing processes are essentially additive too. Non-negativity is then a desirable property which can be exploited for physical/chemical data. We would like to mention other priors, common in some applications or data, and easy to view in the data scatter plots (see Chapters 3 and 5).

Sparsity is a typical property in mass spectra. Effect of sparsity can be nicely viewed in scatter plots and suggest algorithms based on clustering. Moreover, sparsity can also lead to source separation in the underdetermined case, i.e. with fewer sensors than sources. In hyperspectral unmixing, simple algorithms are looking for observations (pixels) without mixture, i.e., related to a pure endmember (unique source). In fact, in such case, the observation reduces to the spectrum of the pure endmember, instead of a weighted mixture of spectra. Another prior, very usual in hyperspectral unmixing, is the sum-to-one assumption. This prior claims that the sum of fractional abundances (i.e., proportions) of endmembers in a pixel is equal to 1. This prior allows to reduce the number of unknowns by one, but implies also interesting geometrical properties in the data scatter plot.

Methods of source separation that we will report in this book are effectively *unsupervised*, but not all actually *blind*. In fact, priors are desirable, both for ensuring solution uniqueness and leading to simpler and efficient algorithms. Sometimes, source separation is said *semi-blind*.

1.3 Statistical Methods for Source Separation

As already discussed, an important class of BSS methods relies on a statistical modeling of the sources, in which they are described either as random variables or stochastic processes, and one assumes that the parameters related to the mixing process are deterministic. Alternatively, when the parameters of the mixing process are also modeled in a statistical fashion, the BSS problem can be solved by formulating a Bayesian inference problem, as will be discussed in Section 1.5.3 and, with more details, in Chapter 4. In this section, we shall focus on the latter formulation, which corresponds to most adopted solution to deal with linear instantaneous BSS models.

A fundamental aspect to understand statistical methods for BSS is to consider its connections with estimation theory, a well-established field of statistics which aims at estimating a set of parameters given a set of samples. In this context, a first fundamental question that arises is related to the conditions that must held so that the problem can be uniquely solved – in technical terms, these conditions are known as identifiability conditions. In the context of BSS, there is a strong result related to identifiability that comes from Darmois–Skitovic theorem [35]. We shall discuss the importance of this result in Section 1.3.1.

Once the identifiability conditions are clarified, BSS methods can be developed by resorting to frameworks that are typical of estimation theory. These approaches include maximum likelihood estimation, information theory, and higher-order statistics, and have led to a general class of methods known as ICA. In Section 1.3.2, we shall review these different strategies to perform ICA.

Finally, when the statistical modeling is based on stochastic processes to represent the sources, a possible way to carry out BSS is to consider formulations that exploit the covariance matrices of the mixtures calculated for different lags or on different sample (time) windows. Such an approach, which is commonly referred to as second-order BSS methods, will be discussed in Section 1.3.3.

1.3.1 Factorization of Independent Sources

As previously discussed, even in its instantaneous linear formulation, the BSS problem is an ill-posed one. In other words, unless additional prior information is considered, the factorization expressed in (1.6) is not unique and, thus, the observed mixtures can be explained by different sources and mixing matrices.

Assuming sources are coming from devices without any relationships, one can assume that, at any time t, the observation of one of the sources, say $s_i(t)$, does not provide any information on the other sources $s_j(t), j \neq i$. Consequently, a very natural constraint that arises in a statistical framework is to consider that the sources can be modeled as independent random variables, which means that:

$$f_{S_1,S_2,\ldots,S_R}(s_1,s_2,\ldots,s_R) = f_{S_1}(s_1) \times f_{S_2}(s_2) \times \ldots \times f_{S_R}(s_R), \tag{1.29}$$

where $f_{S_1,S_2,\ldots,S_R}(s_1,s_2,\ldots,s_R)$ stands for the joint distribution of the sources, whereas $f_{S_r}(s_r)$ corresponds to the marginal distribution of the r-th source. Under such an assumption, BSS can be formulated as a factorization problem where a multidimensional random variable related to the observed mixtures is, in the linear instantaneous case, factorized as linear combinations of independent random variables (the sources). Therefore, a natural question that arises is: When can a multidimensional random variable be uniquely (up to scale and permutation ambiguities) factorized through independent variables?

In order to answer this important question, we here restrict ourselves to the determined linear instantaneous case in the absence of noise. In this situation, the factorization based on independent sources is equivalent to set up a separating matrix $\boldsymbol{B}$ so that the components

$$\boldsymbol{y} = \boldsymbol{B}\boldsymbol{x} \tag{1.30}$$

correspond to the estimated sources which are expected to be independent. The global transformation between $\boldsymbol{s}$ and $\boldsymbol{y}$ can be represented by

$$\boldsymbol{y} = \boldsymbol{G}\boldsymbol{s}, \tag{1.31}$$

where $\boldsymbol{G} = \boldsymbol{B}\boldsymbol{A}$ – as previously discussed, the task of BSS is fulfilled when $\boldsymbol{G}$ is close to a diagonal matrix up to scale and permutation ambiguities.

In view of the discussion above, one can note that if having independent variables $\boldsymbol{y}$ leads to a diagonal $\boldsymbol{G}$, then the factorization of independent sources is enough to ensure source separation. The Darmois–Skitovic theorem [35] provides a first theoretical result related to this factorization and considers the following decomposition:

$$y_1 = \sum_{i=1}^{R} g_{1r} s_r, \tag{1.32}$$

$$y_2 = \sum_{i=1}^{R} g_{2r} s_r. \tag{1.33}$$

In the case of independent random variables $s_1, s_2, \ldots, s_R$, if y_1 and y_2 are independent, then $g_{1k}g_{2k} \neq 0$ if and only if s_k is a Gaussian variable. In other words, if independent components y_1 and y_2 are retrieved, then the mixtures cannot contain common sources, except if these sources are Gaussian distributed. As a consequence, if the sources are modeled as random variables, where no temporal information is considered, then it is not possible to separate Gaussian sources – this is a well-known limitation of ICA methods based on higher (than two) statistics.

A more general application of the Darmois–Skitovic theorem to BSS was addressed by Pierre Comon [2], who provided the conditions of separability to the more general case of P mixtures. In brief terms, Comon showed that adjusting a separating matrix $\boldsymbol{B}$ that provides independent components $\boldsymbol{y}$ leads to source separation if $\boldsymbol{A}$ is full rank and there is at most one Gaussian source. If there are more than two Gaussian sources, say k, one can separate all the sources except the Gaussian sources, which form a subspace of dimension k, in which the Gaussian sources cannot be separated. This important theoretical result has paved the way for the formalization of ICA and its application to BSS. In the sequel, we shall review the main ICA paradigms.

1.3.2 Independent Component Analysis (ICA)

ICA can be seen as a generalization of principal component analysis (PCA), a classical approach in multivariate analysis that is intensively applied in, for instance, chemometrics. In PCA, the goal is to factorize given multidimensional random variables as linear combinations of uncorrelated components. Unfortunately, such a factorization is not enough to ensure source separation and, at best, can solve the problem up an orthogonal matrix ambiguity. In other words, given a set of mixtures, if one applies PCA to this set, then one obtains a set of components $\boldsymbol{y}$ given by:

$$\boldsymbol{y} = \boldsymbol{R}\boldsymbol{s}, \tag{1.34}$$

where $\boldsymbol{R}$ is an orthogonal matrix.

The limitation of PCA discussed above highlights the need for considering statistical independence rather than decorrelation in a BSS context, which justifies the application of ICA. In this framework, the goal is to search for independent sources and can be achieved by means of different strategies, as will be discussed in the sequel. Our discussion will again be restricted to determined linear instantaneous mixing models. Moreover, in ICA, the sources are seen as iid random variables: it means that we do not exploit sample dependencies, but the iid assumption is not required, and ICA works for sources with sample dependencies.

1.3.2.1 ICA by Mutual Information Minimization

A first natural approach to perform ICA is to set up a separation criterion that attains its optimum value if, and only if, the retrieved sources are statistically independent. One possibility here is to consider a measure, common in statistics, the Kullback–Leibler divergence, but also well known in information theory as mutual information. Given a set of random variables represented by $\boldsymbol{y}$, their mutual information is given by:

$$I(\boldsymbol{y}) = \int f_y(\boldsymbol{y}) \log \left(\frac{f_y(\boldsymbol{y})}{\prod_{i=1}^{R} f_{y_i}(y_i)} \right) d\boldsymbol{y}, \tag{1.35}$$

which corresponds to the Kullback–Leibler divergence between the joint distribution of $\boldsymbol{y}$ and the product of their marginal distributions. Alternatively, the mutual information can be written as follows:

$$I(\boldsymbol{y}) = \sum_{i=1}^{R} H(y_i) - H(\boldsymbol{y}), \tag{1.36}$$

where $H(\cdot)$ stands for Shannon's (differential) entropy [37].

It is worth noticing that $I(\boldsymbol{y}) \geq 0$, where the equality achieved if, and only if, the elements of $\boldsymbol{y}$ are statistically independent. Therefore, ICA can be conducted by formulating the following optimization problem:

$$\min_{\boldsymbol{B}} I(\boldsymbol{y}). \tag{1.37}$$

If a gradient descent optimization method is considered, then one obtains the following iterative learning rule for estimating the separating matrix:

$$\boldsymbol{B}^{(k+1)} = \boldsymbol{B}^{(k)} - \mu \left(E\{\Psi_y(\boldsymbol{y})\boldsymbol{x}^T\} - \boldsymbol{B}^{-T} \right), \tag{1.38}$$

where $\Psi_y(\boldsymbol{y})$ denotes the vector of score functions, whose r-th element is given by

$$\psi_{y_i}(y_i) = -\frac{d \log p_{y_i}(y_i)}{dy_i}.$$

One of the main limitations of the mutual information approach for source separation is related to the estimation of the score functions $\psi_{y_i}(y_i)$. Indeed, such an estimation problem is equivalent to that of estimating probability distributions, which is a computationally expensive procedure. Fortunately, as will be discussed in the sequel, there are easier routes to perform source separation in the linear case.

1.3.2.2 ICA by Maximum Likelihood Estimation

The probabilistic model that is often adopted in source separation allows one to tackle the problem of estimating the separating matrix $\boldsymbol{B}$ as a problem of maximum likelihood estimation (MLE). Indeed, if the sources are modeled as a joint random variable of joint distribution given by $p_s(\boldsymbol{s})$, then the distribution of the mixtures is given by

$$f_X(\boldsymbol{x}) = \frac{f_S(\boldsymbol{A}^{-1}\boldsymbol{x})}{|\det \boldsymbol{A}|}. \tag{1.39}$$

Therefore, the distribution of the mixtures carries information on the mixing matrix $\boldsymbol{A}$ and, as a consequence, can be used to estimate $\boldsymbol{A}$ or, equivalently, the separating matrix $\boldsymbol{B}$. Given T samples of the mixtures, $\boldsymbol{x}(1), \boldsymbol{x}(2), \ldots, \boldsymbol{x}(T)$, and by assuming that the sources are iid, then one obtains the following expression:

$$f_{X(1),X(2),\ldots,X(T)}(\boldsymbol{x}(1), \boldsymbol{x}(2), \ldots, \boldsymbol{x}(T)|\boldsymbol{A}) = \prod_{t=1}^{T} \frac{f_{S(t)}(\boldsymbol{A}^{-1}\boldsymbol{x}(t))}{|\det \boldsymbol{A}|}. \tag{1.40}$$

When the samples are fixed (and not random variables) and $\boldsymbol{A}$ is allowed to vary freely, Eq. (1.40) becomes the likelihood function $\mathcal{L}(\boldsymbol{A})$. In MLE, the estimation of $\boldsymbol{A}$ is given by the value that maximizes the likelihood $\mathcal{L}(\boldsymbol{A})$, which, in mathematical terms, can be expressed as the following optimization problem

$$\max_{\boldsymbol{A}} \mathcal{L}(\boldsymbol{A}) = \prod_{t=1}^{T} \frac{f_{S(t)}(\boldsymbol{A}^{-1}\boldsymbol{x}(t))}{|\det \boldsymbol{A}|}. \tag{1.41}$$

Equivalently, in the linear determined case, the MLE framework for estimating the separating matrix is analogously given by:

$$\max_{\boldsymbol{B}} \mathcal{L}(\boldsymbol{B}) = \prod_{t=1}^{T} f_{S(t))}(\boldsymbol{B}\boldsymbol{x}(t))|\det \boldsymbol{B}|. \tag{1.42}$$

To sum up, according to (1.42), MLE requires the following inputs: the probability distribution of the sources and a set of samples of the mixtures. In the case of ICA, the distribution of the sources $p_{\boldsymbol{s}}(\boldsymbol{s})$ can be factorized according to (1.29), since the sources are modeled as independent variables. Such a factorization simplifies the optimization problem expressed in (1.42), which can be solved by the following gradient-based learning:

$$\boldsymbol{B}^{(k+1)} = \boldsymbol{B}^{(k)} - \mu \left(E\{\Psi_{\boldsymbol{s}}(\boldsymbol{y})\boldsymbol{x}^T\} - \boldsymbol{B}^{-T} \right). \tag{1.43}$$

It is worth noticing that the mutual information and maximum likelihood approaches lead to very close learning rules (see Eqs. (1.38) and (1.43)). The difference between these two methods is that, while in MLE one assumes that the distribution of the sources (and thus their score functions) is known beforehand, in the mutual information approach the score functions are related to the estimated sources and must be estimated from data.

At first glance, the strong assumption that the distribution of the sources is known in the MLE approach seems unrealistic *vis-a-vis* the problem of BSS. However, curiously and fortunately, the maximum likelihood approach can achieve source separation even when the assumed distributions for the sources are not equal to the actual ones. In other words, there is a sort of tolerance for mismatches that allows the MLE approach to operate even when one disposes only of a rough estimation of the sources distribution. More details on this topic can be found in [11].

Another source separation paradigm that is closely related to the MLE approach stems from the application of the Infomax principle [8]. Indeed, the resulting learning rule from Infomax is similar to Equ. (1.43). However, the principles behind the MLE and Infomax are different. In the Infomax approach, one searches for the maximization of the mutual information between the mixtures and the outputs of a nonlinear mapping of the mixtures, z_i. The nonlinear mappings ideally are the cumulative density functions of the estimated sources, y_i, so that the z_i's are uniformly distributed in [0,1].

The MLE approach, as a general statistical method, can also be considered for source separation in models other than the linear one. For instance, MLE is often employed to deal with an important class of nonlinear models known as the linear–quadratic model [38].

1.3.2.3 ICA by Kurtosis Maximization

Given that second-order statistics are not enough to provide a BSS framework, both the mutual information and maximum likelihood approaches make use of information contained in source distributions. An alternative way to go beyond second statistics is to resort to an approximation of the sources distribution through higher-order statistics. This can be done by resorting to the formalism of cumulants [11] (Chapter 3).

Among the different BSS strategies involving cumulants, many works considered the particular case of the (normalized) fourth-order marginal cumulant (or kurtosis), which, for a given zero-mean signal $y(t)$ (of variance 1), is given by

$$kurt(y(t)) = E\{y(t)^4\} - 3. \tag{1.44}$$

Typically, the optimization of kurtosis is conducted in the context of blind source extraction (BSE), which can be seen as a particular case of BSS in which one is interested in retrieving a single source. This can be done by setting up an extracting vector $\boldsymbol{w}$ so that:

$$y_1(t) = \boldsymbol{w}^T \boldsymbol{x}(t) \tag{1.45}$$

provides an estimation of one source. It is worth mentioning that BSS can be performed by executing a BSE algorithm several times. However, one must set up a procedure in order to ensure that the different executions of the BSE method do not lead to the same estimated source. This can be done, for instance, by a deflation procedure [7], which, after a given execution of the BSE method, subtracts the contribution of the extracted source from the mixtures.[1]

1 A disadvantage of deflation is that it may lead to an accumulation of errors, since the estimation error associated with a given execution of the extraction algorithms propagates in the next executions.

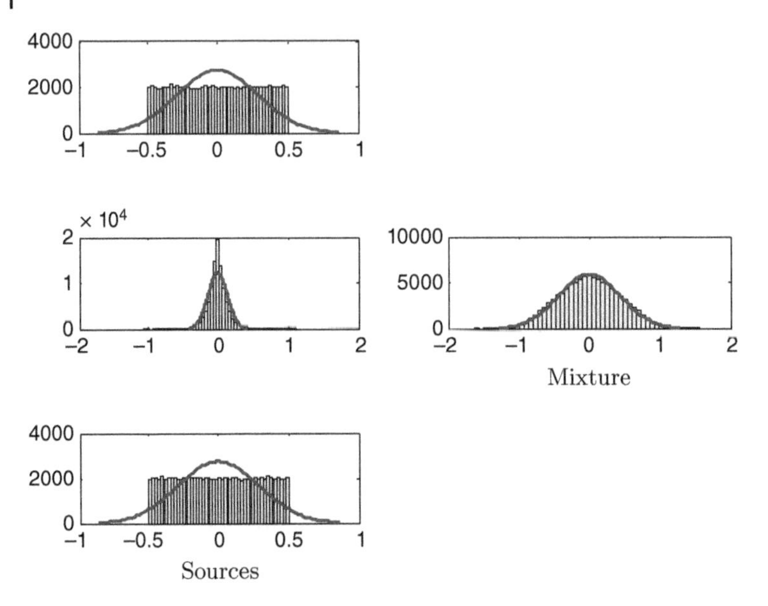

Figure 1.4 Non-gaussianity maximization principle is based on the observation that the mixtures tend to be more Gaussian than the sources.

A possible BSE framework based on kurtosis can be defined by the following optimization problem

$$\max_{\boldsymbol{w}_1} kurt(\mathbf{y}(t))^2 = kurt(\boldsymbol{w}^T\mathbf{z}(t))^2, s.t. ||\mathbf{w}||_2 = 1. \tag{1.46}$$

where $\mathbf{z}(t)$ denotes a whitened version of observed mixtures. The normalization imposed is necessary since the kurtosis is scale variant.

The problem expressed in (1.46) is commonly justified in terms of the non-gaussianity maximization principle. Indeed, as illustrated in Figure 1.4, as a consequence of the central limit theorem, the mixtures tend to be more Gaussian than the sources. Therefore, in the non-gaussianity maximization principle, a given source is estimated by retrieving a signal that maximizes a measure of non-gaussianity. Among the possibilities of measuring non-gaussianity is the kurtosis, since it takes zero for Gaussian variables.

1.3.3 Methods Based on Second-Order Statistics

In classical ICA approaches, the sources are modeled as iid random processes. However, actual signals are seldom iid, which motivated the development of separation approaches that exploit not only spatial diversity but also other types of diversity such as temporal diversity in the context of 1-D signals or the relation between adjacent pixels in images. A possible way to take these types of diversity into account is to consider a framework based on second-order statistics, as will be discussed in the following. Our discussion will focus on 1-D signals, but similar considerations hold for the case of images.

The second-order approach in BSS relies on the correlation matrix. Given the set of sources $\mathbf{s}(t)$, its correlation matrix $\mathbf{R}_{\mathbf{s}}(\tau)$ for a given lag τ is given by

$$\begin{bmatrix} E\{s_1(t)s_1(t+\tau)\} & E\{s_1(t)s_2(t+\tau)\} & \dots & E\{s_1(t)s_R(t+\tau)\} \\ \vdots & \vdots & \ddots & \vdots \\ E\{s_R(t)s_1(t+\tau)\} & E\{s_R(t)s_2(t+\tau)\} & \dots & E\{s_R(t)s_R(t+\tau)\} \end{bmatrix}. \quad (1.47)$$

In classical second-order methods, the central assumption is that the sources are mutually uncorrelated for all lags τ, which means that the correlation matrices $\mathbf{R}_{\mathbf{s}}(\tau)$ are diagonal for all τ.

Given the determined linear mixing model (1.6), one can easily check that the correlation matrices related to the mixtures are given by

$$\mathbf{R}_{\mathbf{x}}(\tau) = \mathbf{A}\mathbf{R}_{\mathbf{s}}(\tau)\mathbf{A}^T. \quad (1.48)$$

Note that, due to the mixing process, the correlation matrices of the observed signals are not diagonal. Therefore, a possible idea to recover the sources is to set up a separating matrix $\mathbf{B}$ such that the correlation matrices $R_{\mathbf{y}}(\tau)$ of $\mathbf{y}(t) = \mathbf{B}\mathbf{x}(t)$ are diagonal. Of course, if this procedure is done for only one matrix, $R_{\mathbf{y}}(0)$, then it will not work, since in this case one is simply (spatially) decorrelating the mixtures, which, as discussed before, is not enough to perform source separation.

In the algorithm for multiple unknown signals extraction (AMUSE) [39], in addition to diagonalize $R_{\mathbf{y}}(0)$, the idea is to diagonalize $R_{\mathbf{y}}(\tau)$ for an additional lag $\tau \neq 0$. Such a strategy works when the eigenvalues of $R_{\mathbf{y}}(\tau)$ for the selected τ are uniquely defined. In this case, the joint diagonalization of 2 matrices leads to an exact (analytical) solution.

A more statistically robust separation can be obtained by jointly diagonalizing $R_{\mathbf{y}}(\tau)$ for several lags τ at the same time. Such a task can be carried out by formulating a problem of approximate[2] joint diagonalization, which can be expressed as follows

$$\min_{\mathbf{B}} J(\mathbf{B}) = \sum_{\tau} off(\mathbf{B}\mathbf{R}_{\mathbf{x}}(\tau)\mathbf{B}^T), \quad (1.49)$$

where the operator $off(\cdot)$ corresponds to the sum of the squares of the terms that do not belong to the matrix diagonal.

---

2 since, generally, there is no longer exact solution like in the simple case of joint diagonalization of 2 matrices

The formulation expressed in (1.49) is the basis of the second-order blind identification (SOBI) algorithm [40]. An interesting feature found in SOBI, as well as in other related methods such as weighted-adjusted SOBI (WASOBI), is that they can separate even Gaussian sources. On the other hand, it is required that the sources have a temporal structure. Moreover, these temporal structures or, equivalently, the related spectra must be different [11] (Chapter 7): it is again the principle of *diversity*.

The source separation formulation based on joint diagonalization can also be considered for dealing with non-stationary sources. This can be done, for instance, by assuming that signals are block non-stationary [6] or by considering the joint diagonalization of time-frequency distributions [41]. In that case, the matrices to jointly diagonalize are computed in different time windows or time-frequency windows. The condition on sources is that they follow independent distribution variations, e.g., the ratio of variance of any pair of sources must not be a constant: it is again the principle of *diversity*.

1.4 Source Separation Problems in Physical–Chemical Sensing

Source separation problems arise in different contexts of physical–chemical sensing. In this section, we shall provide an (non-exhaustive) overview of these problems. We shall discuss which are the priors that are usual in these problems – they comprise: non-negativity and sum-to-one condition, since the sources in these cases are very often related to concentrations.

1.4.1 Material Analysis by Spectroscopy

Spectroscopy is a sensing technique which is frequently used in physical and chemical applications since it provides signals (spectra) related to the vibrational properties of molecules in interaction with an excitation laser light [42, 43]. These properties are directly linked to the molecular composition and structure of the analyzed material. Therefore, this technique is a widely used tool in analytical chemistry for material recognition, characterization, and monitoring during chemical reactions. This section presents an application example presented with more details in [44]. Figure 1.5 shows a set of Raman spectroscopy data obtained from monitoring the crystallization of the carbonate calcium [44]. Depending on the operating conditions (temperature and phase transformation time), a calcium carbonate sample is a mixture of three different phases (calcite, aragonite, or vaterite), which are characterized by three different Raman spectra. It is frequently needed to

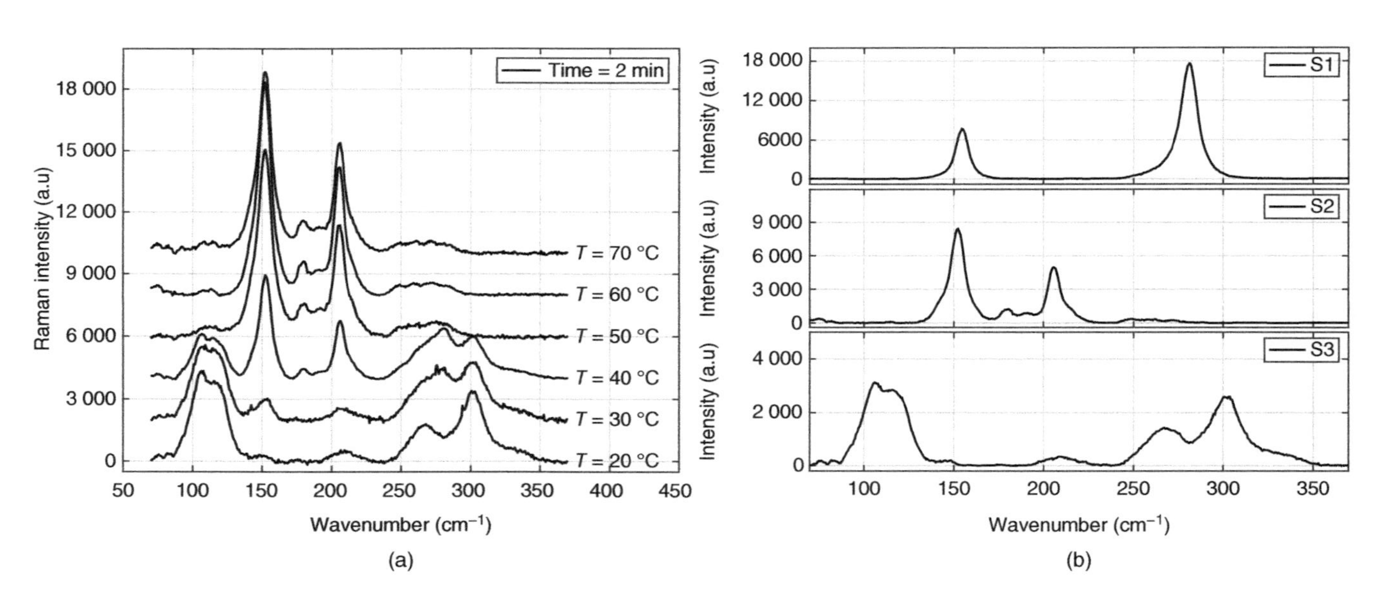

Figure 1.5 Calcium carbonate crystallization: (a) Raman spectra of a carbonate calcium sample after two minutes of phase transformation for different temperature values and (b) Raman spectra (sources) associated to calcite (S1), aragonite (S2), and vaterite (S3).

process spectroscopic signals resulting from the analysis of multicomponent materials. The signal processing in this context aims at identifying the pure components of the materials and at quantifying their abundances. These aims can be formulated as a source separation problem, where the sources signals are identified as the pure component spectra, while the mixing coefficients are linked to the relative abundances of the components [45]. For instance, in the case of absorption spectroscopy and low concentrations of analytes, the linear instantaneous mixing model gives an acceptable approximation according to the Beer–Lambert law [46].

The processing of mixture data in spectroscopy in order to estimate the pure component spectra and their abundances requires specific source separation algorithms that account for the properties of such signals:

- Source separation algorithms based on statistical independence alone fail to give correct estimates. Indeed, spectroscopic signals are mutually correlated since the components are very often made with very similar molecular structures which make their vibrational peaks located around the same wavelengths with a significant overlapping. Consequently, even if some ICA algorithms are designed to account for the non-negativity of the source signals [47], these methods do not ensure the non-negativity of the mixing coefficients.
- The accounting for the non-negativity of the mixing coefficients is possible by addressing the separation in a Bayesian framework [19]. In this case, appropriate distribution functions should be designed to encode the prior information on the spectra and the mixing coefficients. For instance, sparsity is a very important property of the pure component spectra whose exploitation allows to efficiently discriminate the components. Actually, this property can be seen as a hard constraint in the case of mass spectrometry, while in absorption spectroscopy it can be considered as a soft constraint that is introduced using regularization methods.

1.4.2 Hyperspectral Imaging

Hyperspectral imaging corresponds to the measurement of an incident light reflection at the ground surface of an observed scene in several contiguous spectral bands [48]. This imaging technique is used in several applications of physical and chemical sensing (material surface characterization, earth observation, and planetology) [49]. An analysis of the interactions between incident photons and the surface allows to identify the different compounds present in the surface. Despite the high spatial resolution that can be attained by recent imaging devices, the surface area covered by any pixel of the image may contain different components. Therefore, the reflectance

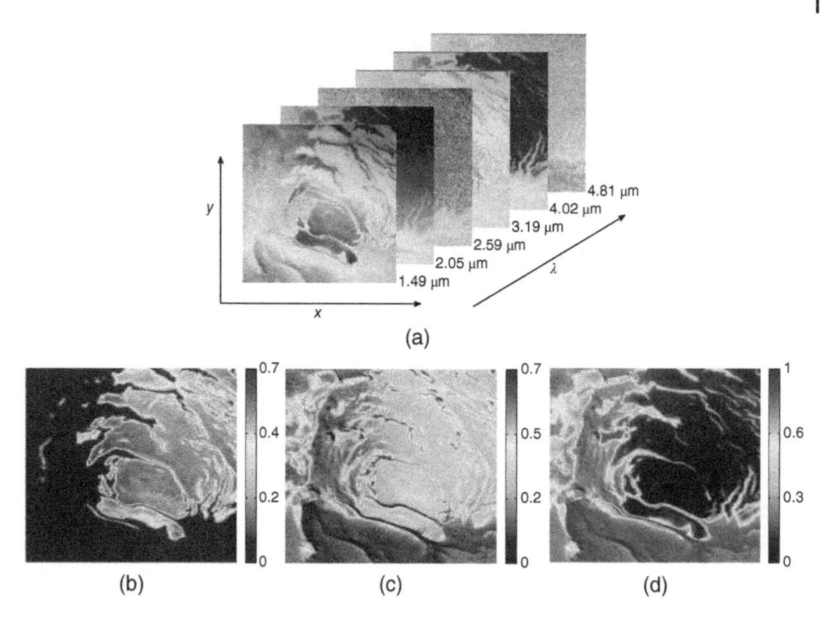

Figure 1.6 Hyperspectral imaging : (a) six hyperspectral images of the south polar cup of Mars planet acquired in six spectral bands during the Mars Express mission. The data cube contains 256 images of 300 × 128 pixels taken in 256 spectral bands. Images (b), (c), and (d) show the spatial distribution (abundance) of three components (CO_2 ice, water ice, and dust).

spectrum in each pixel can be explained by a mixture depending on the individual component reflectance spectra and the abundance of each component in this pixel area [50].

Figure 1.6 gives an example of a hyperspectral image provided by the spectro-imager OMEGA during the Mars Express mission and the obtained abundances of three identified components [51].

Two types of physical mixing at the ground can be observed [13]:

- *Macroscopic mixture.* each pixel is a patchy area made of several pure compounds. This type of mixture, sometimes called "sub-pixel mixture," happens when the spatial resolution is not large enough to observe the complex combination pattern. The total reflectance in this case can be nicely modeled as weighted sum of the pure constituent reflectances. The mixing coefficients (abundance fractions) associated to each pure component are surface proportions inside the pixel:

$$X(x,y,\lambda) = \sum_{r=1}^{R} A_r(x,y) S_r(\lambda) + E(x,y,\lambda), \tag{1.50}$$

where $X(x,y,\cdot)$ is the reflectance spectrum in the pixel of coordinate (x,y), $S_r(\cdot)$ the spectrum of the r-th component, and $A_r(x,y)$ its relative abundance in this pixel. The last term $E(x,y,\cdot)$ is an additive noise corresponding to model approximation and measurement errors.

- *Intimate mixture.* each pixel is made of one single terrain type which is a mixture at less than the typical meanpath scale. The total reflectance in this case will be a nonlinear function of pure constituent reflectances. Providing accurate nonlinear models allowing efficient separation methods is a very challenging task. Most popular models are simplified versions of the Hapke model. The linear–quadratic model is adopted in the case of multiple reflections:

$$R(x,y,\lambda) = \sum_{r=1}^{R} \alpha_r(x,y)\, S_r(\lambda) + \sum_{r=1}^{R-1} \sum_{t=r+1}^{R} \beta_{(r,t)}(x,y) S_r(\lambda) S_t(\lambda), \tag{1.51}$$

and in a more general context one can adopt the post-nonlinear model

$$R(x,y,\lambda) = f_{x,y}\left(\sum_{r=1}^{R} \alpha_r(x,y) S_r(\lambda)\right), \tag{1.52}$$

with nonlinear functions $f_{x,y}(\cdot)$ identified by the separation algorithm [52].

Unmixing hyperspectral data aims at the identification of the observed surface components (endmembers) and the determination of their fractional abundances inside each pixel area. These objectives are formalized as a source separation problem where the pure component spectra are identified as the estimated source signals and the fractional abundances as mixing coefficients. The main constraint in this application is the non-negativity of both the source signals and the mixing coefficients. Therefore, the processing of hyperspectral data can be viewed as non-negative source separation with mixing coefficients that should sum-to-one.

The processing of hyperspectral data using ICA methods is not a reliable approach since the pure component spectra and their spatial distributions are highly correlated mutually. However, ICA methods can be used as a pre-processing method for artifact detection and dimension reduction [51].

1.4.3 Electrochemical Sensor Arrays

In an electrochemical sensor, the transducer mechanism relies on the conversion of chemical information into electrical one. If the electrical information is provided in the form of voltage, then the electrochemical sensor is referred to as a potentiometric sensor.

An important sub-class of potentiometric sensors, known as ion-selective electrode (ISE), has been intensively employed in practice to measure the activity of an ion within liquid solution. A well-known example of ISE is the glass electrode, which is used to measure the pH (activity of the ion H^+). A miniaturized version of the ISE can be built by incorporating a sensitive membrane into a metal–oxide–semiconductor field-effect transistor (MOSFET) device. The resulting device is known as ion-sensitive field-effect transistor (ISFET) and has been applied mostly to measure ions in physiological solutions (such as blood and urine).

Ideally, ISEs and ISFETs should respond exclusively to the ion of interest. However, the sensitive membranes usually lack selectivity, and, as a consequence, the signal acquired by these sensors may be due to a mixture of different ions. Examples of typical interfering ions include NH_4^+ and K^+, which are often used in applications such as water quality surveillance and food industry.

The interference phenomenon that arises in ISEs and ISFETs can be modeled by the Nicolsky–Eisenman (NE) equation [53], which is an empirical extension of the Nernst equation. According to the NE equation, the signal acquired by those devices is given by:

$$x(t) = e + d \log\left(s_1(t) + \sum_{r=2}^{R-1} a_r s_r(t) \right), \tag{1.53}$$

where e and d are parameters associated with physical constants, $s_1(t)$ represents the activity of the target ion, whereas $s_2(t), \ldots, s_R(t)$ denote the activities of the interfering ions. The parameters a_r are called selectivity coefficients and express how strong is the interference of the r-th ion.

A first approach to tackle the interference problem is to develop sensors having low selectivity coefficients. However, such a strategy is costly and may lead to devices that cannot be commercialized in large scale. An alternative approach is to consider an array of sensors that are not necessarily selective. Then, the mixing process that arises due to the interference is dealt with by source separation methods.

A first issue that must be tackled when developing ISE or ISFET arrays based on source separation algorithm is related to the nature of the model described in (1.53). Indeed, linear source separation methods fail in this case as the NE equation is clearly nonlinear, thus requiring the application of nonlinear methods, which are more complex.

A second difficulty is related to the number of samples that are often available in practice. Usually, this number is small and, therefore, source separation methods that require a large number of samples cannot be applied. Moreover, the sources observed in practice can be dependent, as they are connected through a chemical equilibrium or reaction. In these situations,

ICA cannot be applied. Finally, noise can be very harmful in this application, since it can be amplified by the nonlinearities within the process.

The development of separation methods to ISE and ISFET arrays can be simplified by taking into account some prior information. The most straightforward one is the non-negativity of the sources, since they represent ionic activities. Another useful information is related to the strong temporal correlation observed for the sources. Indeed, ion concentrations tend to vary slowly. Finally, in some situations, there are "silent" ions, that is, ions that present an almost null value of concentration. A possible way to incorporate some of these priors can be done by means of a Bayesian approach, as will be discussed in Chapter 4.

1.5 Source Separation Methods for Chemical–Physical Sensing

Several methods have been proposed to address the source separation problem in the context of physical or chemical sensing. A recent survey on these methods can be found in [18, 54]. Each separation method is based on three ingredients: an objective function, the constraints on the source signals and mixing coefficients, and the employed algorithm for their estimation by minimizing the objective function. The objective function is designed in such a way to measure the data fitting while the constraints allow to ensure some desirable properties of the sought sources and mixing coefficients, and finally the optimization algorithm intends to calculate efficiently a solution of the separation problem.

The main constraint in the context of spectral data is the non-negativity of the source signals and the mixing coefficients. Such separation gave rise in chemometrics to a family of problems termed as *multivariate curve resolution* [55, 56] which are solved using various approaches such as *self-modeling curve resolution* (SMCR) [55], *alternating least squares* (ALS) [56], or more recently *non-negative matrix factorization* (NMF) [57]. However, since accounting for non-negativity alone does not lead to unique solution of the non-negative separation problem, additional constraints or assumptions are required to reduce the set of admissible solutions. A discussion on uniqueness conditions and on the calculation of the admissible solutions will be given in Chapter 3 of this book. According to the retained additional constraints and to the way these constraints are accounted for in the separation algorithm, *Bayesian methods* and *geometrical approaches* can be pointed out. Finally, since chemical analysis of material samples can performed simultaneously with respect to several parameters, *tensorial methods* are also used to reduce the set of admissible solutions.

1.5.1 Self-Modeling Curve Resolution

This approach is firstly based on an initial factorization of the data matrix $\boldsymbol{X}$ (whose each row contains observations provided by one sensor) using methods such as factor analysis (FA), principal component analysis (PCA), orthogonal projection approach (OPA), and independent component analysis (ICA),

$$\boldsymbol{X} = \boldsymbol{U}\boldsymbol{V},$$

where $\boldsymbol{U} \in \mathbb{R}^{P\times R}$ and $\boldsymbol{V} \in \mathbb{R}^{R\times T}$. Such preliminary step allows dimension reduction, noise removal, while maximizing the explained information present in the data matrix. In this step, the factorization rank R, which corresponds to the number of sources, should be assessed. This separation approach is historically the first one that was proposed in the chemometrics community to address the curve resolution problem (non-negative source separation), and named *self-modeling curve resolution* (SMCR) [55], in the case of two sources and dimension reduction by PCA. The second step in these methods is based on finding a linear transformation $\boldsymbol{T}$ of the estimated components in such a way to get non-negative sources and non-negative mixing coefficients

$$\begin{cases} \boldsymbol{S} = \boldsymbol{T}\boldsymbol{V}, \\ \boldsymbol{A} = \boldsymbol{U}\boldsymbol{T}^{-1}. \end{cases}$$

A Monte Carlo search procedure was proposed in [58] for finding possible matrices $\boldsymbol{T}$. This approach was generalized to three sources in [59] and more in [60, 61].

However, in practical situations it is needed to find a unique solution (the more plausible one). This can be achieved when *pure variables* exist in the mixture data. The pure variable can be for example wavelengths in spectroscopic measurements, for which the measured signal depends only on one component. In analytical chemistry, most popular methods for determining the pure variables are evolving factor analysis (EFA) [62] and simple-to-use interactive self-modeling analysis (SIMPLISMA) [63].

1.5.2 Non-Negative Matrix Factorization

NMF corresponds to the factorization of a non-negative matrix into the product of two non-negative matrices [64]. NMF algorithms such as those proposed in [20] and [65] allow to solve the source separation problem when both the sources and the mixing coefficients are subject to non-negativity constraints.

Alternating non-negative least squares belong to the family of NMF algorithms when a least squares criterion is used to measure the factorization accuracy [56, 66]. It consists in estimating alternatively the mixing coefficients and the source signals by minimizing a least squares criterion under non-negativity constraints.

$$(\widehat{\boldsymbol{S}}, \widehat{\boldsymbol{A}}) = \underset{\boldsymbol{A},\boldsymbol{S}}{\operatorname{argmin}} \, \|\boldsymbol{X} - \boldsymbol{A}\boldsymbol{S}\|_F^2 \quad \text{subject to } \boldsymbol{S} \geqslant \boldsymbol{0}, \boldsymbol{A} \geqslant \boldsymbol{0}, \tag{1.54}$$

The non-negative least squares algorithm (NNLS), originally proposed in [67] and accelerated in [68], is the most commonly used one. It consists in iteratively solving

$$\widehat{\boldsymbol{S}}^{(k)} = \underset{\boldsymbol{S}}{\operatorname{argmin}} \, \|\boldsymbol{X} - \widehat{\boldsymbol{A}}^{(k-1)}\boldsymbol{S}\|_F^2 \quad \text{subject to } \boldsymbol{S} \geqslant \boldsymbol{0}, \tag{1.55}$$

and

$$\widehat{\boldsymbol{A}}^{(k)} = \underset{\boldsymbol{A}}{\operatorname{argmin}} \, \|\boldsymbol{X} - \boldsymbol{A}\widehat{\boldsymbol{S}}^{(k)}\|_F^2 \quad \text{subject to } \boldsymbol{A} \geqslant \boldsymbol{0}, \tag{1.56}$$

at each iteration $k \geqslant 1$. The initial values $\widehat{\boldsymbol{A}}^{(0)}$ or $\widehat{\boldsymbol{S}}^{(0)}$ are generally specified using similar approaches as in SMCR methods. In addition to non-negativity, additional constraints, such as closure (sum-to-one), unimodality, absence/presence of sources, are accounted for to reduce the set of admissible solutions and to get a meaningful solution, from the chemical/physical point of view [69]. Adding additional constraints require specific constrained least squares optimization algorithms [70].

More generally, the approximate factorization can be performed by minimizing an objective function which is not necessarily the least squares criterion [65]. For instance, various data fitting measures [71], such as Kullback–Leibler, Itakura–Saito, and beta divergences, can be employed to design a factorization algorithm better suited to the considered application.

The convergence of the iterative optimization algorithm proposed in [65] is faster than the ALS method and is numerically advantageous since it does not require any matrix inversion. Note that the NMF algorithm scheme uses multiplicative update equation that are derived using a majorization–minimization principle that will be detailed in Chapter 2 of this book. Accounting for additional constraints in NMF is possible thanks to resorting to penalized estimation algorithms that can be obtained by formulating the problem in a Bayesian estimation framework.

1.5.3 Bayesian Separation Approach

The separation of non-negative sources is an ill-posed inverse problem due to the non-uniqueness of the solution. The resort to a Bayesian estimation is a natural way to encode the available information on the sought quantities in order to get a meaningful solution [72]. In that respect, the available information or additional assumptions on the noise statistics, the source signals, and the mixing coefficients are expressed in terms of probability distribution functions. This statistical modeling of the mixing process allows to derive the likelihood $p(\boldsymbol{X}|\boldsymbol{S},\boldsymbol{A})$ and the posterior distribution $p(\boldsymbol{S},\boldsymbol{A}|\boldsymbol{X})$, thanks to Bayes' rule,

$$p(\boldsymbol{S},\boldsymbol{A}|\boldsymbol{X}) = p(\boldsymbol{X}|\boldsymbol{S},\boldsymbol{A}) \times p(\boldsymbol{S}) \times p(\boldsymbol{A}) \div p(\boldsymbol{X}), \tag{1.57}$$

where the independence between $\boldsymbol{A}$ and $\boldsymbol{S}$ is assumed. The probability distribution functions $p(\boldsymbol{S})$ and $p(\boldsymbol{A})$ allow to represent the available constraints or to encode the assumptions on the source signals and the mixing coefficients, while $p(\boldsymbol{X})$ can be interpreted as a normalization constant. From this posterior density, the estimation of both $\boldsymbol{A}$ and $\boldsymbol{S}$ can be achieved by using various Bayesian estimators [72].

The first step of any Bayesian inference is to encode prior knowledge on the noise statistics, the source signals, and the mixing coefficients by appropriate probability distributions. While the noise is generally assumed to be Gaussian or Poissonian, specific prior distributions should be chosen to encode the constraints. For instance:

- Non-negativity is traduced by truncated Gaussian and Gamma distributions,
- Sum-to-one constraint is encoded using either a truncated Gaussian or the Dirichlet distributions,
- Sparsity is represented by an exponential distribution.

The Gamma distribution is an exponential family distribution which is used for modeling non-negative quantities. Moreover, it exhibits various shapes depending on the value of its two parameters and it has the exponential, the Khi-2, and the Rayleigh densities as particular cases [73].

The formulation of source separation using Bayesian estimation theory has been suggested in [74, 75], while its application to the separation of linear spectral mixture data has been performed in [19, 76, 77]. Bayesian separation methods in the case of linear mixing can be reduced to NMF algorithms, while the Bayesian approach offers a very well-stated framework for addressing the case of nonlinear mixing [28, 52]. Chapter 4 is dedicated to the presentation of the Bayesian source separation approach, its link with NMF methods, and its application in the context of physical/chemical data processing.

1.5.4 Geometrical Approaches

Since the mixture data sets result from a non-negative mixing of non-negative sources, the scatter plot of these mixture signals is contained in the simplicial cone generated by the columns of the mixing matrix. Therefore, the problem of finding the columns of the mixing matrix reduces to the identification of the edges of a minimal cone containing the mixture data matrix. In addition, in the case of mixing coefficients summing to one, the mixture data are distributed on a simplex whose vertices correspond to the source signals. Figure 1.7 gives a graphical illustration of the impact of the mixing coefficients properties on the statistical distribution of the mixture data inside a simplicial cone, in the case of three sources.

In chemical imaging spectrometry (such as Raman, Infrared) and remote sensing by hyperspectral imaging [13], the spectral mixture analysis (also known as spectral unmixing) is often handled using a two-step procedure: the source signals (spectral signatures) estimation and the mixing coefficients (fractional abundances) assessment, respectively. In the first step, the source signals are identified by using an *Endmember Extraction Algorithm* (EEA). See for instance [78] for a recent performance comparison and discussion on some standard EEAs. The most popular EEAs is the N-FINDR algorithm. N-FINDR estimates the source signals by identifying the largest simplex whose vertices are taken from the convex hull of the data. Another popular and faster alternative is the *Vertex Component Analysis*-(VCA)

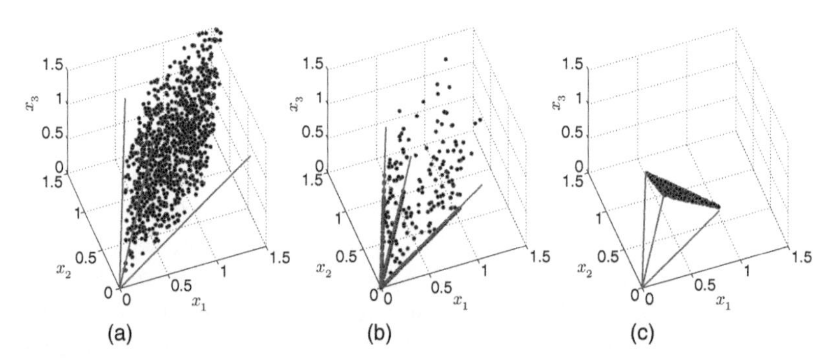

Figure 1.7 Scatter plots of three mixture data for different distributions of the mixing coefficients: (a) uniform distribution, (b) sparse, and (c) uniform constrained to sum-to-one.

method which has been proposed in [79]. It consists in iteratively estimating the vertices of the simplex without calculating the convex hull. A common assumption in VCA and N-FINDR is the existence of pure pixels (pixels composed of a single component) in the observed data. Alternatively, when there is no pure observation for all the sources, the *Minimum Volume Transform* (MVT) finds the smallest simplex that contains all the pixels [80]. On the other hand, there is an increasing interest to the joint estimation of the endmembers and their abundances, which can be formulated as a source separation problem under constraints such as non-negativity and sum-to-one.

1.5.5 Tensor Factorization Methods

Most of the above methods simply exploit one type of diversity: it leads to a two-way array of data, easily represented by a matrix. Above examples consider two-way arrays based on space and time (mixtures of signals), or space and frequency (spectral mixtures) dimensions. But in many chemical experiments, one or more additional diversities can be considered, leading to a three-way (or more generally a multi-way) array of data, which can be represented by a tensor. As an example, in fluorescence spectroscopy, the (measured) fluorescence intensity is dependent on three variables: the fluorescence emission spectrum, the absorbance spectrum, and the relative concentration of compounds. Historically, Parallel Factor Analysis (PARAFAC) or Canonical Polyadic Decomposition (CPD) – inspired from early works on factor analysis in psychometrics and even much earlier works in mathematics [31] – has been intensively studied in chemometrics from 1990s and popularized by R. Bro [21]. Currently, many theoretical contributions and applications are addressing CPD decomposition of three-way arrays. More details of tensorial methods and an application in fluorescence spectroscopy can be found in Chapter 6 of this book.

1.6 Organization of the Book

This chapter was a quick overview of source separation problems and methods. In addition to being a natural introduction to Chapters 2–6 of this book, it is designed for providing a comprehensive display for chemists about the main advanced ideas of source separation in signal processing.

The chapters of this book have been written by global experts of the topics of each chapter. However, for consistency and readability, the main notations are shared in all the chapters. The list of common notations is given in the beginning of this book. Local notations are precised in the beginning of each chapter.

The rest of the book is organized as follows.

As we point out in this chapter, optimization methods, especially with constraints and in non-convex situations, are in the core of many algorithms of source separation. Chapter 2 will introduce the theoretical background which allows to develop efficient algorithms to successfully address these problems.

Chapters 3–6 will focus on four approaches which are well suited to source separation of physical/chemical data. All these chapters follow a similar organization: a methodological part with theoretical results and algorithms, followed by a few examples of various physical/chemical data as illustration.

Since non-negativity of sources as well as mixtures coefficients is a very usual assumption in physical data, Chapter 3 addresses the concept of non-negative matrix factorization (NMF), main theoretical results, some factorization algorithms, and a few examples in spectroscopy.

Chapter 4 presents the Bayesian approach for source separation, which is a very flexible way to take into account priors on sources and on mixing coefficients. It includes principles of Bayesian estimation, algorithms for linear and nonlinear mixtures, and ends with a few examples for spectrometry, ion-sensitive sensor array, and hyperspectral imaging.

Non-negative sources observed under the linear mixing model, special assumptions like "pure pixel" or "sum-to-one" exhibit insightful properties of convex geometry. Chapter 5 will review key theoretical developments and algorithms of convex geometry and pure pixel search for source separation of non-negative sources, inspired by hyperspectral unmixing.

Many chemical data are recorded using more than two diversities. Consequently, experimental data constitute a three-way (or n-way, $n > 2$) array, i.e., a data tensor. Chapter 6 describes the main ideas for tensor decompositions of 3-way arrays and points out the differences with matrix factorization. Illustrations include a few applications based on fluorescence spectroscopy, electrophoresis, or mass spectrogram recordings.

References

1 Jutten, C. and Hérault, J. (1991) Separation of sources, Part I. *Signal Processing*, 24 (1), 1–10.

2 Comon, P. (1994) Independent component analysis, a new concept? *Signal Processing*, 36 (3), 287–314.

3 Cardoso, J.F. and Souloumiac, A. (1993) Blind beamforming for non Gaussian signals. *IEE Proceedings-F*, 140 (6), 362–370.

4 Cardoso, J.F. (1998) Blind signal separation: statistical principles. *Proceedings of the IEEE*, 86 (10), 2009–2025.

5 Pham, D. and Garat, P. (1997) Blind separation of mixture of independent sources through a quasi-maximum likelihood approach. *IEEE Transactions on Signal Processing*, 45 (7), 1712–1725.

6 Pham, D.T. and Cardoso, J.F. (2001) Blind separation of instantaneous mixtures of nonstationary sources. *IEEE Transactions on Signal Processing*, 49 (9), 1837–1848.

7 Delfosse, N. and Loubaton, P. (1995) Adaptive blind separation of independent sources: a deflation approach. *Signal Processing*, 45 (1), 59–83.

8 Bell, A.J. and Sejnowski, T.J. (1995) An information-maximization approach to blind separation and blind deconvolution. *Neural Computation*, 7 (6), 1129–1159.

9 Hyvärinen, A. (1999) Fast and robust fixed-point algorithms for independent component analysis. *IEEE Transactions on Neural Networks*, 10 (3), 626–634.

10 Cichocki, A. and Amari, S. (2002) *Adaptive Blind Signal and Image Processing - Learning Algorithms and Applications*, John Wiley & Sons.

11 Comon, P. and Jutten, C. (eds) (2010) *Handbook of Blind Source Separation: Independent Component Analysis and Applications*, Academic Press.

12 Makino, S., Lee, T.W., and Sawada, H. (2007) *Blind Speech Separation*, Springer Berlin.

13 Bioucas-Dias, J., Plaza, A., Dobigeon, N., Parente, M., Du, Q., Gader, P., and Chanussot, J. (2012) Hyperspectral unmixing overview: geometrical, statistical, and sparse regression-based approaches. *IEEE Journal of Selected Topics in Applied Earth Observations and Remote Sensing*, 5 (2), 354–379.

14 De Lathauwer, L., De Moor, B., and Vandewalle, J. (1994) Blind source separation by higher-order singular value decomposition, in *Proceedings of the 7th European Signal Processing Conference (EUSIPCO'94)*, Edinburgh, UK, pp. 175–178.

15 Makeig, S., Bell, A., Jung, T.P., and Sejnowski, T. (1996) Independent component analysis of electroencephalographic data. *Advances in Neural Information Processing Systems*, pp. 145–151.

16 Cardoso, J.-F. (2010) Precision cosmology with the cosmic microwave background. *IEEE Signal Processing Magazine*, 27 (1), 55–66 10.1109/MSP.2009.934715.

17 Nuzillard, D., Bourg, S., and Nuzillard, J.M. (1998) Model-free analysis of mixtures by {NMR} using blind source separation. *Journal of Magnetic Resonance*, 133 (2), 358–363.

18 Monakhova, Y., Astakhov, S., Mushtakova, S., and Gribov, L. (2011) Methods of the decomposition of spectra of various origin in the analysis of complex mixtures. *Journal of Analytical Chemistry*, 66 (4), 351–362.

19 Moussaoui, S., Carteret, C., Brie, D., and Mohammad-Djafari, A. (2006) Bayesian analysis of spectral mixture data using Markov chain Monte Carlo methods. *Chemometrics and Intelligent Laboratory Systems*, 81 (2), 137–148.

20 Paatero, P. and Tapper, U. (1994) Positive matrix factorization: a nonnegative factor model with optimal utilization of error estimates of data values. *Environmetrics*, 5, 111–126.

21 Bro, R. (1997) PARAFAC. Tutorial and applications. *Chemometrics and Intelligent Laboratory Systems*, 38, 149–171.

22 Widrow, B., Glover, J.R., McCool, J.M., Kaunitz, J., Williams, C.S., Hearn, R.H., Zeidler, J.R., Dong, E. Jr., and Goodlin, R. (1975) Adaptive noise cancelling: principles and applications. *Proceedings of the IEEE*, 63 (12), 1692–1716.

23 Comon, P. (2004) Contrasts, independent component analysis, and blind deconvolution. *International Journal of Adaptive Control and Signal and Processing*, 18 (3), 225–243.

24 Pedersen, M.S., Larsen, J., Kjems, U., and Parra, L.C. (2007) A survey of convolutive blind source separation methods, in *Springer Handbook of Speech Processing*, Springer Press.

25 Emile, B., Comon, P., and Leroux, J. (1998) Estimation of time delays with fewer sensors than sources. *IEEE Transactions on Signal Processing*, 46 (7), 2012–2015.

26 Delmaire, G. and Roussel, G. (2012) Joint estimation decision methods for source localization and restoration in parametric convolution processes. Application to accidental pollutant release. *Digital Signal Processing*, 22 (1), 34–46.

27 Bermejo, S., Jutten, C., and Cabestany, J. (2006) ISFET source separation: foundations and techniques. *Sensors and Actuators B: Chemical*, 113, 222–233.

28 Duarte, L., Jutten, C., and Moussaoui, S. (2009) A Bayesian nonlinear source separation method for smart ion-selective electrode arrays. *IEEE Sensors Journal*, 9 (12), 1763–1771.

29 Heylen, R., Parente, M., and Gader, P. (2014) A review of nonlinear hyperspectral unmixing methods. *IEEE Journal of Selected Topics in Applied Earth Observations and Remote Sensing*, 7 (6), 1844–1868.

30 Niknazar, M., Becker, H., Rivet, B., Jutten, C., and Comon, P. (2014) Blind source separation of underdetermined mixtures of event-related sources. *Signal Processing*, 101, 52–64.

31 Comon, P. (2014) Tensors: a brief introduction. *IEEE Signal Processing Magazine*, 31 (3), 44–53 10.1109/MSP.2014.2298533.

32 Idier, J. (2008) *Bayesian Approach to Inverse Problems*, ISTE Ltd and John Wiley & Sons Inc.

33 Gribonval, R. and Lesage, S. (2006) A survey of sparse component analysis for blind source separation: principles, perspectives, and new challenges, in *ESANN'06 Proceedings-14th European Symposium on Artificial Neural Networks*, d-side publi., pp. 323–330.

34 Cohen, J.E., Comon, P., and Luciani, X. (2016) Correcting inner filter effects, a non multilinear tensor decomposition method. *Chemometrics and Intelligent Laboratory Systems*, 150, 29–40.

35 Darmois, G. (1953) Analyse générale des liaisons stochastiques. *Revue de l'Institut International de Statistique*, 21, 2–8.

36 Cardoso, J.F. (2001) The three easy routes to independent component analysis, contrasts and geometry, in *Proceedings ICA 2001*, pp. 1–6.

37 Cover, T.M. and Thomas, J.A. (1991) *Elements of Information Theory*, Wiley Interscience.

38 Hosseini, S. and Deville, Y. (2004) Blind maximum likelihood separation of a linear-quadratic mixture, in *Proceedings of the 5th International Workshop on Independent Component Analysis and Blind Signal Separation, ICA 2004*, pp. 694–701.

39 Tong, L., Liu, R.W., Soon, V.C., and Huang, Y.F. (1991) Indeterminacy and identifiability of blind identification. *IEEE Transactions on Circuits and Systems*, 38 (5), 499–509.

40 Belouchrani, A., Meraim, K.A., Cardoso, J.F., and Moulines, E. (1997) A blind source separation technique using second order statistics. *IEEE Transactions on Signal Processing*, 45, 434–444.

41 Belouchrani, A. and Amin, M.G. (1998) Blind source separation based on time-frequency signal representations. *IEEE Transactions on Signal Processing*, 46 (11), 2888–2897.

42 Diem, M. (1993) *Introduction to Modern Vibrational Spectroscopy*, Wiley Interscience Publication, Wiley.

43 Hollas, J. (2004) *Modern Spectroscopy*, Wiley.

44 Carteret, C., Dandeu, A., Moussaoui, S., Muhr, H., Humbert, B., and Plasari, E. (2009) Polymorphism studied by lattice phonon Raman spectroscopy and statistical mixture analysis method. *Crystal Growth and Design*, 9, 807–812.

45 Malinowski, E. (2000) *Factor Analysis in Chemistry*, John Willey & Sons, 3rd edn.

46 Ricci, R., Ditzler, M., and Nestor, L. (1994) Discovering the Beer-Lambert law. *Journal of Chemical Education*, 71, 983–985.

47 Plumbley, M.D. (2002) Conditions for non-negative independent component analysis. *IEEE Signal Processing Letters*, 9 (6), 177–180.

48 Chang, C.I. (2007) *Hyperspectral Data Exploitation*, Wiley Interscience.

49 Green, R.O., Eastwood, M.L., Sarture, C.M., Chrien, T.G., Aronsson, M., Chippendale, B.J., Faust, J.A., Pavri, B.E., Chovit, C.J., Solis, M., Olah, M.R., and Williams, O. (1998) Imaging spectroscopy and the airborne visible/infrared imaging (AVIRIS) spectrometer. *Remote Sensing of Environment*, 65, 227–248.

50 Healey, G. and Slater, D. (1999) Models and methods for automated material identification in hyperspectral imagery acquired under unknown illumination and atmospheric conditions. *IEEE Transactions on Geoscience and Remote Sensing*, 37 (6), 2706–2716.

51 Moussaoui, S., Hauksdóttir, H., Schmidt, F., Jutten, C., Chanussot, J., Brie, D., Douté, S., and Benediktsson, J. (2008) On the decomposition of Mars hyperspectral data by ICA and Bayesian positive source separation. *Neurocomputing*, 71 (10-12), 2194–2208.

52 Halimi, A., Altmann, Y., Dobigeon, N., and Tourneret, J.Y. (2011) Nonlinear unmixing of hyperspectral images using a generalized bilinear model. *IEEE Transactions on Geoscience and Remote Sensing*, 49 (11), 4153–4162.

53 Gründler, P. (2007) *Chemical Sensors: An Introduction for Scientists and Engineers*, Springer.

54 De Juan, A., Jaumot, J., and Tauler, R. (2014) Multivariate curve resolution (MCR). Solving the mixture analysis problem. *Analytical Methods*, 6 (14), 4964–4976.

55 Lawton, W.H. and Sylvestre, E. (1971) Self-modeling curve resolution. *Technometrics*, 13, 617–633.

56 Tauler, R., Izquierdo-Ridorsa, A., and Casassas, E. (1993) Simultaneous analysis of several spectroscopic titrations with self-modelling curve resolution. *Chemometrics and Intelligent Laboratory Systems*, 18 (3), 293–300.

57 Sajda, P., Du, S., Brown, T., Stoyanova, R., Shungu, D., Mao, X., and Parra, L. (2004) Nonnegative matrix factorization for rapid recovery of constituent spectra in magnetic resonance chemical shift imaging of the brain. *IEEE Transactions on Medical Imaging*, 23 (12), 1453–1465.

58 Ohta, N. (1973) Estimating absorption bands of component dyes by means of principal component analysis. *Analytical Chemistry*, 45 (3), 553–557.

59 Borgen, O. and Kowalski, B. (1985) An extension of the multivariate component resolution method to three component. *Analytica Chimica Acta*, 174, 1–26.

60 Sasaki, K., Kawata, S., and Minami, S. (1983) Constrained nonlinear method for estimating component spectra from multicomponent mixtures. *Applied Optics*, 22 (22), 3599–3606.

61 Henry, R. and Kim, B. (1990) Extension of self-modeling curve resolution to mixtures of more than three components. Part I. Finding the basic feasible region. *Chemometrics and Intelligent Laboratory Systems*, 8, 205–216.

62 Maeder, M. (1987) Evolving factor analysis for the resolution of overlapping chromatographic peaks. *Analytical Chemistry*, 59, 527–530.

63 Windig, W. and Guilment, J. (1991) Interactive self-modeling mixture analysis. *Analytical Chemistry*, 63, 1425–1432.

64 Thomas, L.B. (1974) Rank factorizations of nonnegative matrices. *SIAM Review*, 16, 393–394.

65 Lee, D. and Seung, H. (1999) Learning the parts of objects by non-negative matrix factorization. *Nature*, 401, 788–791.

66 Paatero, P. (1997) Least squares formulation of robust non-negative factor analysis. *Chemometrics and Intelligent Laboratory Systems*, 37, 23–35.

67 Lawson, C.L. and Hanson, R.J. (1974) *Solving Least-Squares Problems*, Prentice-Hall.

68 Bro, R. and De Jong, S. (1997) A fast non-negativity constrained least squares algorithm. *Journal of Chemometrics*, 11, 393–401.

69 de Juan, A., Heyden, Y.V., Tauler, R., and Massart, D.L. (1997) Assessment of new constraints applied to the alternating least squares method. *Analytica Chimica Acta*, 346, 307–318.

70 Bro, R. and Sidoropoulos, N. (1998) Least squares algorithms under unimodality and non-negativity constraints. *Journal of Chemometrics*, 12, 223–247.

71 Févotte, C. and Idier, J. (2011) Algorithms for nonnegative matrix factorization with the beta-divergence. *Neural Computation*, 23 (9), 2421–2456.

72 Robert, C. (2001) *The Bayesian Choice*, Springer-Verlag, 2nd edn.

73 Stuart, A. and Ord, J. (1994) *Kendall's Advanced Theory of Statistics*, vol. 1: Distribution Theory, John Wiley & Sons, 6th edn.

74 Roberts, S. (1998) Independent component analysis: source assessment and separation, a Bayesian approach. *IEE Proceedings on Vision, Image and Signal Processing*, 145 (3), 149–154.

75 Mohammad-Djafari, A. (1999) A Bayesian approach to source separation, in *Proceedings of the International Workshop on Bayesian Inference and*

Maximum Entropy Methods in Science and Engineering (MaxEnt'1999), pp. 36–46.

76 Ochs, M.F., Stoyanova, R.S., Arias-Mendoza, F., and Brown, T.R. (1999) A new method for spectral decomposition using a bilinear Bayesian approach. *Journal of Magnetic Resonance*, 137, 161–176.

77 Dobigeon, N., Moussaoui, S., Tourneret, J.Y., and Carteret, C. (2009) Bayesian separation of spectral sources under non-negativity and full additivity constraints. *Signal Processing*, 89 (12), 2657–2669.

78 Plaza, A., Martinez, P., Perez, R., and Plaza, J. (2004) A quantitative and comparative analysis of endmember extraction algorithms from hyperspectral data. *IEEE Transactions on Geoscience and Remote Sensing*, 42 (3), 650–663.

79 Nascimento, J.M.P. and Bioucas-Dias, J.M. (2005) Vertex component analysis: a fast algorithm to unmix hyperspectral data. *IEEE Transactions on Geoscience and Remote Sensing*, 43 (4), 898–910.

80 Craig, M. (1994) Minimum-volume transforms for remotely sensed data. *IEEE Transactions on Geoscience and Remote Sensing*, 32 (3), 542–552.

2

Optimization

Emilie Chouzenoux[1] *and Jean-Christophe Pesquet*[2]

[1] *OPIS, Inria Saclay, University Paris-Saclay, Gif-sur-Yvette, France*
[2] *CVN, CentraleSupélec, University Paris Saclay, France*

2.1 Introduction to Optimization Problems

2.1.1 Problem Formulation

In this chapter, we are interested in general problems involving the minimization of a cost function f with respect to a vector $\boldsymbol{u} \in \mathbb{R}^N$ with components $(u_n)_{1 \leq n \leq N}$ representing variables to estimate. In the context of source separation, the unknown vector can often be block-decomposed into $J \geq 1$ blocks having various roles and properties. More specifically, we will decompose $\boldsymbol{u}$ as $(\boldsymbol{u}^{(j)})_{1 \leq j \leq J} \in \mathbb{R}^{N_1} \times \cdots \times \mathbb{R}^{N_J}$, where, for every $j \in \{1, \ldots, J\}$, $\boldsymbol{u}^{(j)} = \left(u_n\right)_{n \in \mathbb{J}_j} \in \mathbb{R}^{N_j}$ and $\mathbb{J}_j$ is the index set of the components corresponding to the j-th block. With this notation, the optimization problem reads

$$\underset{\boldsymbol{u}^{(1)} \in \mathbb{R}^{N_1}, \ldots, \boldsymbol{u}^{(J)} \in \mathbb{R}^{N_J}}{\text{minimize}} \ f(\boldsymbol{u}). \tag{2.1}$$

Function f is usually expressed as a composite objective function consisting of a sum of terms: a data fidelity term describing the existing relations between the observed data and the target variables, and penalization (or regularization) terms modeling the available prior information on the sought variables. Various examples of possible choices for these functions in the context of source separation will be given in Section 2.1.3. In particular, one may want to introduce some constraints on the domain of each block, namely that, for every $j \in \{1, \ldots, J\}$, $\boldsymbol{u}^{(j)}$ belongs to some set $C_j \subset \mathbb{R}^{N_j}$. Problem (2.1) then becomes a constrained optimization problem

$$\underset{\boldsymbol{u}^{(1)} \in C_1, \ldots, \boldsymbol{u}^{(J)} \in C_J}{\text{minimize}} \ f(\boldsymbol{u}). \tag{2.2}$$

Source Separation in Physical-Chemical Sensing, First Edition.
Edited by Christian Jutten, Leonardo Tomazeli Duarte, and Saïd Moussaoui.

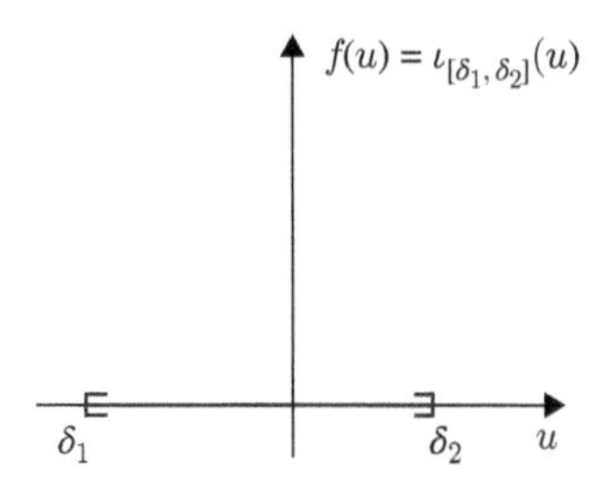

Figure 2.1 Indicator function of the interval $[\delta_1, \delta_2]$

It is important to note that Problem (2.2) actually is an instance of Problem (2.1). Indeed, it suffices to replace the cost function f in (2.1) by the sum $f + \iota_C$, where ι_C is the indicator function of set C:

$$(\forall \boldsymbol{u} \in \mathbb{R}^N) \quad \iota_C(\boldsymbol{u}) = \begin{cases} 0 & \text{if } \boldsymbol{u} \in C \\ +\infty & \text{elsewhere,} \end{cases} \tag{2.3}$$

and C gathers all the constraints on the block variables, i.e. C is the Cartesian product $C_1 \times \cdots \times C_J$ of the J constraint spaces. In a nutshell, allowing a cost function value to be equal to $+\infty$ over some set is a convenient way for constraining its minimizer to be out of this set (Figure 2.1).

With the exception of very specific cases (typically, fully quadratic problems), solving Problem (2.1) requires the design of an algorithm that builds a sequence of iterates $(\boldsymbol{u}_k)_{k\in\mathbb{N}}$ converging to a minimizer $\hat{\boldsymbol{u}}$ of f. There is in the literature a bunch of optimization methods, which differ by their convergence properties, their computational cost, their memory requirements, their convergence speed, and their sensitivity to numerical errors. The goal of this chapter is to help the reader to bring answers to the following questions:

- How to exploit the mathematical properties of each term involved in f? How to handle constraints efficiently? How to deal with non-differentiable terms in f? Which convergence result can be expected if f is non-convex?
- How to reduce the memory requirements of an optimization algorithm? How to avoid large-size matrix inversion?
- What are the benefits of block alternating strategies? What are their convergence guarantees?
- How to accelerate the convergence speed of a first-order (gradient-like) optimization method?

2.1.2 Theoretical Background

2.1.2.1 Convex Functions

Throughout this chapter, all the functions will be assumed to take their values in $\mathbb{R}$, possibly extended to $\mathbb{R} \cup \{+\infty\}$. The domain $\operatorname{dom} f$ of a function

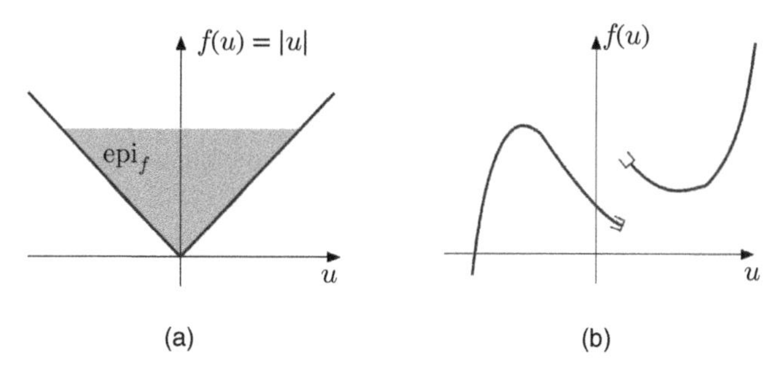

Figure 2.2 Non-lsc function f (a) ; lsc function f (b).

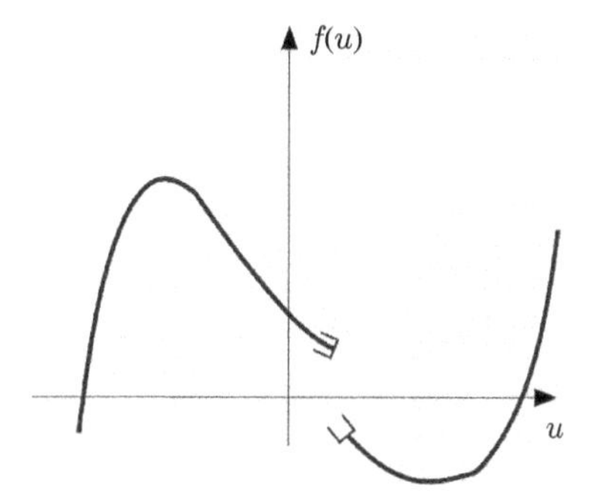

Figure 2.3 Epigraph of the absolute value function.

f from $\mathbb{R}^N$ to $\mathbb{R} \cup \{+\infty\}$ is the subset of $\mathbb{R}^N$ over which this function is finite-valued. When this domain is nonempty, it is said that the function is *proper*. In optimization, one is able to handle discontinuous functions, but most of the time, the involved functions must be assumed to be *lower-semicontinuous* (lsc). A function f from $\mathbb{R}^N$ to $\mathbb{R} \cup \{+\infty\}$ is lower-semicontinuous (Figure 2.2) if its epigraph,

$$\text{epi}(f) = \{(\boldsymbol{u}, \zeta) \in \mathbb{R}^N \times \mathbb{R} \mid f(\boldsymbol{u}) \leq \zeta\},$$

which basically represents the area above the graph of the function (Figure 2.3), is a closed set. It is then easy to check that continuous functions are lower-semicontinuous.

In the rest of this chapter, we will assume, without recalling it explicitly, that all the considered functions are proper and lower-semicontinuous.

Another desirable property of a cost function is its convexity. Recall that a set C is convex if any line segment linking two arbitrary points of C is included in C. A function f from $\mathbb{R}^N$ to $\mathbb{R} \cup \{+\infty\}$ is convex if its epigraph is convex, which is also equivalent to the following inequality:

$$\left(\forall(\boldsymbol{u},\boldsymbol{v})\in(\mathbb{R}^N)^2\right)(\forall\lambda\in]0,1[)\quad f(\lambda\boldsymbol{u}+(1-\lambda)\boldsymbol{v})\leq\lambda f(\boldsymbol{u})+(1-\lambda)f(\boldsymbol{v}).$$

In addition, the function is strictly convex if the above inequality is strict whenever $(\boldsymbol{u},\boldsymbol{v})\in(\operatorname{dom} f)^2$ and $\boldsymbol{u}\neq\boldsymbol{v}$. An important property of a convex function is that any of its local minimizers, i.e. minimizer on a local neighborhood in its domain, is a (global) minimizer over $\mathbb{R}^N$. If the function is strictly convex, such a minimizer, whenever it exists, is unique. These properties are important since many optimization algorithms are only guaranteed to converge to a local minimizer, which may lead to suboptimal performance in practice, when the associated cost function is non-convex.

2.1.2.2 Differentiability and Subdifferentiability

The subdifferential of a function f from $\mathbb{R}^N$ to $\mathbb{R}\cup\{+\infty\}$, at $\boldsymbol{u}\in\mathbb{R}^N$, is defined as

$$\partial f(\boldsymbol{u})=\{\boldsymbol{t}\in\mathbb{R}^N\mid(\forall\boldsymbol{v}\in\mathbb{R}^N)\ f(\boldsymbol{v})\geq f(\boldsymbol{u})+\boldsymbol{t}^{\top}(\boldsymbol{v}-\boldsymbol{u})\}.\tag{2.4}$$

Any vector $\boldsymbol{t}\in\partial f(\boldsymbol{u})$ is called a subgradient of f at $\boldsymbol{u}\in\mathbb{R}^N$. The above definition actually corresponds to the Moreau subdifferential, but other definitions are possible [1], which may be more relevant in the non-convex case. It is worth noting that the affine terms $f(\boldsymbol{u})+\boldsymbol{t}^{\top}(\boldsymbol{v}-\boldsymbol{u})$, $\boldsymbol{t}\in\partial f(\boldsymbol{u})$ actually correspond to a supporting hyperplane crossing the graph of f at $\boldsymbol{u}$, thus generalizing the notion of tangent plane to non-necessarily differentiable functions. In particular, a function may have several subgradients at a given point $\boldsymbol{u}\in\mathbb{R}^N$ or none (in which case, $\partial f(\boldsymbol{u})$ is empty). But, if f is convex and continuous at $\boldsymbol{u}\in\mathbb{R}^N$, then $\partial f(\boldsymbol{u})$ is nonempty. Furthermore, if f is differentiable at $\boldsymbol{u}\in\mathbb{R}^N$ and it is convex, its only subgradient at $\boldsymbol{u}\in\mathbb{R}^N$ is its gradient $\nabla f(\boldsymbol{u})$.

A key property of the Moreau subdifferential is that it allows the characterization of solutions to an optimization problem. Indeed, according to Fermat's rule, $\hat{\boldsymbol{u}}$ is a (global) minimizer of f if and only if $\boldsymbol{0}$ is a subgradient of f at $\hat{\boldsymbol{u}}$. However, if f is differentiable at $\hat{\boldsymbol{u}}$ and non-convex, the condition $\nabla f(\hat{\boldsymbol{u}})=\boldsymbol{0}$ only constitutes a necessary condition for $\hat{\boldsymbol{u}}$ to be a minimizer of f (Figure 2.4).

2.1.3 Examples in the Context of Source Separation

2.1.3.1 Non-negative Matrix Factorization

The problem in non-negative matrix factorization is to decompose a given, possibly noisy, data matrix $\boldsymbol{X}\in\mathbb{R}^{P\times T}$ into a product $\boldsymbol{AS}$, where $\boldsymbol{A}\in\mathbb{R}^{P\times R}$

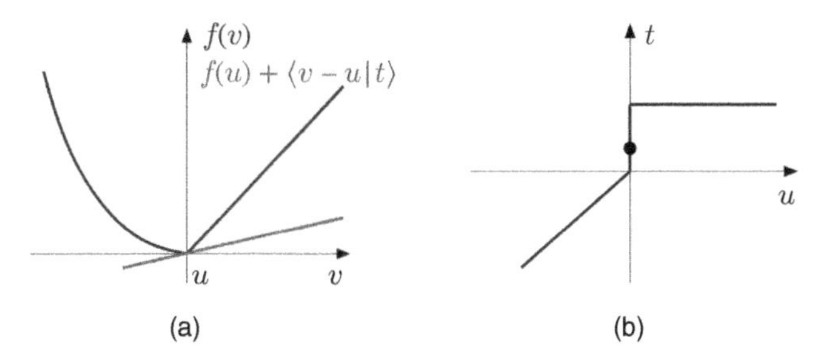

Figure 2.4 Function f and a supporting line of it at $u = 0$ (a). Graph of the subdifferential (b).

Table 2.1 Examples of discrepancy measures in NMF (with $\beta \in \mathbb{R}\backslash\{0,1\}$).

Name	Expression	References
Euclidean	$\frac{1}{2}\|\boldsymbol{X} - \boldsymbol{AS}\|_{\mathrm{F}}^2$	[2]
Kullback–Leibler	$\sum_{p=1}^{P}\sum_{t=1}^{T}(\boldsymbol{AS})_{p,t} - X_{p,t}\log((\boldsymbol{AS})_{p,t})$	[2]
Itakura–Saito	$\sum_{p=1}^{P}\sum_{t=1}^{T}X_{p,t}/(\boldsymbol{AS})_{p,t} + \log((\boldsymbol{AS})_{p,t})$	[3]
β-divergence	$\sum_{p=1}^{P}\sum_{t=1}^{T}\beta^{-1}(\boldsymbol{AS})_{p,t}^{\beta} - (\beta-1)^{-1}X_{p,t}(\boldsymbol{AS})_{p,t}^{\beta-1}$	[4]

and $\boldsymbol{S} \in \mathbb{R}^{R\times T}$ are constrained to have non-negative entries. A common strategy to solve this algebraic problem is to define $\boldsymbol{A}$ and $\boldsymbol{S}$ as the solutions of a constrained optimization problem of the form:

$$\underset{\boldsymbol{A}\geq 0,\boldsymbol{S}\geq 0}{\text{minimize}}\ \Phi(\boldsymbol{AS},\boldsymbol{X}) + \Theta_{\mathrm{a}}(\boldsymbol{A}) + \Theta_{\mathrm{s}}(\boldsymbol{S}), \tag{2.5}$$

where Φ measures the discrepancy between the data $\boldsymbol{X}$ and the sought product $\boldsymbol{AS}$ (see Table 2.1 for some examples), and Θ_{a} (resp. Θ_{s}) introduces some *a priori* information on the unknown matrices.

2.1.3.2 Independent Component Analysis

The problem of ICA consists in retrieving unknown statistically independent components of a vector $\boldsymbol{s}(t)$ which are mixed by an unknown linear operator $\boldsymbol{A}$, leading to an observed vector $\boldsymbol{x}(t)$, at time $t = 1, \ldots, T$. If $\boldsymbol{A}$ is an $R \times R$ invertible matrix, and $\boldsymbol{B} = \boldsymbol{A}^{-1}$ denotes the separating matrix, the likelihood of $\boldsymbol{B}$ is

$$\mathcal{L}(\boldsymbol{B}) = \prod_{t=1}^{T} p_{\boldsymbol{s}(t)}(\boldsymbol{Bx}(t))\,|\det \boldsymbol{B}|, \tag{2.6}$$

where $p_{\boldsymbol{s}(t)}$ is the probability density function of the source vector at time t. For simplicity, let us now assume that the vectors $(\boldsymbol{s}(t))_{1\leq t\leq T}$ are independent and identically distributed. The maximum likelihood estimator of $\boldsymbol{B}$ is then obtained by minimizing the neg-log-likelihood of $\boldsymbol{B}$, that is

$$f(\boldsymbol{B}) = \sum_{t=1}^{T} \sum_{r=1}^{R} \varphi_r(\boldsymbol{B}, \boldsymbol{x}(t)) - T \log|\det \boldsymbol{B}|, \tag{2.7}$$

where for every $r = 1, \ldots, R$, $\varphi_r(\boldsymbol{B}, \cdot) = -\log p_{s_r}\left([\boldsymbol{B}\cdot]_r\right)$ and p_{s_r} is the probability density function of the r-th component of the source vector. Note that, even if these probability density functions are log-concave, f is a non-convex function due to the last logarithmic term. In addition, several global minimizers may exist, especially when some of the source component probability distributions are equal (permutation ambiguity) or even (sign ambiguity). When the source distributions are unknown, a criterion similar to (2.7) can be minimized where some approximations to the probability distributions of the sources are employed [5].

A commonly used technique in ICA is to perform a prewhitening of the observations which aims at computing a matrix $\boldsymbol{W}$ such that, for every $t = 1, \ldots, T, \boldsymbol{y}(t) = \boldsymbol{W}\boldsymbol{x}(t)$ has an identity correlation matrix $\mathbb{E}\{\boldsymbol{y}(t)\boldsymbol{y}(t)^{\top}\}$. We have then $\boldsymbol{s}(t) = \boldsymbol{Q}\boldsymbol{y}(t)$ where $\boldsymbol{Q}$ is a matrix which remains to be estimated. If the sources are assumed to be zero-mean with unit variance, then this matrix is orthogonal and the neg-log-likelihood reduces to

$$\boldsymbol{Q} \mapsto \sum_{t=1}^{T} \sum_{r=1}^{R} \varphi_r(\boldsymbol{Q}, \boldsymbol{y}(t)) \tag{2.8}$$

This simplification comes at the expense of the orthogonality constraint which needs to be imposed on the sought matrix. The resulting cost function to be minimized on the space of $R \times R$ matrices reads

$$f(\boldsymbol{Q}) = \sum_{t=1}^{T} \sum_{r=1}^{R} \varphi_r(\boldsymbol{Q}, \boldsymbol{y}(t)) + \iota_{\mathcal{O}}(\boldsymbol{Q}), \tag{2.9}$$

where $\mathcal{O}$ is the non-convex closed set of orthogonal matrices.

2.1.3.3 Tensor Decomposition

Source separation problems can often be recast either directly, or after some preprocessing (e.g. by computing higher-order cumulants of the observations), under a tensorial form [6]. One important problem in such tensor analysis is the Canonical Polyadic Decomposition (CPD) which consists in decomposing a tensor $\mathcal{T}$ of order D in a sum of R one-rank tensors, i.e.

$$\mathcal{T} = \sum_{r=1}^{R} \boldsymbol{a}^{(r,1)} \otimes \cdots \otimes \boldsymbol{a}^{(r,d)}, \tag{2.10}$$

where R is the tensor rank and, for every $r = 1, \ldots, R$ and $d = 1, \ldots, D$, $\boldsymbol{a}^{(r,d)}$ is a P_d-dimensional vector. More generally, one may be interested in approximating a tensor $\mathcal{T}$ by a sum corresponding to the right-hand side summation in (2.10) where R is then less than or equal to the tensor rank. It is now known that this problem is ill-posed, and that it is mandatory to impose constraints [7] on the sought vectors $(\boldsymbol{a}^{(r,d)})_{1 \le r \le R, 1 \le d \le D}$, which are related to properties of the underlying physical model. For example, the components of these vectors can sometimes be assumed to belong to $\mathbb{R}_+$. Due to possible symmetry properties, some of these vectors can also be constrained to be equal. Finally, to limit indeterminacies, some of these vectors can be normalized by setting their norms to 1. All these constraints mean that $\boldsymbol{a} = (\boldsymbol{a}^{(r,d)})_{1 \le r \le R, 1 \le d \le D}$ has to belong to a given constraint set C. The resulting constrained CPD problem can therefore be reformulated as the minimization of

$$f(\boldsymbol{a}) = \Phi\left(\mathcal{T} - \sum_{r=1}^{R} \boldsymbol{a}^{(r,1)} \otimes \cdots \otimes \boldsymbol{a}^{(r,d)}\right) + \iota_C(\boldsymbol{a}), \tag{2.11}$$

where Φ is a given distance measure. The latter function is often chosen equal to the squared Frobenius norm. Because of the multilinearity of model (2.10), even if Φ is a convex function and C is a convex set, the resulting optimization problem is non-convex.

2.1.4 Chapter Outline

In Section 2.2, we will introduce the Majorization–Minimization (MM) principle which will be our main guideline for all the approaches presented in this chapter. General recipes for building a good surrogate majorant to a given cost function will be discussed. Then, we will focus on quadratic MM methods, which will be shown to cover a wide range of existing algorithms for minimizing both smooth functions (e.g. MM subspace algorithms) and non-smooth ones (e.g. the Variable Metric Forward-Backward algorithm).

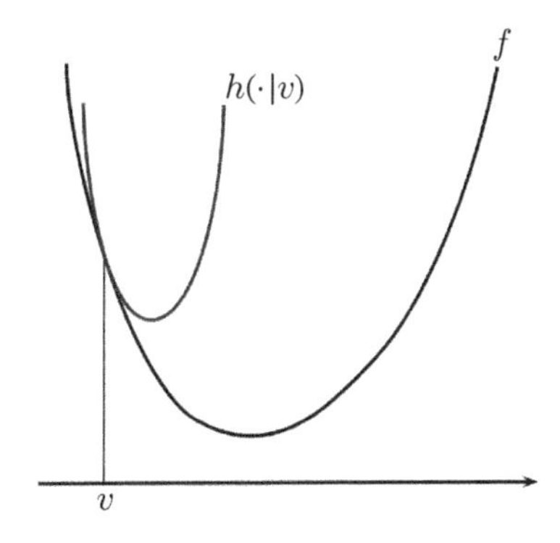

Figure 2.5 Graph of function f and a majorant function $h(\cdot \mid v)$ of f at v

In the last part of this section, we will present block-coordinate approaches which now play a prominent role in the solution of large-scale optimization problems. Section 2.3 will be dedicated to primal-dual algorithms which have gained much popularity in the last years. These methods will be introduced from the viewpoint of Lagrange duality. Both proximal approaches and interior point methods will be described. We conclude this chapter by presenting in Section 2.4 an example of signal restoration, arising in spectroscopy, and various optimization approaches to resolve it (Figure 2.5).

2.2 Majorization–Minimization Approaches

Majorization–Minimization (MM) [8], also known as optimization transfer [9], auxiliary function method [10], quadratic approximation [11], surrogate minimization [12], or successive upper-bound minimization [13], is at the core of many optimization algorithms for large-scale signal/image processing. In particular, it appears that most of the optimization methods that are useful to solve source separation problems (gradient descent schemes, NMF algorithms, proximal methods, etc.) can be rewritten under this generic framework, which motivates us for making a detailed focus on MM algorithms, along with their convergence guarantees, and their applicability on specific examples inspired from the context of source separation.

2.2.1 Majorization–Minimization Principle

The main idea behind MM methods lies in the concept of *majorant function*, defined below.

Definition 2.1 *Let* $f : \mathbb{R}^N \to \mathbb{R} \cup \{+\infty\}$. *Let* $\boldsymbol{v} \in \mathbb{R}^N$. $h(\cdot|\boldsymbol{v}) : \mathbb{R}^N \to \mathbb{R} \cup \{+\infty\}$ *is a majorant function of* f *at* $\boldsymbol{v}$ *if*

$$(\forall \boldsymbol{u} \in \mathbb{R}^N) \qquad f(\boldsymbol{u}) \leq h(\boldsymbol{u}|\boldsymbol{v}), \qquad \text{and} \qquad f(\boldsymbol{v}) = h(\boldsymbol{v}|\boldsymbol{v}).$$

The MM algorithm consists of solving the minimization problem

$$\underset{\boldsymbol{u}\in\mathbb{R}^N}{\text{minimize}}\, f(\boldsymbol{u}), \tag{2.12}$$

by alternating between two steps:

(1) **M**ajorize the criterion at current iterate with a *majorant function*,
(2) **M**inimize the majorant function to define the next iterate.

A simple instance of MM algorithm for minimizing function f over $\mathbb{R}^N$ reads:

$$\begin{aligned}&\text{For}\ \ k = 0, 1, \dots\\ &\left\lfloor \boldsymbol{u}_{k+1} \in \underset{\boldsymbol{u}\in\mathbb{R}^N}{\text{Argmin}}\ \ h(\boldsymbol{u}|\boldsymbol{u}_k)\right.\end{aligned} \tag{2.13}$$

where $h(\cdot|\boldsymbol{u}_k)$ is a majorant function for f at $\boldsymbol{u}_k$. By construction, we have the following inequalities:

$$(\forall k \in \mathbb{N})\quad f(\boldsymbol{u}_{k+1}) \underset{\text{Maj.}}{\leq} h(\boldsymbol{u}_{k+1}|\boldsymbol{u}_k) \underset{\text{Min.}}{\leq} h(\boldsymbol{u}_k|\boldsymbol{u}_k) = f(\boldsymbol{u}_k).$$

We thus see that the sequence $\big(f(\boldsymbol{u}_k)\big)_{k\in\mathbb{N}}$ is decreasing. The construction of an MM algorithm requires to define a strategy for building majorant functions, as well as a strategy for minimizing (or at least, decreasing) them, with the aim to guarantee the convergence of the sequence $(\boldsymbol{u}_k)_{k\in\mathbb{N}}$ to a solution to (2.12).

2.2.2 Majorization Techniques

In this section, we recall some important properties from [9, 12] that may be useful to construct a suitable majorant function for a given criterion f.

Property 2.1 *Let* $f_1 : \mathbb{R}^N \to \mathbb{R} \cup \{+\infty\}$, $f_2 : \mathbb{R}^N \to \mathbb{R} \cup \{+\infty\}$ *and* $\boldsymbol{v} \in \mathbb{R}^N$. *Assume that* $h_1(\cdot|\boldsymbol{v}) : \mathbb{R}^N \to \mathbb{R} \cup \{+\infty\}$ *is a majorant function of* f_1 *at* $\boldsymbol{v}$, *and* $h_2(\cdot|\boldsymbol{v}) : \mathbb{R}^N \to \mathbb{R} \cup \{+\infty\}$ *is a majorant function of* f_2 *at* $\boldsymbol{v}$. *Then,*

(i) $h_1(\cdot|\boldsymbol{v}) + h_2(\cdot|\boldsymbol{v})$ *is a majorant function of* $f_1 + f_2$ *at* $\boldsymbol{v}$.
(ii) If, for all $\boldsymbol{u} \in \mathbb{R}^N$, $f_1(\boldsymbol{u}) \geq 0$ *and* $f_2(\boldsymbol{u}) \geq 0$, *then* $h_1(\cdot|\boldsymbol{v})h_2(\cdot|\boldsymbol{v})$ *is a majorant function of* $f_1 f_2$ *at* $\boldsymbol{v}$.
(iii) If $\phi : \mathbb{R} \to \mathbb{R} \cup \{+\infty\}$ *is an increasing function, then* $\phi(h_1(\cdot|\boldsymbol{v}))$ *is a majorant function of* $\phi(f_1)$ *at* $\boldsymbol{v}$.

The definition of the subdifferential in (2.4) yields the following useful property.

Property 2.2 *Let* $f : \mathbb{R}^N \to \mathbb{R} \cup \{+\infty\}$ *and* $\boldsymbol{v} \in \mathbb{R}^N$. *If* f *is concave (i.e.* $-f$ *is convex) on* $\mathbb{R}^N$, *then a majorant function for* f *at* $\boldsymbol{v}$ *is*

$$(\forall \boldsymbol{u} \in \mathbb{R}^N)\quad h(\boldsymbol{u}|\boldsymbol{v}) = f(\boldsymbol{v}) + \boldsymbol{t}^\top(\boldsymbol{u} - \boldsymbol{v}),$$

where $(-\boldsymbol{t}) \in \partial(-f)(\boldsymbol{v})$, *where* ∂f *denotes the subdifferential of* f.

Note that the above Property 2.2 is at the core of DC (difference of convex) programming algorithms for minimizing the difference of two convex functions [14]. At each iteration of such methods, the concave part of the criterion is majorized at the current point by a linear function, while the convex part is majorized by a simpler convex (typically quadratic) function. The resulting majorant function is then minimized, according to an MM scheme.

Quadratic majorant functions can be derived by making use of second-order properties of f.

Property 2.3

(i) *Assume that f is β-Lipschitz differentiable on $\mathbb{R}^N$, i.e. there exists $\beta > 0$ such that*

$$(\forall(\boldsymbol{u},\boldsymbol{v}) \in \mathbb{R}^N \times \mathbb{R}^N) \quad \|\nabla f(\boldsymbol{v}) - \nabla f(\boldsymbol{u})\| \leq \beta\|\boldsymbol{u} - \boldsymbol{v}\|.$$

Then a majorant function for f at $\boldsymbol{v}$ is

$$(\forall \boldsymbol{u} \in \mathbb{R}^N) \quad h(\boldsymbol{u}|\boldsymbol{v}) = f(\boldsymbol{v}) + (\nabla f(\boldsymbol{v}))^\top(\boldsymbol{u} - \boldsymbol{v}) + \frac{\mu}{2}\|\boldsymbol{u} - \boldsymbol{v}\|^2,$$

where $\mu \in [\beta, +\infty[$.

(ii) *If f is twice differentiable on $\mathbb{R}^N$ with Hessian $\nabla^2 f$, then a majorant function for f at $\boldsymbol{v}$ is*

$$(\forall \boldsymbol{u} \in \mathbb{R}^N) \quad h(\boldsymbol{u}|\boldsymbol{v}) = f(\boldsymbol{v}) + (\nabla f(\boldsymbol{v}))^\top(\boldsymbol{u} - \boldsymbol{v}) + \frac{1}{2}\|\boldsymbol{u} - \boldsymbol{v}\|_{\boldsymbol{A}}^2,$$

where $\boldsymbol{A} \in \mathbb{R}^{N\times N}$ is a positive semidefinite matrix such that, for every $\boldsymbol{u} \in \mathbb{R}^N$, $\boldsymbol{A} - \nabla^2 f(\boldsymbol{u})$ is positive semidefinite and $\|\boldsymbol{u}\|_{\boldsymbol{A}} = \sqrt{\boldsymbol{u}^\top\boldsymbol{A}\boldsymbol{u}}$ denotes the weighted Euclidean norm of $\boldsymbol{u}$ within the metric induced by $\boldsymbol{A}$.

Example 2.4 *Consider Problem (2.12), where f is β-Lipschitz differentiable on $\mathbb{R}^N$. Then, Property 2.3(i) holds, and, for this particular choice of majorizing function (by setting $\mu = \gamma^{-1}$), the MM algorithm (2.13) reads:*

$$\begin{array}{l} \boldsymbol{u}_0 \in \mathbb{R}^N \\ \gamma \in]0, \beta^{-1}[\\ \text{For } k = 0, 1, \ldots \\ \lfloor\ \boldsymbol{u}_{k+1} = \boldsymbol{u}_k - \gamma\nabla f(\boldsymbol{u}_k) \end{array} \tag{2.14}$$

The above algorithm identifies with the steepest descent algorithm *(or* gradient algorithm*) with fixed stepsize. Its convergence properties can be found in [15], for the over-relaxed case when $\gamma \in]0, 2\beta^{-1}[$. The advantage of this method is its simplicity, but this may come at the expense of a slow convergence rate. The convergence speed of the steepest descent algorithm can be much improved by introducing a variable stepsize associated to a linesearch strategy [15, Chap.3]. More sophisticated strategies, based on preconditioning, or subspace acceleration will be discussed in Section 2.2.3.*

Example 2.5 *Consider Problem (2.12), where f is a convex lsc function on $\mathbb{R}^N$. For all $\boldsymbol{v} \in \mathbb{R}^N$, a strongly convex majorant function of f can be derived by setting*

$$(\forall \boldsymbol{u} \in \mathbb{R}^N) \quad h(\boldsymbol{u}|\boldsymbol{v}) = f(\boldsymbol{u}) + \frac{1}{2}\|\boldsymbol{u} - \boldsymbol{v}\|_{\boldsymbol{A}}^2,$$

with $\boldsymbol{A} \in \mathbb{R}^{N\times N}$ a positive definite matrix. The minimizer of the later majorant function is unique and defines the so-called proximity operator *of f at $\boldsymbol{v}$ within the metric induced by $\boldsymbol{A}$:*

$$\mathrm{prox}_f^{\boldsymbol{A}}(\boldsymbol{v}) = \underset{\boldsymbol{u}\in\mathbb{R}^N}{\mathrm{argmin}}\, f(\boldsymbol{u}) + \frac{1}{2}\|\boldsymbol{u} - \boldsymbol{v}\|_{\boldsymbol{A}}^2 \tag{2.15}$$

with the simplified notation $\mathrm{prox}_f^{\boldsymbol{I}} \equiv \mathrm{prox}_f$. The proximity operator benefits from many interesting properties [16].[1] *In particular, if $\hat{\boldsymbol{u}}$ is a minimizer of f, we have the fixed-point relation:*

$$\hat{\boldsymbol{u}} = \mathrm{prox}_f^{\boldsymbol{A}}(\hat{\boldsymbol{u}}),$$

which is at the core of the convergence analysis of the proximal point algorithm *given below:*

$$\begin{array}{l} \boldsymbol{u}_0 \in \mathbb{R}^N \\ \textit{For}\;\; k = 0, 1, \ldots \\ \left\lfloor \;\boldsymbol{u}_{k+1} = \mathrm{prox}_f^{\boldsymbol{A}}(\boldsymbol{u}_k). \right. \end{array} \tag{2.16}$$

Let $\boldsymbol{A} = \gamma^{-1}\boldsymbol{I}$, with $\gamma \in]0, +\infty[$. According to the definition of the proximity operator, the main iteration (2.16) can be rewritten as:

$$\boldsymbol{u}_{k+1} = \boldsymbol{u}_k - \gamma \boldsymbol{t}_k, \qquad \text{with} \qquad \boldsymbol{t}_k \in \partial f(\boldsymbol{u}_{k+1}), \tag{2.17}$$

which (2.17) can be viewed as an implicit subgradient scheme. The main advantage of this implicit scheme is that it allows the stepsize γ to be constant along the iterations, whereas the stepsize is required to decrease to zero in the case of the standard (explicit) subgradient algorithm [17].

Example 2.6 *Let us consider the resolution of Problem (2.12) where*

$$(\forall \boldsymbol{u} \in \mathbb{R}^N) \quad f(\boldsymbol{u}) = \frac{1}{2}\|\boldsymbol{H}\boldsymbol{u} - \boldsymbol{z}\|_2^2 + \lambda \sum_{n=1}^{N} \log(\eta + |u_n|) \tag{2.18}$$

with $\boldsymbol{H} \in \mathbb{R}^{M\times N}$, $\boldsymbol{z} \in]0, +\infty[^M$, $(\lambda, \eta) \in]0, +\infty[^2$. Function $u \mapsto \log(\eta + u)$ is concave on $[0, +\infty[$, so according to Property 2.2, for every $\boldsymbol{v} \in \mathbb{R}^N$, a tangent majorant function of (2.18) is

$$(\forall \boldsymbol{u} \in \mathbb{R}^N) \quad h(\boldsymbol{u}|\boldsymbol{v}) = \frac{1}{2}\|\boldsymbol{H}\boldsymbol{u} - \boldsymbol{z}\|_2^2 + \lambda \sum_{n=1}^{N} \frac{|u_n|}{\eta + |v_n|} + C(\boldsymbol{v}) \tag{2.19}$$

1 See also http://proximity-operator.net

where $C(\boldsymbol{v})$ is a constant such that $h(\boldsymbol{v}|\boldsymbol{v}) = f(\boldsymbol{v})$. The application of the MM algorithm to the majorizing approximation (2.19) leads to the so-called iterative reweighted ℓ_1 (IRL1) algorithm for sparse signal reconstruction [18–21]. This name comes from the fact that function (2.19) reads as a least squares term, penalized by a weighted ℓ_1 norm, whose weights depend on the entries of $\boldsymbol{v}$.

The following property is at the core of iterative reweighted least-squares algorithms [22–26], and of half quadratic methods [27–33] for the minimization of penalized quadratic functions.

Property 2.7 *Let $f : \mathbb{R} \to \mathbb{R}$ be a continuous function such that, for every $u \in \mathbb{R}$, $f(u) = \psi(|u|)$, where:*

(i) ψ is differentiable on $]0, +\infty[$,
(ii) $\psi(\sqrt{\cdot})$ is concave on $]0, +\infty[$,
(iii) ψ is increasing on $]0, +\infty[$.

Then, for all $v \in \mathbb{R}^*$,

$$(\forall u \in \mathbb{R}) \quad f(u) \le f(v) + \dot{f}(v)(u - v) + \frac{1}{2}\omega(|v|)(u - v)^2$$

where, for every $\xi \in]0, +\infty[$, $\omega(\xi) = \dot{\psi}(\xi)/\xi$ and $\dot{f}$ denotes the derivative of the scalar function f.

Example 2.8 *Let us focus on the search for a minimizer on $\mathbb{R}^N$ of the non-smooth function $f : \mathbb{R}^N \to \mathbb{R}$ defined as:*

$$(\forall \boldsymbol{u} \in \mathbb{R}^N) \quad f(\boldsymbol{u}) = \sum_{s=1}^{S} \|\boldsymbol{L}_s \boldsymbol{u} - \mathbf{c}_s\| \tag{2.20}$$

with $(\forall s \in \{1, \dots, S\})$ $\boldsymbol{L}_s \in \mathbb{R}^{P_s \times N}$, $\mathbf{c}_s \in \mathbb{R}^{P_s}$. Applying Property 2.7 with f the absolute value function, yields, for every $v \in \mathbb{R}^*$,

$$(\forall u \in \mathbb{R}) \quad |u| \le |v| + \mathrm{sign}(v)(u - v) + \frac{1}{2|v|}(u - v)^2 = \frac{u^2}{2|v|} + \frac{|v|}{2}.$$

A direct consequence is that, for all $\boldsymbol{v}$ belonging to the set

$$C = \left\{ \boldsymbol{v} \in \mathbb{R}^N \mid (\forall s \in \{1, \dots, S\})\ \boldsymbol{L}_s \boldsymbol{v} - \mathbf{c}_s \neq 0 \right\},$$

the quadratic function

$$(\forall \boldsymbol{u} \in \mathbb{R}^N) \quad h(\boldsymbol{u}|\boldsymbol{v}) = \frac{1}{2} \sum_{s=1}^{S} \frac{\|\boldsymbol{L}_s \boldsymbol{u} - \mathbf{c}_s\|^2}{\|\boldsymbol{L}_s \boldsymbol{v} - \mathbf{c}_s\|} + \|\boldsymbol{L}_s \boldsymbol{v} - \mathbf{c}_s\|,$$

is a majorant function for f at $\boldsymbol{v}$. This leads to an MM algorithm for minimizing f usually known as the iterative reweighted least-squares *(IRLS) algorithm. A difficulty is that IRLS needs to be stopped as soon as an iterate does not belong to C (since the majorant function would not be defined anymore). This constitutes a difficulty to prove the convergence of the method. In particular, one must ensure that the algorithm does not stop before reaching the convergence. An analysis of IRLS convergence properties is available in specific cases, namely if for every $s \in \{1, \dots, S\}$, $\boldsymbol{L}_s = \boldsymbol{I}$, IRLS identifying with the Weiszfeld algorithm [24, 34], or when for every $s \in \{1, \dots, S\}$, $P_s = 1$, $\boldsymbol{L}_s = \boldsymbol{e}_s^\top$, with $\boldsymbol{e}_s$ the s-th canonical basis vector of $\mathbb{R}^N$, and f contains an additional quadratic term [22]. Note that, even if the iterates $(\boldsymbol{u}_k)_{k\in\mathbb{N}}$ stay in C, numerical issues may arise if, for some k, some component of $(\|\boldsymbol{L}_s\boldsymbol{u}_k - \boldsymbol{c}_s\|)_{1\le s\le S}$ becomes close to zero. A common strategy adopted in the literature to avoid this problem is to replace the expression of the majorant function, by a smooth approximation of it, defined on the whole space $\mathbb{R}^N$. For instance, [13, 35, 36] define, for every $\boldsymbol{v} \in \mathbb{R}^N$, the smooth majorant function*

$$(\forall \boldsymbol{u} \in \mathbb{R}^N) \quad \tilde{h}(\boldsymbol{u}|\boldsymbol{v}) = \frac{1}{2}\sum_{s=1}^{S} \frac{\|\boldsymbol{L}_s\boldsymbol{u} - \boldsymbol{c}_s\|^2}{\sqrt{\|\boldsymbol{L}_s\boldsymbol{v} - \boldsymbol{c}_s\|^2 + \eta^2}} + C(\boldsymbol{v}),$$

where $C(\boldsymbol{v})$ is constant with respect to $\boldsymbol{u}$ such that $f(\boldsymbol{v}) = \tilde{h}(\boldsymbol{v}|\boldsymbol{v})$, and $\eta > 0$ is a smoothing parameter. It can be easily shown that this modified IRLS algorithm identifies with an MM algorithm for minimizing a smooth approximation to (2.20), given by

$$(\forall \boldsymbol{u} \in \mathbb{R}^N) \quad \tilde{f}(\boldsymbol{u}) = \sum_{s=1}^{S} \sqrt{\|\boldsymbol{L}_s\boldsymbol{u} - \boldsymbol{c}_s\|^2 + \eta^2} - \eta.$$

In practice, parameter η is chosen close to zero, so that the minimizers of f and $\tilde{f}$ almost coincide, while numerical issues are avoided.

We finish this section by stating a property resulting from Jensen's inequality which is recalled below:

Lemma 2.9 *Let $\psi : \mathbb{R} \to \mathbb{R} \cup \{+\infty\}$ be a convex function and let $\omega = (\omega_n)_{1\le n\le N} \in [0, +\infty[^N$ be such that $\sum_{n=1}^N \omega_n = 1$. Then,*

$$(\forall (u_1, \dots, u_N) \in \mathbb{R}^N) \quad \psi\left(\sum_{n=1}^{N} \omega_n u_n\right) \le \sum_{n=1}^{N} \omega_n \psi(u_n).$$

Property 2.10 *Let $\psi : \mathbb{R} \to \mathbb{R} \cup \{+\infty\}$ be a convex function.*

(i) $\left(\forall (\boldsymbol{u}, \boldsymbol{v}, \boldsymbol{c}) \in (]0, +\infty[^N)^3\right) \quad \psi(\boldsymbol{c}^\top\boldsymbol{u}) \le \sum_{n=1}^N \frac{c_n v_n}{\boldsymbol{c}^\top\boldsymbol{v}} \psi\left(\frac{\boldsymbol{c}^\top\boldsymbol{v}}{v_n} u_n\right).$

(ii) Let $\boldsymbol{\omega} \in]0,+\infty[^N$ *such that* $\sum_{n=1}^{N} \omega_n = 1$. *We have*

$$(\forall(\boldsymbol{u},\boldsymbol{v},\boldsymbol{c}) \in \left(]-\infty,+\infty[^N\right)^3) \quad \psi(\boldsymbol{c}^\top \boldsymbol{u}) \leq \sum_{n=1}^{N} \omega_n \psi\left(\tfrac{c_n}{\omega_n}(u_n - v_n) + \boldsymbol{c}^\top \boldsymbol{v}\right).$$

This property is useful in the derivation of several optimization algorithms in the context of signal/image restoration.

Example 2.11 *Let us consider the resolution of Problem (2.12) where*

$$(\forall \boldsymbol{u} \in \mathbb{R}^N) \quad f(\boldsymbol{u}) = \sum_{m=1}^{M} \psi((\boldsymbol{H}\boldsymbol{u})_m; z_m), \tag{2.21}$$

$\boldsymbol{H} \in \mathbb{R}^{M\times N}$, $\boldsymbol{z} \in [0,+\infty[^M$, *and* ψ *is the Kullback-Leibler divergence defined as*

$$(\forall u \in \mathbb{R})(\forall z \in [0,+\infty[) \quad \psi(u;z) = \begin{cases} u - z\log(u), & \text{if } \ u > 0 \text{ and } z > 0 \\ 0 & \text{if } \ u = 0 \text{ and } z = 0 \\ +\infty, & \text{elsewhere.} \end{cases} \tag{2.22}$$

Function f is convex but its minimizers have no closed form expression. A strategy is then to build a majorant function of f, which is simple to minimize. The Richardson–Lucy algorithm [37] consists in applying the MM method (2.13) where the majorant function is given by

$$(\forall \boldsymbol{u} \in]0,+\infty[^N) \quad h(\boldsymbol{u}|\boldsymbol{v}) = \sum_{n=1}^{N} a_n u_n - b_n(\boldsymbol{v})\log(u_n) + c_n(\boldsymbol{v}),$$

with, for every $n \in \{1,\dots,N\}$,

$$\begin{aligned} a_n &= [\boldsymbol{H}^\top \mathbf{1}]_n \\ b_n(\boldsymbol{v}) &= [\boldsymbol{H}^\top(\boldsymbol{z} \oslash \boldsymbol{H}\boldsymbol{v})]_n v_n \\ c_n(\boldsymbol{v}) &= [\boldsymbol{H}^\top(\boldsymbol{z} \boxdot \log(\boldsymbol{H}\boldsymbol{v}) \oslash \boldsymbol{H}\boldsymbol{v})]_n v_n - [\boldsymbol{H}^\top(\boldsymbol{z} \oslash \boldsymbol{H}\boldsymbol{v})]_n v_n \log(v_n) \end{aligned} \tag{2.23}$$

(consequence of Property 2.10(i)). Hereabove, $\boxdot$ *(resp.* $\oslash$*) denotes the component-wise product (resp. division) between matrices of the same size. This majorant function is easy to optimize since it is separable with respect to* $\boldsymbol{u}$. *Its minimization leads to the following multiplicative update rule:*

$$\begin{array}{l} \boldsymbol{u}_0 \in]0,+\infty[^N \\ \textit{For } \ k = 0,1,\dots \\ \left\lfloor \begin{array}{l} \textit{For } \ n = 1,2,\dots,N \\ \left\lfloor u_{n,k+1} = u_{n,k} \dfrac{[\boldsymbol{H}^\top(\boldsymbol{z} \oslash \boldsymbol{H}\boldsymbol{u}_k)]_n}{a_n}. \right. \end{array} \right. \end{array} \tag{2.24}$$

The convergence of the iterations (2.24) has been studied for instance in [38]. Note that extensions of the Richardson–Lucy algorithm to the case of a penalized KL criterion are available (see, for instance, [39]).

2.2.3 Quadratic MM Methods

The numerical efficiency of the MM algorithm relies on the use of simple majorant functions providing good approximations to f and whose minimizer is simple to compute. An intuitive choice is thus to choose quadratic majorant functions, which leads to the MM quadratic algorithm [11]. Half-quadratic methods [30–33] are a particular case of this method.

2.2.3.1 Quadratic MM Algorithm

Assume that, for every $\boldsymbol{v} \in \mathbb{R}^N$, there exists a positive definite symmetric matrix $\boldsymbol{A}(\boldsymbol{v}) \in \mathbb{R}^{N\times N}$ such that the quadratic function

$$(\forall \boldsymbol{u} \in \mathbb{R}^N) \quad h(\boldsymbol{u}|\boldsymbol{v}) = f(\boldsymbol{v}) + (\boldsymbol{u}-\boldsymbol{v})^\top \nabla f(\boldsymbol{v}) + \frac{1}{2}\|\boldsymbol{u}-\boldsymbol{v}\|^2_{\boldsymbol{A}(\boldsymbol{v})} \tag{2.25}$$

is a majorant function of f at $\boldsymbol{v}$. The MM quadratic algorithm then reads:

$$\begin{array}{l} \boldsymbol{u}_0 \in \mathbb{R}^N \\ \text{For } k = 0, 1, \ldots \\ \left\lfloor \begin{array}{l} \theta_k \in]0, 2[\\ \boldsymbol{u}_{k+1} = \boldsymbol{u}_k - \theta_k \boldsymbol{A}(\boldsymbol{u}_k)^{-1} \nabla f(\boldsymbol{u}_k). \end{array} \right. \end{array} \tag{2.26}$$

The parameters $(\theta_k)_{k\in\mathbb{N}}$ act as stepsizes. For $\theta_k \equiv 1$, we recover the basic MM algorithm. An illustration of the role of the stepsize is illustrated in the scalar case in Figure 2.6. The simple Algorithm (2.26), which can be viewed as a preconditioned gradient algorithm, for the specific class of preconditioners $(\boldsymbol{A}(\boldsymbol{u}_k)^{-1})_{k\in\mathbb{N}}$ [40, Sec.1.3], enjoys the following convergence properties:

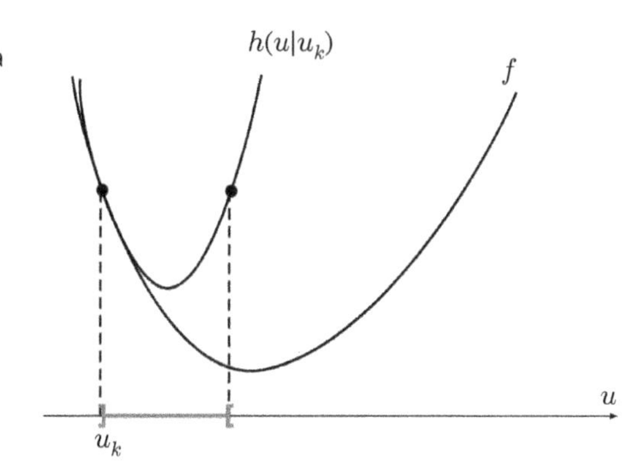

Figure 2.6 MM quadratic algorithm. Function f and a quadratic majorizing function $h(\cdot|u_k)$. The gray interval depicts associated values for u_{k+1} for $\theta_k \in]0, 2[$. The choice $\theta_k = 1$ corresponds to choose u_{k+1} as the minimizer of $h(\cdot|u_k)$

Theorem 2.12 *Let $f : \mathbb{R}^N \to \mathbb{R}$ be a differentiable function such that $\lim_{\|\boldsymbol{u}\| \to +\infty} f(\boldsymbol{u}) = +\infty$. Assume that there exists $(\underline{\nu}, \overline{\nu}) \in]0, +\infty[^2$ such that $(\forall k \in \mathbb{N})$ $\underline{\nu} \boldsymbol{I} \preceq \boldsymbol{A}(\boldsymbol{u}_k) \preceq \overline{\nu} \boldsymbol{I}$,[2] and $(\underline{\theta}, \overline{\theta}) \in]0, +\infty[^2$ such that, $(\forall k \in \mathbb{N})$ $\underline{\theta} \leq \theta_k \leq 2 - \overline{\theta}$. Then, the following statements hold:*

(1) As $k \to +\infty$, $\nabla f(\boldsymbol{u}_k) \to 0$ and $f(\boldsymbol{u}_k) \searrow f(\tilde{\boldsymbol{u}})$,[3] where $\tilde{\boldsymbol{u}}$ is a critical point of f.
(2) If f is convex, any sequential cluster point of $(\boldsymbol{u}_k)_{k \in \mathbb{N}}$ is a minimizer of f.
(3) If f is strictly convex, then $\boldsymbol{u}_k \to \hat{\boldsymbol{u}}$ where $\hat{\boldsymbol{u}}$ is the unique minimizer of f.

Example 2.13 *Algorithm (2.26) has been applied to the problem of chromatogram baseline correction and noise reduction in [41]. Chromatogram measurements $\boldsymbol{y} \in \mathbb{R}^N$ are modeled as $\boldsymbol{y} = \boldsymbol{u} + \boldsymbol{v} + \boldsymbol{w}$ with $\boldsymbol{u} \in \mathbb{R}^N$ the unknown signal of interest, $\boldsymbol{v} \in \mathbb{R}^N$ an unknown smooth background also called baseline, and $\boldsymbol{w} \in \mathbb{R}^N$ a noise term. The goal is to retrieve $\boldsymbol{u}$ under some prior assumptions on $\boldsymbol{u}$ and $\boldsymbol{v}$. In [41], the authors proposed the following penalized cost function*

$$(\forall \boldsymbol{u} \in \mathbb{R}^N) \quad f(\boldsymbol{u}) = \frac{1}{2} \|\boldsymbol{H}(\boldsymbol{y} - \boldsymbol{u})\|^2 + \sum_{i=0}^{M} \lambda_i \phi(\boldsymbol{D}_i \boldsymbol{u}), \tag{2.27}$$

with $\boldsymbol{H}$ a suitable high-pass filter, and, for every $i = 0, \ldots, M$, $\boldsymbol{D}_i$ the i-th order difference operator, $\lambda_i > 0$ and ϕ a smooth even potential function. The minimization of f is performed by Algorithm (2.26), leading to so-called Baseline Estimation And Denoising with Sparsity (BEADS) method. An illustration of result obtained by BEADS on real chromatogram data is displayed in Figure 2.7.

Example 2.14 *Let us consider the resolution of Problem (2.12) where*

$$(\forall \boldsymbol{u} \in \mathbb{R}^N) \quad f(\boldsymbol{u}) = \frac{1}{2} \|\boldsymbol{H}\boldsymbol{u} - \boldsymbol{z}\|^2, \tag{2.28}$$

$\boldsymbol{H} \in \mathbb{R}^{M \times N}$, $\boldsymbol{z} \in \mathbb{R}^M$. According to Property 2.10(ii), for every $\boldsymbol{v} \in \mathbb{R}^N$, a separable majorant function of f at $\boldsymbol{v}$ is:

$$(\forall \boldsymbol{u} \in \mathbb{R}^N) \quad h(\boldsymbol{u}|\boldsymbol{v}) = f(\boldsymbol{v}) + \nabla f(\boldsymbol{v})^\top (\boldsymbol{u} - \boldsymbol{v}) + \frac{1}{2} \sum_{n=1}^{N} a_n (u_n - v_n)^2,$$

with, for every $n \in \{1, \ldots, N\}$,

$$a_n = [|\boldsymbol{H}|^\top |\boldsymbol{H}| \; \boldsymbol{1}]_n,$$

---

2 We use $\preceq$ to define the Loewner partial ordering on $\mathbb{R}^{N \times N}$: For every $(\boldsymbol{A}, \boldsymbol{B}) \in \mathbb{R}^{N \times N} \times \mathbb{R}^{N \times N}$, $\boldsymbol{A} \preceq \boldsymbol{B}$ if and only if, for every $\boldsymbol{u} \in \mathbb{R}^N$, $\boldsymbol{u}^\top \boldsymbol{A} \boldsymbol{u} \leq \boldsymbol{u}^\top \boldsymbol{B} \boldsymbol{u}$.
3 We use the shorter notation $\searrow$ to indicate monotonically decreasing convergence.

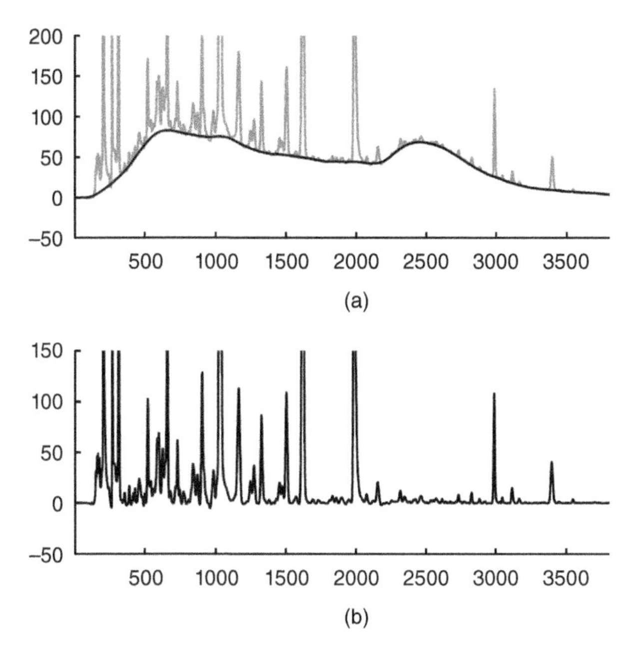

Figure 2.7 Baseline correction using BEADS. Baselines estimated in (a) and estimated peaks (obtained by subtraction of estimated baseline from data) in (b).

where $|\boldsymbol{H}|$ *denotes a matrix with same size than* $\boldsymbol{H}$*, whose entries are equal to the absolute value of those of* $\boldsymbol{H}$*. In that case, Algorithm (2.30) becomes equivalent to the* simultaneous algebraic reconstruction technique *(SART) [42] for tomography reconstruction:*

$$
\begin{array}{l}
\boldsymbol{u}_0 \in \mathbb{R}^N \\
\textit{For} \;\; k = 0, 1, \ldots \\
\left\lfloor \begin{array}{l}
\textit{For} \;\; n = 1, 2, \ldots, N \\
\left\lfloor u_{n,k+1} = u_{n,k} - \frac{[\boldsymbol{H}^\top(\boldsymbol{H}\boldsymbol{u}_k - \boldsymbol{z})]_n}{a_n}. \right.
\end{array} \right.
\end{array}
\tag{2.29}
$$

Note that the separable form of the quadratic majorant function in that case leads to a fully parallelizable algorithm, in the sense that each component of $\boldsymbol{u}$ *is updated in a parallel manner. Generalizations of the SART method to the minimization of a smoothed regularized least squares criterion can be found for instance in [43, 44].*

Remark 2.1 In the context of large-scale optimization, the exact inversion of the majorant matrix involved at each iteration of Algorithm (2.26) can be computationally intensive. In practice, one can instead solve approximately the associated linear system using a linear solver, such as the conjugate gradient algorithm. A nice feature of Algorithm (2.26) is that its convergence properties are not modified if we replace the update by

$$\begin{aligned}&\text{For } k = 0, 1, \ldots\\&\left\lfloor\begin{array}{l}\theta_k \in]0, 2[\\ \boldsymbol{u}_{k+1} = \boldsymbol{u}_k - \theta_k \boldsymbol{d}_k.\end{array}\right.\end{aligned} \tag{2.30}$$

where $\boldsymbol{d}_k$ results from $J_k \geq 1$ iterations of a conjugate gradient algorithm (possibly preconditioned) applied to the system $\boldsymbol{A}(\boldsymbol{u}_k)\boldsymbol{d} = \nabla f(\boldsymbol{u}_k)$ [45]. Surprisingly, the best performance in practice for the modified Algorithm (2.30) is achieved for small values of J_k (typically, $J_k = 5$).

Remark 2.2 Let us consider the minimization of a twice differentiable, strictly convex function $f : \mathbb{R}^N \to \mathbb{R}$, by using the *Newton algorithm*:

$$\begin{aligned}&\text{For } k = 0, 1, \ldots\\&\lfloor\ \boldsymbol{u}_{k+1} = \boldsymbol{u}_k - (\nabla^2 f(\boldsymbol{u}_k))^{-1} \nabla f(\boldsymbol{u}_k).\end{aligned} \tag{2.31}$$

The strict convexity of f is however not sufficient to guarantee the convergence of Algorithm (2.31), nor to ensure the monotonicity of the sequence $(f(\boldsymbol{u}_k))_{k\in\mathbb{N}}$, which makes the Newton method with unit stepsize inadequate for general use [11]. The convergence of the algorithm can actually be ensured, according to Theorem 2.12, if, for every $\boldsymbol{u} \in \mathbb{R}^N$, the following majorization holds:

$$f(\boldsymbol{u}) \leq f(\boldsymbol{v}) + (\boldsymbol{u} - \boldsymbol{v})^\top \nabla f(\boldsymbol{v}) + \frac{1}{2} \|\boldsymbol{u} - \boldsymbol{v}\|^2_{\nabla^2 f(\boldsymbol{v})}. \tag{2.32}$$

However, this assumption is rarely satisfied in practice. Three practical strategies to secure the convergence of Algorithm (2.31) are discussed in [15, Chap.6], [40, Sec.1.4]: (i) the introduction of a backtracking linesearch ensuring that the Newton step satisfies some sufficient decrease condition, (ii) the addition of a definite positive matrix, possibly varying along the iterations, to the Hessian matrix in order to satisfy a majorizing condition similar to (2.32), or (iii) the combination of the previous technique with a trust region approach that check the majorization inequality only within a neighborhood of the current iterate.

2.2.3.2 Half-Quadratic MM Algorithms

Half-quadratic methods [27–33] can be interpreted as MM quadratic algorithms for minimizing on $\mathbb{R}^N$ the following class of functions:

$$(\forall \boldsymbol{u} \in \mathbb{R}^N) \quad f(\boldsymbol{u}) = \frac{1}{2} \|\boldsymbol{H}\boldsymbol{u} - \boldsymbol{z}\|^2 + \sum_{s=1}^{S} \psi(\|\boldsymbol{L}_s \boldsymbol{u} - \boldsymbol{c}_s\|) \tag{2.33}$$

with $\boldsymbol{H} \in \mathbb{R}^{M\times N}$, $\boldsymbol{z} \in \mathbb{R}^{M}$, $\psi : \mathbb{R} \mapsto \mathbb{R}$, and $(\forall s \in \{1,\dots,S\})$ $\boldsymbol{L}_s \in \mathbb{R}^{N\times P_s}$, $\boldsymbol{c}_s \in \mathbb{R}^{P_s}$. Two particular choices for matrix $\boldsymbol{A}(\cdot)$ for the majorant defined (2.25) are employed in this context, depending on the assumptions on ψ.

The first strategy [31] relies on the assumption that ψ has a β-Lipschitzian gradient on $\mathbb{R}$. Let us define

$$(\forall \boldsymbol{v} \in \mathbb{R}^N) \quad \boldsymbol{A}(\boldsymbol{v}) = \boldsymbol{H}^\top\boldsymbol{H} + \mu\boldsymbol{L}^\top\boldsymbol{L}, \quad \mu \geq \beta, \tag{2.34}$$

with $\boldsymbol{L} = \left[\boldsymbol{L}_1^\top \mid \dots \mid \boldsymbol{L}_S^\top\right]^\top \in \mathbb{R}^{P\times N}$, $P = \sum_{s=1}^{S} P_s$. Then, according to Property 2.3(i), (2.25) is a majorant function of (2.33).

The second strategy [30] assumes that ψ is differentiable on $]0,+\infty[$, $\psi(\sqrt{\cdot})$ is concave, $(\forall x \in [0,+\infty[)$, $\dot{\psi}(x) \geq 0$, and $\lim_{\substack{x\to 0\\ x>0}} \left(\omega(x) = \frac{\dot{\psi}(x)}{x}\right) \in \mathbb{R}$. Then, Property 2.7 implies that (2.25) is a majorant function of (2.33), if

$$(\forall \boldsymbol{v} \in \mathbb{R}^N) \quad \boldsymbol{A}(\boldsymbol{v}) = \boldsymbol{H}^\top\boldsymbol{H} + \boldsymbol{L}^\top \mathrm{Diag}\left(\left(\omega(\|\boldsymbol{L}_s\boldsymbol{v} - \boldsymbol{c}_s\|)\boldsymbol{1}\right)_{1\leq s\leq S}\right)\boldsymbol{L}. \tag{2.35}$$

The later approach leads to a better convergence rate in theory [27, 33], but its computational cost may be higher. Indeed, in this case, the majorant matrix (2.35), as well as its inverse, must be recomputed at each iteration, which can be prohibitive in the context of large-scale problems. Figure 2.8 illustrates on a simple scalar example the two different approaches for half-quadratic majorant construction.

The term *half-quadratic* stems from the fact that the application of the MM quadratic algorithm to the minimization of (2.33), with majorant matrix (2.34) or (2.35) can be written as an alternating minimization algorithm:

$$\begin{aligned}
&\text{For } k = 0,1,\dots\\
&\left\lfloor
\begin{array}{l}
\boldsymbol{b}_k \in \underset{\boldsymbol{b}\in D}{\mathrm{Argmin}}\ \Phi(\boldsymbol{u},\boldsymbol{b})\\
\overline{\boldsymbol{u}}_k \in \underset{\boldsymbol{u}\in\mathbb{R}^N}{\mathrm{Argmin}}\ \Phi(\boldsymbol{u},\boldsymbol{b})\\
\boldsymbol{u}_{k+1} = \theta_k\overline{\boldsymbol{u}}_k + (1-\theta_k)\boldsymbol{u}_k, \qquad \theta_k \in]0,2[,
\end{array}
\right.
\end{aligned} \tag{2.36}$$

where $\Phi : \mathbb{R}^N \times D \to (\mathbb{R}\cup\{+\infty\}) \times (\mathbb{R}\cup\{+\infty\})$ is such that for every $\boldsymbol{b} \in D$, $\boldsymbol{u} \mapsto \Phi(\boldsymbol{u},\boldsymbol{b})$ is a *quadratic* function.

2.2.3.3 Subspace Acceleration Strategy

In the context of large-scale optimization, the minimization of a quadratic majorant function at each iteration may become untractable. The main idea of subspace acceleration is to restrict the minimization space to a subspace spanned by a small number of vectors, instead of minimizing the majorant

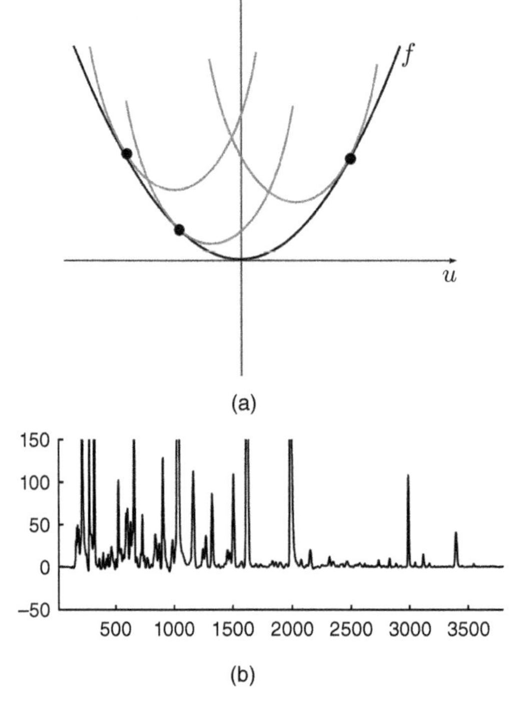

Figure 2.8 Half-quadratic majorant function. Function $f(u) = \sqrt{1 + u^2}$ and quadratic majorizing functions $h(\cdot|v)$ for different values of the tangency point v, built according to the strategy (2.34) (a) and (2.35) (b). All the majorizing functions obtained from (2.34) share the same curvature, corresponding to the maximum curvature of f (here, obtained at $u = 0$). In contrast, the majorizing functions resulting from (2.35) are even functions, with curvature depending on the tangency point v.

over the whole space $\mathbb{R}^N$. The general form of an MM subspace algorithm is thus

$$(\forall k \in \mathbb{N}) \quad \boldsymbol{u}_{k+1} \in \underset{\boldsymbol{u} \in \operatorname{span}\left\{ \boldsymbol{d}_k^1, \boldsymbol{d}_k^2, \ldots, \boldsymbol{d}_k^{M_k} \right\}}{\operatorname{Argmin}} \quad h(\boldsymbol{u}|\boldsymbol{u}_k),$$

where, for every $k \in \mathbb{N}$, $M_k \geq 1$. Equivalently, we obtain:

$$\begin{array}{l} \boldsymbol{u}_0 \in \mathbb{R}^N \\ \text{For } k = 0, 1, \ldots \\ \left\lfloor \begin{array}{l} \text{Choose } \boldsymbol{D}_k \in \mathbb{R}^{N \times M_k}, \\ \boldsymbol{s}_k \in \underset{\boldsymbol{s} \in \mathbb{R}^{M_k}}{\operatorname{Argmin}} \; h\left(\boldsymbol{u}_k + \boldsymbol{D}_k \boldsymbol{s} \middle| \, \boldsymbol{u}_k \right), \\ \boldsymbol{u}_{k+1} = \boldsymbol{u}_k + \boldsymbol{D}_k \boldsymbol{s}_k, \end{array} \right. \end{array} \tag{2.37}$$

where, for every $k \in \mathbb{N}$, $\boldsymbol{D}_k = \left[\boldsymbol{d}_k^1 \mid \boldsymbol{d}_k^2 \mid \ldots \mid \boldsymbol{d}_k^{M_k} \right] \in \mathbb{R}^{N \times M_k}$ gathers some search directions (see [46, Tab.1] for examples) and $\boldsymbol{s}_k \in \mathbb{R}^{M_k}$ is a multi-dimensional stepsize allowing to optimally combined these directions. By adopting a subspace spanned by a small number of directions, one can expect to reduce the computational cost of the algorithm.

Assume that, for every $\boldsymbol{v} \in \mathbb{R}^N$, there exists a strongly positive self-adjoint operator $\boldsymbol{A}(\boldsymbol{v})$ such that the quadratic function (2.25) is a majorant function of f at $\boldsymbol{v}$. The MM subspace quadratic algorithm reads as follows, where $\dagger$ denotes the pseudo-inverse operation:

$$
\begin{aligned}
&\boldsymbol{u}_0 \in \mathbb{R}^N \\
&\text{For } k = 0, 1, \ldots \\
&\left\lfloor
\begin{aligned}
&\text{Choose } \boldsymbol{D}_k \in \mathbb{R}^{M_k}, \\
&\boldsymbol{s}_k = -\left(\boldsymbol{D}_k^\top \boldsymbol{A}(\boldsymbol{u}_k)\boldsymbol{D}_k\right)^\dagger \left(\boldsymbol{D}_k^\top \nabla f(\boldsymbol{u}_k)\right), \\
&\boldsymbol{u}_{k+1} = \boldsymbol{u}_k + \boldsymbol{D}_k \boldsymbol{s}_k.
\end{aligned}
\right.
\end{aligned}
\tag{2.38}
$$

A convergence result of Algorithm (2.38) similar to Theorem 2.12 has been established in [46], under the assumption that, for every $k \in \mathbb{N}$, the gradient direction $\nabla f(\boldsymbol{u}_k)$ belongs to the vector subspace spanned by the columns of $\boldsymbol{D}_k$. This result has been extended to the case of non-convex functions [47], where the convergence of the iterates of Algorithm (2.38) to a critical point of the objective function is proved. Note that Algorithm (2.38) has been extended to the online case when only a stochastic approximation of the criterion is employed at each iteration [48, 49].

The MM subspace algorithm shows very good performance in practice, when compared with nonlinear conjugate gradient algorithms [50], low memory BFGS strategy [51], and also with graph-cut based discrete optimization methods, as well as proximal algorithms [46, 52, 53]. All the related works illustrate the fact that the choice of the subspace has a major impact on the practical convergence speed of Algorithm (2.38) (see, for instance [46, Section 5], [47, Section 5.1]). In particular, it seems that the worse performance is obtained in the case of a gradient-like algorithm, i.e. when, for every $k \in \mathbb{N}$, $M_k = 1$ and $\boldsymbol{D}_k = -\nabla f(\boldsymbol{u}_k)$. On the opposite, when the search subspace is the full space, Algorithm (2.38) becomes equivalent to the MM quadratic method mentioned in Section 2.2.3.1. This method can converge in a very small number of iterations, but each of them requires the inversion of an $N \times N$ matrix (see (2.26)), which may have a high computational cost. A good compromise is obtained for the memory gradient subspace [54], spanned by the current gradient and the previous direction, leading to the so-called MM Memory Gradient (3MG) algorithm. These practical observations are confirmed by the theoretical analysis of the rate of convergence of Algorithm (2.38) [55]. Extension of the method to the resolution of constrained optimization problems is presented in [56].

2.2.4 Variable Metric Forward–Backward Algorithm

Let us now focus on the minimization of $f = f_1 + f_2$ with f_1 differentiable and f_2 convex non-necessarily differentiable (for instance, a regularization term enforcing sparsity). Assume that, for every $\boldsymbol{v} \in \mathbb{R}^N$, there exists a positive definite matrix $\boldsymbol{A}(\boldsymbol{v}) \in \mathbb{R}^{N\times N}$ such that the quadratic function

$$(\forall \boldsymbol{u} \in \mathbb{R}^N) \quad h(\boldsymbol{u}|\boldsymbol{v}) = f_1(\boldsymbol{v}) + \nabla f_1(\boldsymbol{v})^\top(\boldsymbol{u} - \boldsymbol{v}) + \frac{1}{2}\|\boldsymbol{u} - \boldsymbol{v}\|^2_{\boldsymbol{A}(\boldsymbol{v})}$$

is a majorant function of f_1 at $\boldsymbol{v}$. The variable metric forward–backward (VMFB) algorithm [57, 58] reads

$$\begin{array}{l} \boldsymbol{u}_0 \in \mathbb{R}^N \\ \text{For } k = 0, 1, \ldots \\ \left\lfloor \begin{array}{l} \theta_k \in]0, 2[, \\ \boldsymbol{u}_{k+1} = \operatorname{prox}_{f_2}^{\theta_k^{-1}\boldsymbol{A}(\boldsymbol{u}_k)} \left(\boldsymbol{u}_k - \theta_k \boldsymbol{A}(\boldsymbol{u}_k)^{-1}\nabla f_1(\boldsymbol{u}_k)\right), \end{array} \right. \end{array} \tag{2.39}$$

where the proximity operator has been defined in (2.15). Algorithm (2.39) can actually be viewed as a relaxed form of an MM algorithm. Let $\theta_k \equiv 1$. According to the definition of the proximity operator,

$$\begin{aligned} \boldsymbol{u}_{k+1} &= \underset{\boldsymbol{u}\in\mathbb{R}^N}{\operatorname{argmin}} \ \frac{1}{2}\|\boldsymbol{u} - \boldsymbol{u}_k + \boldsymbol{A}(\boldsymbol{u}_k)^{-1}\nabla f_1(\boldsymbol{u}_k)\|^2_{\boldsymbol{A}(\boldsymbol{u}_k)} + f_2(\boldsymbol{u}) \\ &= \underset{\boldsymbol{u}\in\mathbb{R}^N}{\operatorname{argmin}} \ (\boldsymbol{u} - \boldsymbol{u}_k)^\top \nabla f_1(\boldsymbol{u}_k) + \frac{1}{2}\|\boldsymbol{u} - \boldsymbol{u}_k\|^2_{\boldsymbol{A}(\boldsymbol{u}_k)} + f_2(\boldsymbol{u}) \\ &= \underset{\boldsymbol{u}\in\mathbb{R}^N}{\operatorname{argmin}} \ h(\boldsymbol{u}|\boldsymbol{u}_k) + f_2(\boldsymbol{u}), \end{aligned}$$

which identifies with an MM algorithm to minimize f.

A convergence result similar to Theorem 2.12 can be stated for the VMFB algorithm, under the assumption that f_2 is convex, with non-empty domain, and continuous on its domain. An extension of the convergence properties of VMFB in the context of non-convex optimization can be found in [58].

Remark 2.3 Assume that f_1 is β-Lipschitz differentiable. According to Property 2.3(i), a valid choice is $\boldsymbol{A}(\boldsymbol{u}_k) \equiv \beta^{-1}\boldsymbol{I}$. In that case, VMFB algorithm becomes equivalent to the forward–backward algorithm [59] given below

$$\begin{array}{l} \boldsymbol{u}_0 \in \mathbb{R}^N \\ \text{For } k = 0, 1, \ldots \\ \left\lfloor \begin{array}{l} \theta_k \in]0, 2\beta^{-1}[, \\ \boldsymbol{u}_{k+1} = \operatorname{prox}_{\theta_k f_2}\left(\boldsymbol{u}_k - \theta_k \nabla f_1(\boldsymbol{u}_k)\right). \end{array} \right. \end{array} \tag{2.40}$$

The iterative soft thresholding algorithm (ISTA) [60] is a special case of this algorithm in the context of sparse signal restoration when f_2 is a (possibly weighted) ℓ_1 norm.

Remark 2.4 There exist plenty of properties for the proximity operator, so that a number of functions have a closed form expression for their prox [16].[4]

4 See also the web repository http://proximity-operator.net/.

Moreover, if f is a separable function, i.e.

$$(\forall \boldsymbol{u} \in \mathbb{R}^N) \quad f(\boldsymbol{u}) = \sum_{n=1}^{N} f_n(u_n), \tag{2.41}$$

with $(f_n)_{1 \leq n \leq N}$ functions defined on $\mathbb{R}$, then

$$\text{prox}_f^{\text{Diag}\{\boldsymbol{a}\}}(\boldsymbol{u}) = \left(\text{prox}_{a_n^{-1} f_n}(u_n)\right)_{1 \leq n \leq N}. \tag{2.42}$$

Note that the later property is the cornerstone of the so-called *separable surrogate functionals* algorithm proposed in [43] in the context of sparse signal reconstruction, which identifies with a VMFB algorithm with a diagonal metric.

When the form of the function and/or the metric is more involved, one must resort to an iterative strategy in order to compute the proximity step. Attention must be paid to this problem since the overall computation cost of the optimization method becomes then strongly dependent on the efficiency of the subiterations implemented for computing the proximity operator. A very efficient algorithm for performing the inner loop is the dual forward–backward algorithm [61] (sometimes also known as the dual ascent algorithm) which simply consists in applying the above forward–backward algorithm to the dual of the proximal problem. Note that several recent works [62–65] have proposed block-alternating, parallel, or distributed versions of this method, in order to reduce its computational and memory cost in the context of large-scale optimization problems.

Example 2.15 *Assume that*

$$(\forall \boldsymbol{u} \in \mathbb{R}^N) \quad f_2(\boldsymbol{u}) = \iota_C(\boldsymbol{u}), \tag{2.43}$$

where C is a closed non-empty convex subset of $\mathbb{R}^N$. Then, Algorithm (2.39) becomes equivalent to the scaled projected gradient algorithm

$$\begin{array}{l} \boldsymbol{u}_0 \in \mathbb{R}^N \\ \text{For } k = 0, 1, \ldots \\ \left\lfloor \begin{array}{l} \theta_k \in]0, 2[, \\ \tilde{\boldsymbol{u}}_k = \boldsymbol{u}_k - \theta_k \boldsymbol{A}(\boldsymbol{u}_k)^{-1} \nabla f_1(\boldsymbol{u}_k), \\ \boldsymbol{u}_{k+1} = \underset{\boldsymbol{v} \in C}{\text{argmin}} \ \|\boldsymbol{v} - \tilde{\boldsymbol{u}}_k\|_{\boldsymbol{A}(\boldsymbol{u}_k)}. \end{array} \right. \end{array} \tag{2.44}$$

The convergence of Algorithm (2.44) has been studied, for instance, in [66, 67] under various assumptions on the metric matrices and the stepsize values. Note that the standard projected gradient [68] is recovered if the preconditioning matrices are scaled versions of the identity matrix.

Example 2.16 *The recent work [69, 70] presents an application of the variable metric forward–backward approach in the context of mass spectrometry (MS). MS is a fundamental technology of analytical chemistry. In the context of protein analysis, the MS spectra* $\boldsymbol{y} \in \mathbb{R}^M$ *can be written as* $\boldsymbol{y} = \boldsymbol{K}\boldsymbol{u} + \boldsymbol{w}$ *with* $\boldsymbol{K} \in \mathbb{R}^{M \times N}$ *the MS averaging dictionary,* $\boldsymbol{u} \in \mathbb{R}^N$ *a sparse positive-valued signal to be estimated, and* $\boldsymbol{w}$ *a noise term. The dictionary size* N *is simply equal to* $N = MZ$*, with* M *the grid size for the mass and* Z *the number of charges to be explored. The estimation of* $\boldsymbol{u}$ *can be performed by minimizing* $f = f_1 + f_2$ *with* f_1 *a smooth term favoring the sparsity of* $\boldsymbol{u}$ *and* $f_2(\boldsymbol{u}) = \iota_C(\boldsymbol{u})$ *with*

$$C = \left\{\boldsymbol{u} \in \mathbb{R}^N \mid \|\boldsymbol{K}\boldsymbol{u} - \boldsymbol{y}\| \leq \xi \text{ and } \boldsymbol{u} \geq 0\right\} \tag{2.45}$$

where $\xi > 0$ *depends of the noise level. The authors in [69, 70] introduced the so-called SPOQ ((Smoothed p-Over-q) penalty:*

$$f_1(\boldsymbol{u}) = \log\left(\frac{\left(\ell_{p,\alpha}^p(\boldsymbol{u}) + \beta^p\right)^{1/p}}{\ell_{q,\eta}(\boldsymbol{u})}\right), \tag{2.46}$$

where $\beta > 0$, $\ell_{p,\alpha}$ *and* $\ell_{q,\eta}$ *are smoothed versions of* ℓ_p *and* ℓ_q *norms parametrized by* $(\alpha, \eta) \in]0, +\infty[^2$*. SPOQ penalty consists of a non-convex Lipschitz differentiable surrogate for* ℓ_p*-over-*ℓ_q *quasi-norm/norm ratios, with the ability to enforce sparsity property while preserving the signal scale. An accelerated version of Algorithm (2.39) is then proposed to minimize efficiently the resulting non-convex non-smooth objective function* f*, where the majorization condition is relaxed, using a trust-region based strategy. Figure 2.9 illustrates the results obtained when using SPOQ procedure to restore synthetic MS spectra.*

2.2.5 Block-Coordinate MM Algorithms

2.2.5.1 General Principle

As emphasized in Section 2.1, most of the objective functions encountered in the context of source separation involve blocks of variables (very often, two blocks) with different significance from the application standpoint, which makes the resulting optimization problems intrinsically adapted to block alternating minimization methods. Such methods can also be very appealing in the context of large-scale problems when the whole variable cannot be stored entirely in the local memory of the machine, so that only its update subsets can be loaded and updated at each iteration of the algorithm. The aim of this section is to show how MM algorithms can be combined with block alternating schemes.

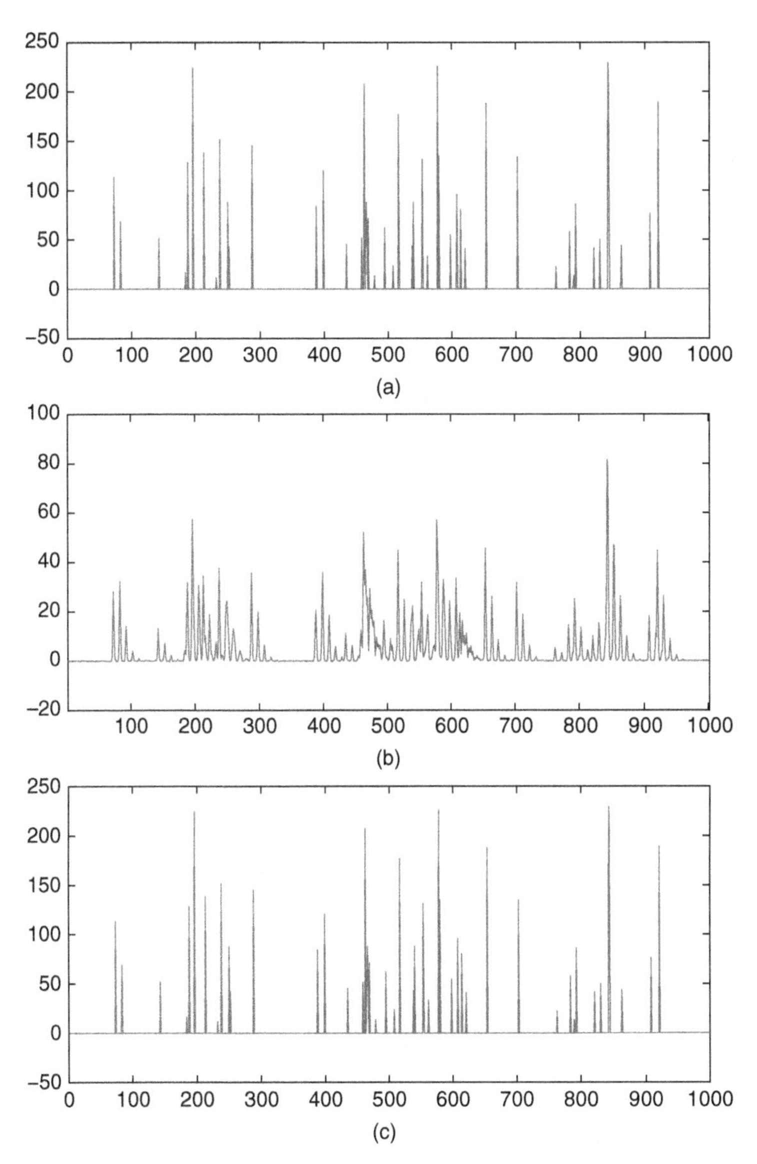

Figure 2.9 Original sparse signal (a), observed measurements (b), and restored signal (c) using SPOQ with $p = 0.75$ and $q = 2$, for a synthetic dataset with $M = 1000$ and mono-charged profile, i.e. $Z = 1$.

For every $\boldsymbol{v} \in \mathbb{R}^N$, for every $j \in \{1, \dots, J\}$, assume that there exists a majorant function $h_j(\cdot|\boldsymbol{v}) : \mathbb{R}^{N_j} \to \mathbb{R}$, which majorizes the restriction of f to its j-th block at $\boldsymbol{v}^{(j)}$, i.e. $h_j(\boldsymbol{v}^{(j)}|\boldsymbol{v}) = f(\boldsymbol{v})$ and,

$$(\forall \boldsymbol{s} \in \mathbb{R}^{N_j}) \qquad h_j(\boldsymbol{s}|\boldsymbol{v}) \leq f(\boldsymbol{v}^{(1)}, \dots, \boldsymbol{v}^{(j-1)}, \boldsymbol{s}, \boldsymbol{v}^{(j+1)}, \dots, \boldsymbol{v}^{(J)}).$$

Then, a block-coordinate MM algorithm for minimizing f reads:

$$\begin{array}{l} \boldsymbol{u}_0 \in \mathbb{R}^N \\ \text{For } k = 0, 1, \dots \\ \left\lfloor \begin{array}{l} \text{Select } j_k \in \{1, \dots, J\}, \\ \boldsymbol{u}_{k+1}^{(j_k)} = \underset{\boldsymbol{s} \in \mathbb{R}^{N_{j_k}}}{\operatorname{argmin}} \; h_j(\boldsymbol{s}|\boldsymbol{u}_k), \\ \boldsymbol{u}_{k+1}^{(\bar{j}_k)} = \boldsymbol{u}_k^{(\bar{j}_k)}. \end{array} \right. \end{array} \tag{2.47}$$

where $\bar{j}_k = \{1, \dots, J\} \setminus \{j_k\}$. When $J = 1$, one recovers the standard MM algorithm (2.13). For $J > 1$, at each iteration $k \in \mathbb{N}$, $j_k \in \{1, \dots, J\}$ can be chosen according to:

- the *cyclic* rule:

$$(\forall k \in \mathbb{N}) \quad j_k - 1 = k \bmod (J),$$

- a *quasi-cyclic* rule:
  There exists a constant $K \geq J$ such that, for every $k \in \mathbb{N}$,

$$\{1, \dots, J\} \subset \{j_k, \dots, j_{k+K-1}\},$$

- a *random* rule:
  For every $k \in \mathbb{N}$, j_k is a realization of a random variable, for example uniformly distributed on $\{1, \dots, J\}$.

The convergence properties of the general scheme (2.47), also known as *block successive upper-bound minimization* [13], have been studied in [71, 72] under various assumptions on the function f, on the majorant function sequence, and on the block selection rule. Discussions regarding the practical implementation of Algorithm (2.47) can be found for instance in [73] in the context of image reconstruction and in [12] in the context of machine learning.

2.2.5.2 Block-Coordinate Quadratic MM Algorithm

Let us now describe a block-coordinate version of the quadratic MM algorithm (2.26). For every $\boldsymbol{v} \in \mathbb{R}^N$, for every $j \in \{1, \dots, J\}$, assume that there exists a symmetric positive definite matrix $\boldsymbol{A}_j(\boldsymbol{v}) \in \mathbb{R}^{N_j \times N_j}$ such that the quadratic function

$$(\forall \boldsymbol{s} \in \mathbb{R}^{N_j}) \; h_j(\boldsymbol{s}|\boldsymbol{v}) = f(\boldsymbol{v}) + (\boldsymbol{s} - \boldsymbol{v}^{(j)})^\top \nabla_j f(\boldsymbol{v}) + \frac{1}{2} \|\boldsymbol{s} - \boldsymbol{v}^{(j)}\|^2_{\boldsymbol{A}_j(\boldsymbol{v})}$$

is a majorant function at $\boldsymbol{v}^{(j)}$ of the restriction of f to its j-th block. Then, the block-coordinate MM quadratic algorithm reads:

$$\left\lfloor\begin{array}{l}\boldsymbol{u}_0 \in \mathbb{R}^N \\ \text{For } k = 0, 1, \dots \\ \quad \text{Select } j_k \in \{1, \dots, J\}, \\ \quad \theta_k \in]0, 2[, \\ \quad \boldsymbol{u}_{k+1}^{(j_k)} = \boldsymbol{u}_k^{(j_k)} - \theta_k \boldsymbol{A}_{j_k}(\boldsymbol{u}_k)^{-1} \nabla_{j_k} f(\boldsymbol{u}_k), \\ \quad \boldsymbol{u}_{k+1}^{(\overline{j}_k)} = \boldsymbol{u}_k^{(\overline{j}_k)}. \end{array}\right. \tag{2.48}$$

For $J = 1$, we recover Algorithm (2.26). For $J > 1$, the convergence properties of Algorithm (2.48), also known as *coordinate-ascent paraboloidal surrogate* method [73], depend on the block selection rule. For example, in the case of a deterministic quasi-cyclic rule, a convergence result similar to Theorem 2.12 holds, under the assumption that there exists $(\underline{\nu}, \overline{\nu}) \in]0, +\infty[^2$ such that $(\forall k \in \mathbb{N})\ \underline{\nu}\boldsymbol{I} \preceq \boldsymbol{A}_{j_k}(\boldsymbol{u}_k) \preceq \overline{\nu}\boldsymbol{I}$.

Example 2.17 *Let us consider the problem of minimizing the following function:*

$$(\forall \boldsymbol{u} \in \mathbb{R}^N) \quad f(\boldsymbol{u}) = \frac{1}{2}\|\boldsymbol{H}\boldsymbol{u} - \boldsymbol{z}\|^2 + \sum_{n=1}^{N} \sqrt{1 + u_n^2}, \tag{2.49}$$

with $\boldsymbol{H} \in \mathbb{R}^{M \times N}$ and $\boldsymbol{z} \in \mathbb{R}^M$. According to Property (2.7), for every $\boldsymbol{v} \in \mathbb{R}^N$,

$$(\forall \boldsymbol{u} \in \mathbb{R}^N) \quad f(\boldsymbol{u}) \leq f(\boldsymbol{v}) + \nabla f(\boldsymbol{v})^\top(\boldsymbol{u} - \boldsymbol{v}) + \frac{1}{2}(\boldsymbol{u} - \boldsymbol{v})^\top \boldsymbol{A}(\boldsymbol{v})(\boldsymbol{u} - \boldsymbol{v}), \tag{2.50}$$

with

$$(\forall \boldsymbol{v} \in \mathbb{R}^N) \quad \boldsymbol{A}(\boldsymbol{v}) = \boldsymbol{H}^\top \boldsymbol{H} + \text{Diag}\{(1 + v_n^2)_{1 \leq n \leq N}\}^{-1/2}. \tag{2.51}$$

Consequently, for every $\boldsymbol{v} \in \mathbb{R}^N$, a majorant approximation of the restriction of f to the n-th component of $\boldsymbol{u}$, i.e. $u \mapsto f([v_1, \dots, v_{n-1}, u, v_n, \dots, v_N]^\top)$, is:

$$(\forall \boldsymbol{u} \in \mathbb{R}^N) \quad h_n(u_n | \boldsymbol{v}) = f(\boldsymbol{v}) + \nabla_n f(\boldsymbol{v})(u_n - v_n) + \frac{a_n(\boldsymbol{v})}{2}(u_n - v_n)^2, \tag{2.52}$$

with

$$(\forall \boldsymbol{v} \in \mathbb{R}^N)(\forall n \in \{1, \dots, N\}) \quad a_n(\boldsymbol{v}) = [\boldsymbol{H}^\top \boldsymbol{H}]_{n,n} + (1 + v_n^2)^{-1/2}, \tag{2.53}$$

and $\nabla_n f(\boldsymbol{v})$ the n-th component of $\nabla f(\boldsymbol{u})$. We can deduce the following alternating algorithm for minimizing f, referred to as the coordinate ascent with parabola surrogate *in [73]:*

$$\left\lfloor\begin{array}{l}\boldsymbol{u}_0 \in \mathbb{R}^N \\ \textit{For } k = 0, 1, \ldots \\ \left\lfloor\begin{array}{l} \text{Select } n \in \{1, \ldots, N\}, \\ \theta_k \in]0, 2[, \\ u_{n,k+1} = u_{n,k} - \theta_k a_n(\boldsymbol{u}_k)^{-1} \nabla_n f(\boldsymbol{u}_k), \\ (\forall n' \in \{1, \ldots, N\} \setminus \{n\}) \quad u_{n',k+1} = u_{n',k}. \end{array}\right.\end{array}\right. \tag{2.54}$$

2.2.5.3 Block-Coordinate VMFB Algorithm

The VMFB algorithm (2.39) can also be extended to a block alternating version of it. Consider the minimization of $f = f_1 + f_2$ with f_1 differentiable and f_2 convex non-necessarily differentiable. Assume additionally that f_2 is block-separable, i.e.

$$\left(\forall \boldsymbol{u} = (\boldsymbol{u}^{(j)})_{1 \leq j \leq J}\right) \qquad f_2(\boldsymbol{u}) = \sum_{j=1}^{J} f_{2,j}(\boldsymbol{u}^{(j)}).$$

For every $\boldsymbol{v} \in \mathbb{R}^N$, for every $j \in \{1, \ldots, J\}$, assume that there exists a symmetric positive definite matrix $\boldsymbol{A}_j(\boldsymbol{v}) \in \mathbb{R}^{N_j \times N_j}$ such that the quadratic function

$$(\forall \boldsymbol{u}^{(j)} \in \mathbb{R}^{N_j})\ h_j(\boldsymbol{u}^{(j)} | \boldsymbol{v}) = f_1(\boldsymbol{v}) + (\boldsymbol{u}^{(j)} - \boldsymbol{v}^{(j)})^\top \nabla_j f_1(\boldsymbol{v}) + \frac{1}{2} \|\boldsymbol{u}^{(j)} - \boldsymbol{v}^{(j)}\|^2_{\boldsymbol{A}_j(\boldsymbol{v})}$$

is a majorant function at $\boldsymbol{v}^{(j)}$ of the restriction of f_1 to its j-th block. Then, the block-coordinate VMFB (BC-VMFB) algorithm reads:

$$\left\lfloor\begin{array}{l}\boldsymbol{u}_0 \in \mathbb{R}^N \\ \text{For } k = 0, 1, \ldots \\ \left\lfloor\begin{array}{l} \text{Select } j_k \in \{1, \ldots, J\}, \\ \theta_k \in]0, 2[, \\ \boldsymbol{u}^{(j_k)}_{k+1} = \operatorname{prox}^{\theta_k^{-1} \boldsymbol{A}_{j_k}(\boldsymbol{u}_k)}_{f_{2,j_k}} \left(\boldsymbol{u}^{(j_k)}_k - \theta_k \boldsymbol{A}_{j_k}(\boldsymbol{u}_k)^{-1} \nabla_{j_k} f_1(\boldsymbol{u}_k)\right), \\ \boldsymbol{u}^{(\bar{j}_k)}_{k+1} = \boldsymbol{u}^{(\bar{j}_k)}_k. \end{array}\right.\end{array}\right. \tag{2.55}$$

Like its non-alternating version, this algorithm can be understood as a relaxed MM algorithm. Its convergence properties have been analyzed in the non-convex case in [74]. In the convex case, the convergence properties are the same as for the VMFB algorithm, assuming that the update rule is quasi-cyclic.

Example 2.18 *Let us consider the minimization of*

$$\underset{\boldsymbol{A} \in \mathbb{R}^{P \times R}, \boldsymbol{S} \in \mathbb{R}^{R \times T}}{\text{minimize}}\ F(\boldsymbol{A}, \boldsymbol{S}) = \frac{1}{2} \|\boldsymbol{AS} - \boldsymbol{X}\|^2_{\mathrm{F}} + \iota_{\mathcal{A}}(\boldsymbol{A}) + \iota_{\mathcal{S}}(\boldsymbol{S}), \tag{2.56}$$

where $\mathcal{A} = [a_{\min}, a_{\max}]^{P\times R}$, with $0 < a_{\min} \le a_{\max}$, and $\mathcal{S} = [s_{\min}, s_{\max}]^{R\times T}$, with $0 < s_{\min} \le s_{\max}$. In order to apply Algorithm (2.55), we need to define, for a given $(\boldsymbol{A}', \boldsymbol{S}') \in \mathcal{A}\times\mathcal{S}$, quadratic majorants of $F(\cdot, \boldsymbol{S}')$ (resp. $F(\boldsymbol{A}', \cdot)$). Since matrices $\boldsymbol{A}' \in \mathcal{A}$ and $\boldsymbol{S}' \in \mathcal{S}$ have positive elements, we can derive the following quadratic majorant functions (consequence of Lemma 2.10)

$$(\forall \boldsymbol{A} \in \mathcal{A})\ F(\boldsymbol{A}, \boldsymbol{S}') \le F(\boldsymbol{A}', \boldsymbol{S}') + \operatorname{tr}\left((\boldsymbol{A} - \boldsymbol{A}')\nabla_1 F(\boldsymbol{A}', \boldsymbol{S}')^\top\right) + \frac{1}{2}\operatorname{tr}\left(((\boldsymbol{A} - \boldsymbol{A}') \boxdot (\boldsymbol{A}'(\boldsymbol{S}'\boldsymbol{S}'^\top)) \oslash \boldsymbol{A}')(\boldsymbol{A} - \boldsymbol{A}')^\top\right), \tag{2.57}$$

$$(\forall \boldsymbol{S} \in \mathcal{S})\ F(\boldsymbol{A}', \boldsymbol{S}) \le F(\boldsymbol{A}', \boldsymbol{S}') + \operatorname{tr}\left((\boldsymbol{S} - \boldsymbol{S}')\nabla_2 F(\boldsymbol{A}', \boldsymbol{S}')^\top\right) + \frac{1}{2}\operatorname{tr}\left(((\boldsymbol{S} - \boldsymbol{S}') \boxdot (\boldsymbol{A}'^\top\boldsymbol{A}'\boldsymbol{S}') \oslash \boldsymbol{S}')(\boldsymbol{S} - \boldsymbol{S}')^\top\right), \tag{2.58}$$

where $\operatorname{tr}(\cdot)$ denotes the trace operator. Then, the BC-VMFB algorithm, using a cyclic rule, reads:

$$\begin{array}{l} \boldsymbol{A}_0 \in \mathcal{A},\ \boldsymbol{S}_0 \in \mathcal{S} \\ \textit{For}\ \ k = 0, 1, \ldots \\ \left\lfloor \begin{array}{l} \tilde{\boldsymbol{A}}_k = \boldsymbol{A}_k - (\boldsymbol{A}_k \oslash (\boldsymbol{A}_k\boldsymbol{S}_k\boldsymbol{S}_k^\top)) \boxdot \nabla_1 F(\boldsymbol{A}_k, \boldsymbol{S}_k), \\ \boldsymbol{A}_{k+1} = \mathrm{P}_{\mathcal{A}}\left(\tilde{\mathrm{A}}_k\right) = \min\left(\max\left(\tilde{\mathrm{A}}_k, a_{\min}\right), a_{\max}\right), \\ \tilde{\boldsymbol{S}}_k = \boldsymbol{S}_k - (\boldsymbol{S}_k \oslash (\boldsymbol{A}_{k+1}^\top\boldsymbol{A}_{k+1}\boldsymbol{S}_k)) \boxdot \nabla_2 F(\boldsymbol{A}_{k+1}, \boldsymbol{S}_k), \\ \boldsymbol{S}_{k+1} = \mathrm{P}_{\mathcal{S}}\left(\tilde{\mathrm{S}}_k\right) = \min\left(\max\left(\tilde{\mathrm{S}}_k, s_{\min}\right), s_{\max}\right). \end{array} \right. \end{array} \tag{2.59}$$

Let us formulate a few remarks about the above algorithm:

- *The projections onto sets $\mathcal{A}$ and $\mathcal{S}$ are here computed in the standard Euclidean metric. This is a direct consequence of the separability of the majorant functions, and of the simple form of the constraint sets. More complex constraints on the sought solution (e.g. sum-to-one, sparsity) can be included at the price of modifying the underlying metric in the projection steps (see for example [75]).*
- *Algorithm (2.59) is reminiscent from the alternating minimization algorithm proposed in [2] in the context of NMF with Euclidean distance, the main difference being the strict positivity constraint on the components of the involved matrices, introduced in the projection step. This constraint allows us to guarantee the convergence of the iterates of the algorithm to a local minimizer of F, according to the convergence theorem from [74].*
- *A cyclic rule has been used in Algorithm (2.59) for updating matrices **A** and **S**. In fact, the convergence of the algorithm still holds when using a quasi-cyclic rule, i.e. when each block is updated at least once every $K \ge 0$ iteration. A particularly interesting strategy is to adopt an inner looping approach, where **A** (resp. **S**) is updated several times before updating **S** (resp. **A**). Such a technique has been applied for instance in [76, 77] in the context of blind deconvolution.*

2.3 Primal-Dual Methods

In this section, we present algorithms making use of primal-dual formulations for dealing with the minimization of a composite cost function of the form

$$(\forall \boldsymbol{u} \in \mathbb{R}^N) \qquad f(\boldsymbol{u}) = f_1(\boldsymbol{u}) + f_2(\boldsymbol{C}\boldsymbol{u}) \tag{2.60}$$

where $f_1 : \mathbb{R}^N \to \mathbb{R} \cup \{+\infty\}$, $f_2 : \mathbb{R}^M \to \mathbb{R} \cup \{+\infty\}$, and $\boldsymbol{C} \in \mathbb{R}^{M\times N}$. Subsequently, functions f_1 and f_2 will be assumed to be convex so as to facilitate our presentation. The presented algorithms can however be applied to the non-convex case, even if convergence guaranties become hazardous.

2.3.1 Lagrange Duality

By introducing an auxiliary variable $\boldsymbol{v}$, the minimization of f can be reformulated as the minimization of a separable sum of functions subject to the linear constraint $\boldsymbol{v} = \boldsymbol{C}\boldsymbol{u}$, that is

$$\underset{\boldsymbol{u}\in\mathbb{R}^N,\ \boldsymbol{v}\in\mathbb{R}^M}{\text{minimize}} f_1(\boldsymbol{u}) + f_2(\boldsymbol{v}) + \iota_{\{0\}}(\boldsymbol{C}\boldsymbol{u} - \boldsymbol{v}), \tag{2.61}$$

with $\iota_{\{0\}}$ is the indicator function of the singleton $\{0\}$. A well-known strategy in optimization to address constrained problems is to resort to the Lagrange multiplier method. This classical technique makes use of the Lagrange function defined as

$$(\forall \boldsymbol{u} \in \mathbb{R}^N)\ \left(\forall (\boldsymbol{v}, \boldsymbol{w}) \in (\mathbb{R}^M)^2\right) \quad \mathcal{L}(\boldsymbol{u}, \boldsymbol{v}, \boldsymbol{w}) = f_1(\boldsymbol{u}) + f_2(\boldsymbol{v}) + \boldsymbol{w}^\top(\boldsymbol{C}\boldsymbol{u} - \boldsymbol{v}), \tag{2.62}$$

where the parameter $\boldsymbol{w}$ is the so-called Lagrange multiplier vector. More precisely, it can be shown that under some mild qualification conditions (for example, the intersection of the interior of the domain of f_2 and the image of the domain of f_1 by $\boldsymbol{C}$ is nonempty), if $(\hat{\boldsymbol{u}}, \hat{\boldsymbol{v}}, \hat{\boldsymbol{w}})$ is a saddle point of $\mathcal{L}$, then $\hat{\boldsymbol{u}}$ is a minimizer of f. Recall that such a saddle point is defined as follows:

$$(\forall \boldsymbol{u} \in \mathbb{R}^N)\ \left(\forall (\boldsymbol{v}, \boldsymbol{w}) \in (\mathbb{R}^M)^2\right) \quad \mathcal{L}(\hat{\boldsymbol{u}}, \hat{\boldsymbol{v}}, \boldsymbol{w}) \le \mathcal{L}(\hat{\boldsymbol{u}}, \hat{\boldsymbol{v}}, \hat{\boldsymbol{w}}) \le \mathcal{L}(\boldsymbol{u}, \boldsymbol{v}, \hat{\boldsymbol{w}}). \tag{2.63}$$

It is characterized by the Karush–Kuhn–Tucker (KKT) optimality conditions:

$$\begin{cases} -\boldsymbol{C}^\top\hat{\boldsymbol{w}} \in \partial f_1(\hat{\boldsymbol{u}}) \\ \hat{\boldsymbol{w}} \in \partial f_2(\hat{\boldsymbol{v}}) \\ \boldsymbol{C}\hat{\boldsymbol{u}} = \hat{\boldsymbol{v}}. \end{cases} \tag{2.64}$$

2.3.2 Alternating Direction Method of Multipliers

2.3.2.1 Basic Form

The saddle point property of the Lagrangian suggests to find a minimizer of the cost function f by alternating between a minimization step with respect to the primal variables $\boldsymbol{u}$ and $\boldsymbol{v}$, and a maximization step with respect to the dual variable $\boldsymbol{w}$. However, in order to obtain good convergence properties of the resulting algorithm, the Lagrange function must be modified as follows:

$$(\forall \boldsymbol{u} \in \mathbb{R}^N)\ \left(\forall (\boldsymbol{v}, \boldsymbol{z}) \in (\mathbb{R}^M)^2\right) \quad \tilde{\mathcal{L}}(\boldsymbol{u}, \boldsymbol{v}, \boldsymbol{z}) = \mathcal{L}(\boldsymbol{u}, \boldsymbol{v}, \gamma \boldsymbol{z}) + \frac{\gamma}{2}\|\boldsymbol{C}\boldsymbol{u} - \boldsymbol{v}\|^2, \tag{2.65}$$

where, for convenience, we have performed a variable change for the Lagrange multiplier by setting $\boldsymbol{w} = \gamma \boldsymbol{z}$ with $\gamma \in]0, +\infty[$. Due to the additional quadratic term, $\tilde{\mathcal{L}}$ is called an augmented Lagrangian. It can be noticed that $\tilde{\mathcal{L}}$ is a majorant function of $\mathcal{L}$, which coincides with it when the constraint $\boldsymbol{v} = \boldsymbol{C}\boldsymbol{u}$ is satisfied. The proposed optimization algorithm then generates a sequence $(\boldsymbol{u}_k, \boldsymbol{v}_k, \boldsymbol{z}_k)_{k \geq 1}$ as follows:

$$\begin{array}{l} \boldsymbol{v}_0 \in \mathbb{R}^M, \boldsymbol{z}_0 \in \mathbb{R}^M \\ \text{For } k = 0, 1, \ldots \\ \left\lfloor \begin{array}{l} \boldsymbol{u}_{k+1} = \underset{\boldsymbol{u} \in \mathbb{R}^N}{\operatorname{argmin}}\ \tilde{\mathcal{L}}(\boldsymbol{u}, \boldsymbol{v}_k, \boldsymbol{z}_k) \\ \boldsymbol{v}_{k+1} = \underset{\boldsymbol{v} \in \mathbb{R}^N}{\operatorname{argmin}}\ \tilde{\mathcal{L}}(\boldsymbol{u}_{k+1}, \boldsymbol{v}, \boldsymbol{z}_k) \\ \boldsymbol{z}_{k+1} \text{ such that } \tilde{\mathcal{L}}(\boldsymbol{u}_{k+1}, \boldsymbol{v}_{k+1}, \boldsymbol{z}_{k+1}) \geq \tilde{\mathcal{L}}(\boldsymbol{u}_{k+1}, \boldsymbol{v}_{k+1}, \boldsymbol{z}_k). \end{array} \right. \end{array} \tag{2.66}$$

If the last step is performed through a gradient ascent step with stepsize $1/\gamma$, we obtain the Alternating Direction Method of Multipliers (ADMM) [78–81], which reads

$$\begin{array}{l} \boldsymbol{v}_0 \in \mathbb{R}^M, \boldsymbol{z}_0 \in \mathbb{R}^M \\ \text{For } k = 0, 1, \ldots \\ \left\lfloor \begin{array}{l} \boldsymbol{u}_{k+1} = \underset{\boldsymbol{u} \in \mathbb{R}^N}{\operatorname{argmin}}\ \frac{1}{2}\|\boldsymbol{C}\boldsymbol{u} - \boldsymbol{v}_k + \boldsymbol{z}_k\|^2 + \frac{1}{\gamma} f_1(\boldsymbol{u}) \\ \boldsymbol{s}_k = \boldsymbol{C}\boldsymbol{u}_{k+1} \\ \boldsymbol{v}_{k+1} = \operatorname{prox}_{\frac{f_2}{\gamma}}\left(\boldsymbol{z}_k + \boldsymbol{s}_k\right) \\ \boldsymbol{z}_{k+1} = \boldsymbol{z}_k + \boldsymbol{s}_k - \boldsymbol{v}_{k+1}. \end{array} \right. \end{array} \tag{2.67}$$

The convergence of the sequence $(\boldsymbol{u}_k)_{k \in \mathbb{N}}$ to a minimizer of the cost function f is then secured provided that $\boldsymbol{C}$ has full column rank, that is $\boldsymbol{C}^\top \boldsymbol{C}$ is invertible. Note that, by duality arguments, ADMM is strongly related to another famous algorithm in convex optimization, the Douglas–Rachford algorithm [82, 83].

2.3.2.2 Minimizing a Sum of More Than Two Functions

In practice, one may be interested in more involved cost functions of the form:

$$(\forall \boldsymbol{u} \in \mathbb{R}^N) \qquad f(\boldsymbol{u}) = \sum_{j=1}^{J} f_j(\boldsymbol{C}_j \boldsymbol{u}), \tag{2.68}$$

where, for every $j = 1, \dots, J$, f_j is a convex function from $\mathbb{R}^{M_j}$ to $\mathbb{R} \cup \{+\infty\}$, and $\boldsymbol{C}_j \in \mathbb{R}^{M_j \times N}$. One might think of a direct extension of ADMM to this context by introducing auxiliary variables $\boldsymbol{v}_j = \boldsymbol{C}_j \boldsymbol{u}$, but one has to be very cautious in doing so, since convergence guaranties may be lost [84]. A better approach may consist in resorting to consensus-like techniques leading to parallel forms of ADMM [85]. Such an efficient algorithm is the Parallel ProXimal Algorithm (PPXA) which was designed in [86]. An extended version of this algorithm (PPXA+) [87] is described next:

$$\begin{array}{l}
(\boldsymbol{y}_{0,j})_{1 \le j \le J} \in \mathbb{R}^N, \boldsymbol{u}_0 = (\sum_{j=1}^J \boldsymbol{C}_j^\top \boldsymbol{C}_j)^{-1} \sum_{j=1}^J \boldsymbol{C}_j^\top \boldsymbol{y}_{0,j} \\
\text{For } k = 0, 1, \dots \\
\left\lfloor \begin{array}{l}
\boldsymbol{v}_{k,j} = \text{prox}_{\gamma f_j}(\boldsymbol{y}_{k,j}), \quad j = 1, \dots, J \\
\boldsymbol{c}_k = (\sum_{j=1}^{J} \boldsymbol{C}_j^\top \boldsymbol{C}_j)^{-1} \sum_{j=1}^{J} \boldsymbol{C}_j^\top \boldsymbol{v}_{k,j} \\
\boldsymbol{y}_{k+1,j} = \boldsymbol{y}_{k,j} + \lambda \left(\boldsymbol{C}_j (2\boldsymbol{c}_k - \boldsymbol{u}_k) - \boldsymbol{v}_{k,j} \right), \quad j = 1, \dots, J \\
\boldsymbol{u}_{k+1} = \boldsymbol{u}_k + \lambda (\boldsymbol{c}_k - \boldsymbol{u}_k),
\end{array} \right.
\end{array} \tag{2.69}$$

where $\gamma \in]0, +\infty[$ and $\lambda \in]0, 2[$ are two parameters of the algorithm. The use of this algorithm requires that $\sum_{j=1}^J \boldsymbol{C}_j^\top \boldsymbol{C}_j$ be invertible. One of the key additional advantages of PPXA+ is that many operations can be performed in parallel on J processors.

Example 2.19 *An illustration of the great performance of Algorithm (2.69) can be found in [88], in the context of signal restoration from DOSY NMR measurements. The application context for this modality is explained in detail in Section 2.4, but we introduce it briefly here for the sake of clarity. We are given measurements $\boldsymbol{y} \in \mathbb{R}^M$, related to the sought DOSY (Diffusion Order SpectroscopY) spectra $\overline{\boldsymbol{u}} \in \mathbb{R}^N$ through the linear relation $\boldsymbol{y} = \boldsymbol{K}\overline{\boldsymbol{u}} + \boldsymbol{w}$, with $\boldsymbol{K} \in \mathbb{R}^{M \times N}$ the observation matrix associated to the DOSY physical model, and $\boldsymbol{w} \in \mathbb{R}^M$ a noise vector. The authors propose to solve the inverse problem of retrieving an estimate of $\overline{\boldsymbol{u}}$ from $\boldsymbol{y}$ and $\boldsymbol{K}$, by minimizing (2.68) where $J = 2$, $\boldsymbol{C}_1 = \boldsymbol{K}$, $f_1 = \iota_C$ with*

$$C = \left\{ \boldsymbol{z} \in \mathbb{R}^M \mid \|\boldsymbol{z} - \boldsymbol{y}\| \le \xi \right\}, \tag{2.70}$$

$\xi > 0$, $\boldsymbol{C}_2 = \boldsymbol{I}_N$ and f_2 is an hybrid penalty combining an entropy term and an ℓ_1 regularization [89]. The application of Algorithm (2.69) leads to

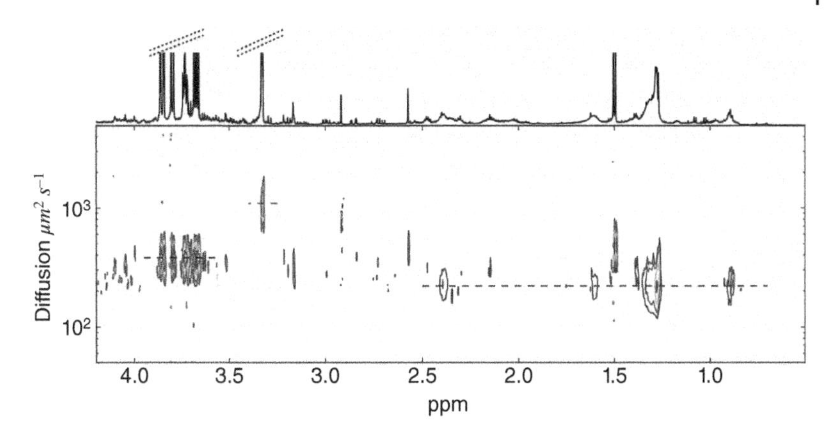

Figure 2.10 Result of PALMA method on an NMR DOSY experiment performed on a brown algae methanol/water extract.

the so-called PALMA algorithm, standing for "Proximal Algorithm for L1 combined with MAxent prior." An example of result obtained by PALMA on the DOSY NMR analysis of crude ethanolic plant extract obtained from brown algae is provided in Figure 2.10.

2.3.3 Primal-Dual Proximal Algorithms

It can be observed that the computation of $\boldsymbol{u}_{k+1}$ at the k-th iteration of ADMM is generally non-explicit. Even when $J = 2$, PPXA+ does not present this shortcoming, it requires a matrix inversion for computing each variable $\boldsymbol{c}_k$. When the involved matrices do not have a simple structure and they are of large size, such inversion has a high computational cost. A number of primal-dual proximal algorithms have been proposed to circumvent this difficulty [90–93], which are applicable to possibly non-smooth optimization problems. A simple way for deriving one of the most popular primal-dual proximal algorithms consists in starting from (2.67) and to replace the update of $\boldsymbol{u}_{k+1}$ by the following semi-implicit subgradient step:

$$\boldsymbol{u}_{k+1} = \boldsymbol{u}_k - \tau\gamma\left(\boldsymbol{C}^{\top}(\boldsymbol{C}\boldsymbol{u}_k - \boldsymbol{v}_k + \boldsymbol{z}_k) + \frac{1}{\gamma}\boldsymbol{t}_k\right), \tag{2.71}$$

where $\boldsymbol{t}_k$ is a subgradient of f_1 at $\boldsymbol{u}_{k+1}$ and τ is a positive stepsize. This is also equivalent to

$$\boldsymbol{y}_k = \gamma(\boldsymbol{C}\boldsymbol{u}_k - \boldsymbol{v}_k + \boldsymbol{z}_k), \tag{2.72}$$

$$\boldsymbol{u}_{k+1} = \operatorname{prox}_{\tau f_1}\left(\boldsymbol{u}_k - \tau\boldsymbol{C}^{\top}\boldsymbol{y}_k\right). \tag{2.73}$$

Now, reverting the order of the updates of $\boldsymbol{v}_k$ and $\boldsymbol{z}_k$ with respect to the original form of ADMM and performing some algebra lead to

$$\begin{aligned}
&\boldsymbol{u}_0 \in \mathbb{R}^N, \boldsymbol{y}_0 \in \mathbb{R}^M \\
&\text{For } k = 0, 1, \ldots \\
&\left\lfloor \begin{array}{l}
\boldsymbol{u}_{k+1} = \operatorname{prox}_{\tau f_1}\left(\boldsymbol{u}_k - \tau \boldsymbol{C}^\top \boldsymbol{y}_k\right) \\
\boldsymbol{d}_k = \boldsymbol{y}_k + \gamma \boldsymbol{C}(2\boldsymbol{u}_{k+1} - \boldsymbol{u}_k) \\
\boldsymbol{y}_{k+1} = \boldsymbol{d}_k - \gamma \operatorname{prox}_{\frac{f_2}{\gamma}}\left(\frac{\boldsymbol{d}_k}{\gamma}\right).
\end{array}\right.
\end{aligned} \tag{2.74}$$

The convergence of $(\boldsymbol{u}_k)_{k\in\mathbb{N}}$ to a minimizer of f can then be shown if $\tau\gamma|||\boldsymbol{C}|||^2 \leq 1$, where we recall that $|||\cdot|||$ denotes the spectral norm. Let us emphasize that no matrix inversion is required in the previous algorithm. Various extensions of this framework are possible. In particular, other forms of primal-dual proximal algorithms can be obtained [92]. It is also possible to add in the original criterion a Lipschitz-continuous term, which is addressed in the algorithm through its gradient [94–97]. Variants of this algorithm have been proposed so as to minimize composite function of the form (2.68) [92]. Block-alternating implementations of primal-dual proximal algorithms, along with their convergence properties, can be found for instance in [98, 99].

Example 2.20 *The primal-dual proximal algorithm is applied in [100] to mass spectrometry. Following a framework similar to the one of Example 2.16, the authors consider now the ℓ_1 norm penalty function $f_1(\boldsymbol{u}) = \sum_{n=1}^{N} |u_n|$. The resulting optimization problem is then convex non-differentiable, and thus Algorithm 2.74 is particularly well suited. Example of results obtained for the analysis of a real MS dataset measured on a Brucker Solarix 15 T, FT-ICR instrument with an ESI source, is provided in Figure 2.11. The considered sample was constituted of 3 μM of the peptide EVEALEKKVAALESKVQALEKKVEALEHG-NH2 (C140H240N38O45) in its trimer form within 50 mM of NH4OAc, acquired in native conditions. Despite the very large size of the problem, the processing time was of about 108 min on a standard laptop.*

2.3.4 Primal-Dual Interior Point Algorithm

Let us consider the optimization problem

$$\underset{\boldsymbol{u}\in C}{\text{minimize}}\ f_1(\boldsymbol{u}), \tag{2.75}$$

subject to the linear equality constraints

$$C = \{\boldsymbol{u} \in \mathbb{R}^N \mid \boldsymbol{C}\boldsymbol{u} + \rho \in]0, +\infty[^N\}, \tag{2.76}$$

f_1 is a convex twice differentiable function, $\boldsymbol{C} \in \mathbb{R}^{M\times N}$, and $\rho = (\rho_m)_{1\leq m\leq M} \in \mathbb{R}^M$. The existence of a minimizer $\hat{\boldsymbol{u}}$ will be subsequently assumed. Interior

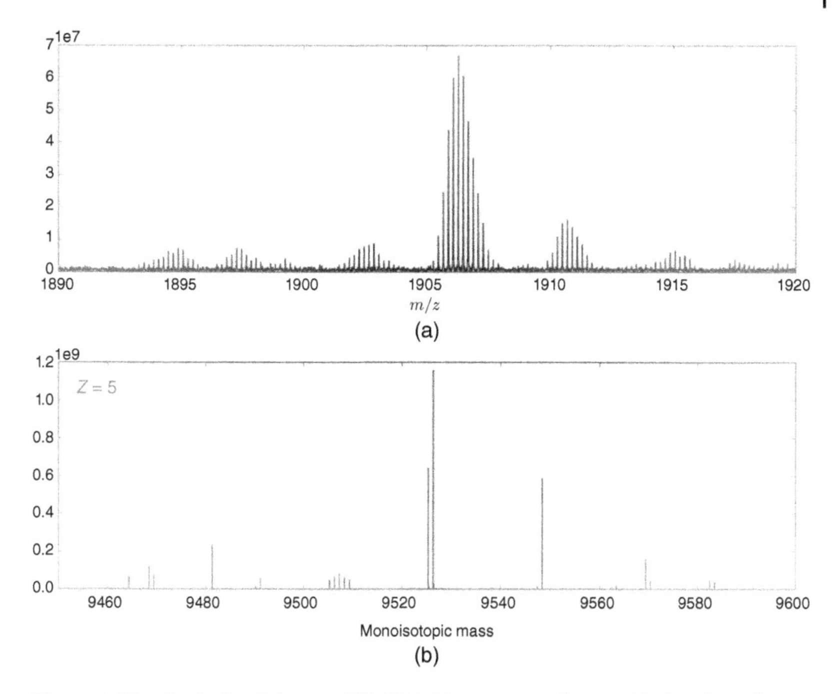

Figure 2.11 Analysis of the real FT-ICR-MS spectrum of a peptide in trimer form: (a) zoom on the acquired data; (b) recovered spectrum for the charge $z = 5$, using Algorithm 2.74 for a mass grid size of $M = 8130981$ and total charge number of $Z = 5$.

point methods, introduced in [101], solve the constrained optimization problem (2.75) by introducing a sequence of unconstrained optimization subproblems:

$$(\forall k \in \mathbb{N}) \quad \underset{\boldsymbol{u} \in \mathbb{R}^N}{\text{minimize}}\, f_{\mu_k}(\boldsymbol{u}) = f_1(\boldsymbol{u}) + \mu_k\, b(\boldsymbol{u}) \tag{2.77}$$

for positive barrier parameter values $(\mu_k)_{k \in \mathbb{N}}$ decaying to 0. The auxiliary function b, called *Barrier function*, penalizes the closeness to the constraint boundaries and maintains the iterates inside the strict interior of the constrained domain. The most widely used auxiliary function is the logarithmic barrier

$$(\forall \boldsymbol{u} \in \mathbb{R}^N) \quad b(\boldsymbol{u}) = f_2(\boldsymbol{C}\boldsymbol{u}), \tag{2.78}$$

where

$$(\forall \boldsymbol{v} = (v_m)_{1 \le m \le M} \in \mathbb{R}^M)$$
$$f_2(\boldsymbol{v}) = \begin{cases} -\sum_{m=1}^{M} \log(v_m + \rho_m) & \text{if } \boldsymbol{v} \in]0, +\infty[^M \\ +\infty & \text{otherwise.} \end{cases} \tag{2.79}$$

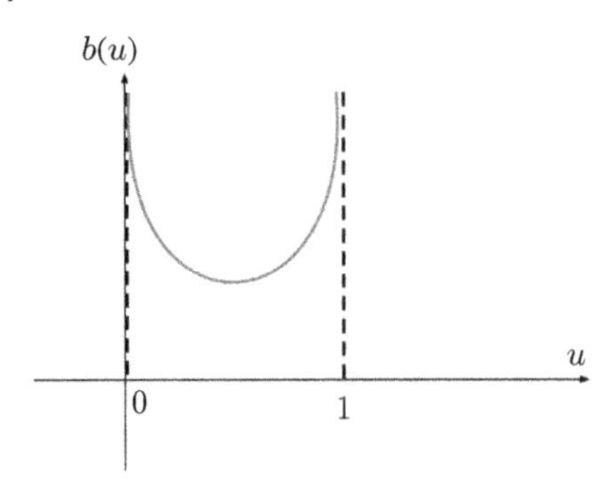

Figure 2.12 Logarithmic barrier function $b(u) = -\log(u) - \log(1-u)$ associated to the constraint $u \in [0, 1]$.

For every $\mu \in]0, +\infty[$, this choice makes the penalized criterion f_μ unbounded at the boundary of the feasible region so that its minimizers $\hat{\boldsymbol{u}}_\mu$ fulfill strictly the constraints. An example is presented in Figure 2.12.

Although various classical interior-point methods can be envisaged to solve Problem (2.77), a primal-dual approach can be followed by reformulating the problem under the form (2.61). The resulting primal-dual interior-point methods simultaneously estimate the primal variables $\boldsymbol{u}$ and the dual Lagrange multiplier vector $\lambda = -\boldsymbol{w}$ associated with the constraints [102]. The joint estimation of primal and dual variables is performed through Karush–Kuhn–Tucker (KKT) optimality conditions optimality conditions (2.64):

$$\begin{cases} \nabla f_1(\hat{\boldsymbol{u}}_\mu) - \boldsymbol{C}^\mathrm{T}\hat{\lambda}_\mu = \boldsymbol{0} \\ \hat{\boldsymbol{\Lambda}}_\mu(\boldsymbol{C}\hat{\boldsymbol{u}}_\mu + \rho) = \boldsymbol{\mu} \\ \boldsymbol{C}\hat{\boldsymbol{u}}_\mu + \rho > \boldsymbol{0}. \end{cases} \tag{2.80}$$

where $\hat{\boldsymbol{\Lambda}}_\mu = \mathrm{Diag}\{\hat{\lambda}_\mu\}$ and $\boldsymbol{\mu} = \mu[1, \dots, 1]^\mathrm{T} \in]0, +\infty[^M$.

Numerically, a sequence of optimization problems is solved for a sequence of penalization (also called perturbation) parameters $(\mu_k)_{k\in\mathbb{N}}$ converging to 0. At each iteration $k \in \mathbb{N}$ of the algorithm, a pair of primal-dual variables $(\boldsymbol{u}_{k+1}, \lambda_{k+1})$ is firstly computed from an approximate solution to (2.80) through a Newton algorithm step on the equality conditions, in association with a linesearch strategy on a merit function incorporating some barrier terms, allowing to ensure the inequality condition [10, Chap.11]. The update strategy is then given by

$$(\boldsymbol{u}_{k+1}, \lambda_{k+1}) = (\boldsymbol{u}_k - \alpha_k \boldsymbol{c}_k, \lambda_k - \alpha_k \boldsymbol{d}_k), \tag{2.81}$$

where α_k is the stepsize and $(\boldsymbol{c}_k, \boldsymbol{d}_k)$ are the primal and dual Newton directions. Finally, the penalization parameter μ_{k+1} is updated in order to ensure the algorithm convergence. Based on the iterative scheme (2.81), several primal-dual interior point methods have been proposed in the literature, each of them calling for its own strategy for the computation of the primal-dual directions, the derivation of a suitable stepsize, and the update of the perturbation parameter (see [102, 103] for a review). We focus here on the iterative scheme introduced in [104, 105]. Note that the algorithm described above is reminiscent from the one from [106], with the introduction of novel tools based on the MM principle aiming at accelerating the practical convergence, as well as reducing the computational cost per iteration.

2.3.4.1 Primal-Dual Directions

Newton directions $(\boldsymbol{c}_k, \boldsymbol{d}_k)$ are obtained by solving the linear system

$$\begin{bmatrix} \nabla^2 f_1(\boldsymbol{u}_k) & -\boldsymbol{C}^\top \\ \boldsymbol{\Lambda}_k \boldsymbol{C} & \mathrm{Diag}\{\boldsymbol{C}\boldsymbol{u}_k + \rho\} \end{bmatrix} \begin{bmatrix} \boldsymbol{c}_k \\ \boldsymbol{d}_k \end{bmatrix} = \boldsymbol{r}(\boldsymbol{u}_k, \lambda_k, \mu_k), \tag{2.82}$$

where $\boldsymbol{\Lambda}_k = \mathrm{Diag}\{\lambda_k\}$ and $\boldsymbol{r}(\boldsymbol{u}, \lambda, \mu)$ is the primal-dual residual defined as

$$\boldsymbol{r}(\boldsymbol{u}, \lambda, \mu) = \begin{pmatrix} \boldsymbol{r}^{\mathrm{prim}}(\boldsymbol{u}, \lambda) \\ \boldsymbol{r}^{\mathrm{dual}}(\boldsymbol{u}, \lambda, \mu) \end{pmatrix} = \begin{pmatrix} \nabla f_1(\boldsymbol{u}) - \boldsymbol{C}^\top \lambda \\ \boldsymbol{\Lambda}(\boldsymbol{C}\boldsymbol{u} + \rho) - \mu \end{pmatrix}. \tag{2.83}$$

As pointed out in [107, 108], the primal-dual matrix in the left side of (2.82) suffers from ill-conditioning as soon as some $[\boldsymbol{C}\boldsymbol{u}_k + \rho]_m$ or some λ_m gets closer to zero. Moreover, this matrix is not guaranteed to be symmetric nor definite positive [103, 109], so that the linear system (2.82) is difficult to solve. Therefore, rather than solving directly (2.82), variable substitution is used [106, 110]. From the second equation of (2.82), one calculates the dual direction according to

$$\begin{aligned} \boldsymbol{d}_k &= \mathrm{Diag}\{\boldsymbol{C}\boldsymbol{u}_k + \rho\}^{-1} \left(\boldsymbol{\Lambda}_k(\boldsymbol{C}\boldsymbol{u}_k + \rho) - \mu_k - \boldsymbol{\Lambda}_k \boldsymbol{C}\boldsymbol{c}_k\right) \\ &= \lambda_k - \mathrm{Diag}\{\boldsymbol{C}\boldsymbol{u}_k + \rho\}^{-1} \left(\mu_k + \boldsymbol{\Lambda}_k \boldsymbol{C}\boldsymbol{c}_k\right). \end{aligned} \tag{2.84}$$

Then, the primal direction is obtained by solving the linear system

$$\boldsymbol{H}_k \boldsymbol{c}_k = \boldsymbol{g}_k \tag{2.85}$$

with

$$\boldsymbol{H}_k = \nabla^2 f_1(\boldsymbol{u}_k) + \boldsymbol{C}^\top \mathrm{Diag}\{\boldsymbol{C}\boldsymbol{u}_k + \rho\}^{-1} \boldsymbol{\Lambda}_k \boldsymbol{C} \tag{2.86}$$

$$\boldsymbol{g}_k = \nabla f_1(\boldsymbol{u}_k) - \boldsymbol{C}^\top \mathrm{Diag}\{\boldsymbol{C}\boldsymbol{u}_k + \rho\}^{-1} \mu_k. \tag{2.87}$$

The computation cost of the primal-dual interior point algorithm is almost exclusively due to the resolution of the primal system (2.85). It is shown in [111] that an approximate solution of this system is sufficient to ensure the convergence of the method. Several solutions exist to calculate such approximation (see [112] and references therein). For instance, [104] proposes to perform an approximate solution using a preconditioned bi-conjugate gradient algorithm based on an incomplete LU factorization of matrix $\boldsymbol{H}_k$. Moreover, [105] emphasizes that, when $\boldsymbol{H}_k$ has a block-diagonal matrix, the resolution of (2.85) reduces to the resolution of a family of linear systems of small size and such structure is well suited for parallel computing as the blocks can be processed independently. When $\boldsymbol{H}_k$ is not easily invertible, a separable quadratic majorization of it can be employed, and an MM algorithm can be applied to solve (2.85). More precisely, let us remark that solving (2.85) is equivalent to the resolution of a quadratic minimization problem of the form

$$\underset{\boldsymbol{c}\in\mathbb{R}^M}{\text{minimize}}\ \frac{1}{2}\boldsymbol{c}^\top \boldsymbol{H}_k \boldsymbol{c} - \boldsymbol{g}_k^\top \boldsymbol{c}. \tag{2.88}$$

Let $\boldsymbol{B}_k$ be a symmetric positive definite matrix such that $\boldsymbol{H}_k \preceq \boldsymbol{B}_k$, and whose inverse is simple to compute (for instance, when $\boldsymbol{B}_k$ is a block diagonal matrix). Then, the solution of (2.88) is computed thanks to the following iterations:

$$\begin{aligned} &\boldsymbol{c}_{k,0} = 0,\\ &\text{For } j = 0, 1, \ldots\\ &\left\lfloor \boldsymbol{c}_{k,j+1} = \boldsymbol{c}_{k,j} - \boldsymbol{B}_k^{-1}\left(\boldsymbol{H}_k \boldsymbol{c}_{k,j} - \boldsymbol{g}_k\right), \right. \end{aligned} \tag{2.89}$$

until the fulfillment of the following stopping criterion on the primal residual: $\|\boldsymbol{H}_k \boldsymbol{d}_{k,j} - \boldsymbol{g}_k\| \leqslant \mu_k \|\boldsymbol{g}_k\|$ where μ_k is the barrier parameter. The benefit of this method has been illustrated in [113] through an example of spectral unmixing. Note that this MM-like approach is not only faster, it is also better suited for parallel implementation [114].

2.3.4.2 Linesearch

At the k-th iteration, the stepsize value α_k must be chosen so as to ensure the convergence of the algorithm and the fulfillment of the inequalities of the perturbed KKT system (2.80). The convergence study of the primal-dual algorithm presented in [106] requests that α_k ensures a sufficient decrease of the primal-dual merit function Ψ_{μ_k}, defined, for every $\boldsymbol{u} \in \mathbb{R}^N$ and $\lambda \in \mathbb{R}^M$, as

$$\Psi_{\mu_k}(\boldsymbol{u}, \lambda) = f_{\mu_k}(\boldsymbol{u}) + \lambda^\top(\boldsymbol{C}\boldsymbol{u} + \rho) - \mu_k \sum_{m=1}^{M} \log(\lambda_m [\boldsymbol{C}\boldsymbol{u} + \rho]_m), \tag{2.90}$$

$$= \mathcal{L}(\boldsymbol{u}, \boldsymbol{C}\boldsymbol{u}, -\lambda) - \mu_k \sum_{m=1}^{M} \log \lambda_m. \tag{2.91}$$

The sufficient decrease is assessed using the Armijo condition,

$$\psi_{\mu_k}(\alpha_k) - \psi_{\mu_k}(0) \leqslant \sigma\, \alpha_k \nabla \psi_{\mu_k}(0) \quad \text{with} \quad \sigma \in]0, 1/2[, \tag{2.92}$$

where, for every $\alpha \in [0, +\infty[$, $\psi_{\mu_k}(\alpha) \triangleq \Psi_{\mu_k}(\boldsymbol{u}_k - \alpha \boldsymbol{c}_k, \boldsymbol{\lambda}_k - \alpha \boldsymbol{d}_k)$. One can note that (2.91) contains two logarithmic barrier functions enforcing the fulfillment of the KKT inequalities, the positivity of the Lagrange multipliers $(\lambda_m)_{1\leq m\leq M}$ being a straightforward consequence of (2.80). The presence of the barrier function may cause the inefficiency of general purpose backtracking line search methods for finding a stepsize satisfying (2.92) and, thus, the slowdown of the algorithm convergence. Several strategies have been proposed to override this issue [115–117]. In particular, the majorization–minimization strategy from [116], relying on the construction of a majorizing function made of a quadratic term and a logarithmic barrier term, was shown to lead to good practical performance through numerical applications to 2D nuclear magnetic resonance signal reconstruction under positivity constraints [118], and to sparse signal deconvolution [117].

2.3.4.3 Penalization Parameter Update

The convergence analysis of interior point methods requires that the sequence $(\mu_k)_{k\in\mathbb{N}}$ tends to 0 as k tends to infinity. An efficient update strategy for the choice of the barrier parameter is the μ-criticity rule introduced in [119]:

$$(\forall k \in \mathbb{N}) \quad \mu_k = \theta \frac{\delta_k}{M}, \tag{2.93}$$

where $\delta_k = (\boldsymbol{C}\boldsymbol{u}_k + \boldsymbol{\rho})^\top \boldsymbol{\lambda}_k$ is the duality gap and $\theta \in]0, 1[$. The barrier parameter is updated as soon as the primal and dual directions fulfill the inner stopping rule [110, 120]:

$$\|\boldsymbol{r}^{\text{prim}}(\boldsymbol{u}_k, \boldsymbol{\lambda}_k)\|_\infty \leqslant \eta^{\text{prim}} \mu_k \text{ and } \frac{\delta_k}{M} \leqslant \eta^{\text{dual}} \mu_k, \tag{2.94}$$

with $\eta^{\text{prim}} > 0$ and $\eta^{\text{dual}} \in]1, 1/\theta[$.

2.3.4.4 Resulting Algorithm

The main steps of the proposed optimization method are summarized below:

$$\begin{array}{l}
\boldsymbol{u}_0 \in \mathbb{R}^N \quad \text{s.t.} \quad \boldsymbol{C}\boldsymbol{u}_0 + \boldsymbol{\rho} \in]0, +\infty[^M, \ \boldsymbol{\lambda}_0 \in]0, +\infty[^M, \ \mu_0 \in]0, +\infty[. \\
\text{For } k = 0, 1, \dots \\
\left\lfloor \begin{array}{l}
\text{If (2.94) holds} \\
\lfloor \ \text{Update} \quad \mu_k \quad \text{according to (2.93)} \\
\text{Calculate} \quad \boldsymbol{c}_k \quad \text{by solving (approximately) (2.85)} \\
\text{Deduce} \quad \boldsymbol{d}_k \quad \text{from (2.84)} \\
\text{Find} \quad \alpha_k > 0 \quad \text{satisfying (2.92)} \\
\text{Update} \quad (\boldsymbol{u}_{k+1}, \boldsymbol{\lambda}_{k+1}) \quad \text{according to (2.81).}
\end{array} \right.
\end{array} \tag{2.95}$$

The convergence properties of Algorithm (2.95) have been studied in [105, 106, 111], under the assumption that the linear constraint set is nonempty and bounded. In particular, it is shown that if f_1 is strictly convex, then the sequence $(\boldsymbol{u}_k)_{k\in\mathbb{N}}$ converges to the unique solution of (2.75).

Example 2.21 *The primal-dual interior point algorithm (2.95) has been applied to two-dimensional nuclear magnetic resonance in [118]. Classical nuclear magnetic resonance (NMR) experiments analyze the spin relaxation process independently, either in terms of longitudinal or transverse relaxation, leading to the so-called T_1 and T_2 relaxation spectra. Joint measurements with respect to both relaxation parameters allow to build two-dimensional distribution which is of high interest for chemical structure determination [121]. Experimental data consist of a series of discrete noisy samples $\boldsymbol{Y} \in \mathbb{R}^{m_1 \times m_2}$ such that $\boldsymbol{Y} = \boldsymbol{K}_1 \boldsymbol{U} \boldsymbol{K}_2^\top + \boldsymbol{W}$ with $\boldsymbol{U} \in \mathbb{R}^{N_1 \times N_2}$ the sought distribution, $\boldsymbol{K}_1$, $\boldsymbol{K}_2$ matrices associated to the acquisition model, and $\boldsymbol{W}$ a noise term. T_1-T_2 NMR reconstruction aims at estimating $\boldsymbol{U}$ given $\boldsymbol{Y}$, $\boldsymbol{K}_1$, $\boldsymbol{K}_2$. This problem can be solved by minimizing a penalized least-squares term:*

$$f_1(\boldsymbol{U}) = \frac{1}{2}\|\boldsymbol{K}_1 \boldsymbol{U} \boldsymbol{K}_2^\top - \boldsymbol{Y}\|_F^2 + \lambda\|\boldsymbol{U}\|_F^2 \tag{2.96}$$

with $\lambda > 0$, subject to the positivity constraint $\boldsymbol{U} \geq 0$. Figure 2.13 presents numerical results of 2D spectra obtained from the analysis of an apple matter sample ($m_1 = 50$, $m_2 = 10000$, $N_1 = 300$, $N_2 = 300$), using Algorithm (2.95) [118].

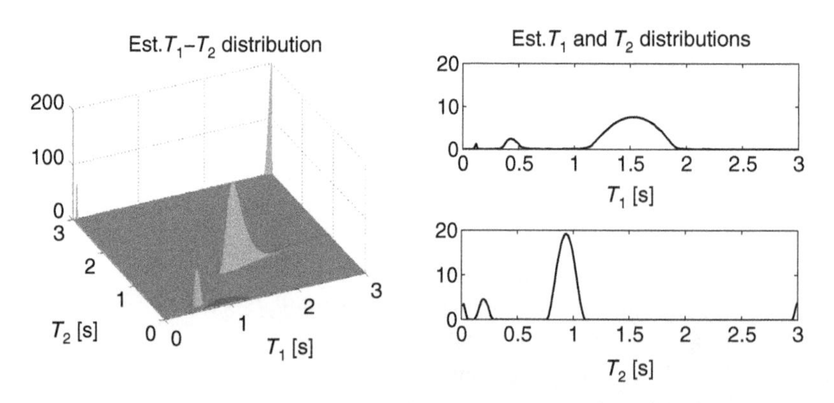

Figure 2.13 Estimated T_1-T_2 NMR distributions from real data (apple) using Algorithm (2.95).

2.4 Application to NMR Signal Restoration

This section illustrates the applicability of the presented optimization tools for signal restoration in NMR. The measurement of diffusion by NMR is used in various application fields (agroalimentary sector, pharmaceutical industry, ecology) to analyze the properties of complex chemical mixtures in order to determine their molecular structure and dynamics [88, 121, 122]. After the immersion of the matter in a strong magnetic field, all the nuclear spins align to an equilibrium state along the field orientation. The application of a short magnetic pulse, i.e. the pulsed field gradient, in resonance with the spin motion disturbs the spin orientation. NMR aims at analyzing the process which corresponds to the re-establishment of the spin into its equilibrium state. During the DOSY experiment [123], a series of measurements is acquired for different pulsed field gradient strengths. The data are then analyzed with the aim to separate different species according to their diffusion coefficient.

The DOSY NMR data $\boldsymbol{y} = (y^{(m)})_{1 \leq m \leq M} \in \mathbb{R}^M$ gathers the results of M experiments characterized by a vector $\boldsymbol{t} = (t^{(m)})_{1 \leq m \leq M} \in \mathbb{R}^M$ related to the pulsed field gradient strength and to the acquisition time. The relation between $\boldsymbol{y}$ and $\boldsymbol{t}$ can be expressed as the following Laplace transform:

$$(\forall m \in \{1, \dots, M\}) \quad y^{(m)} = \int \chi(T) \exp(-t^{(m)} T) dT, \tag{2.97}$$

where $\chi(T)$ is the unknown diffusion distribution. The problem is then to reconstruct $\chi(T)$ on the sampled grid $\boldsymbol{T} = (T^{(n)})_{1 \leq n \leq N}$, from the measurements $\boldsymbol{y}$. After discretization and appropriate normalization, the observation model reads

$$\boldsymbol{y} = \boldsymbol{K}\overline{\boldsymbol{u}} + \boldsymbol{w}, \tag{2.98}$$

where $\boldsymbol{K} \in \mathbb{R}^{M \times N}$ is given by

$$(\forall m \in \{1, \dots, M\})(\forall n \in \{1, \dots, N\}) \qquad K^{(m,n)} = \exp(-T^{(n)} t^{(m)}), \tag{2.99}$$

$\overline{\boldsymbol{u}} \in \mathbb{R}^N$ is the sought signal related to $\left(\chi(T^{(n)})\right)_{1 \leq n \leq N}$ (up to a scaling factor depending on the discretization grid), and $\boldsymbol{w} \in \mathbb{R}^M$ is a perturbation noise.

We propose here to find an estimate $\hat{\boldsymbol{u}} \in \mathbb{R}^N$ of $\overline{\boldsymbol{u}}$ by solving the following minimization problem, requiring the knowledge of $\boldsymbol{K}$ and $\boldsymbol{y}$:

$$\underset{\boldsymbol{u} \in \mathbb{R}^N}{\text{minimize}} \ \frac{1}{2} \|\boldsymbol{K}\boldsymbol{u} - \boldsymbol{y}\|^2 + \beta g(\boldsymbol{u}) \tag{2.100}$$

where $g \in \Gamma_0(\mathbb{R}^N)$ denotes a regularization term and $\beta > 0$ is a regularization parameter. Note that, in practice, a very large number of DOSY

NMR acquisitions (typically, 10^4) are conduced for various settings of the pulsed gradient field, so that Problem (2.100) must be solved many times, which motivates the search for a fast minimization algorithm [88]. In the following, we present optimization solutions, for various choices of function g, in the line of the work published in [89]. For illustration purpose, we will consider a test example with $\overline{\boldsymbol{u}}$ of size $N = 256$, and $\boldsymbol{y}$ of size $M = 50$, both represented in Figure 2.14. The noise $\boldsymbol{w}$ is a realization of a zero-mean white Gaussian noise, with standard deviation equalling to $10^{-2}y_0 \approx 0.34$. The grid $\boldsymbol{T} = (T^{(n)})_{1\leq n\leq N}$ is sampled following a logarithmic rule, within the interval $[T_{\min}, T_{\max}] = [1, 10^3]$ $\mu m^2\ s^{-1}$, while $\boldsymbol{t} = (t^{(m)})_{1\leq m\leq M}$ is regularly sampled on $[t_{\min}, t_{\max}] = [0, 1.5]$ seconds. The quality of the results obtained by the different tested restoration approaches will be assessed quantitatively by

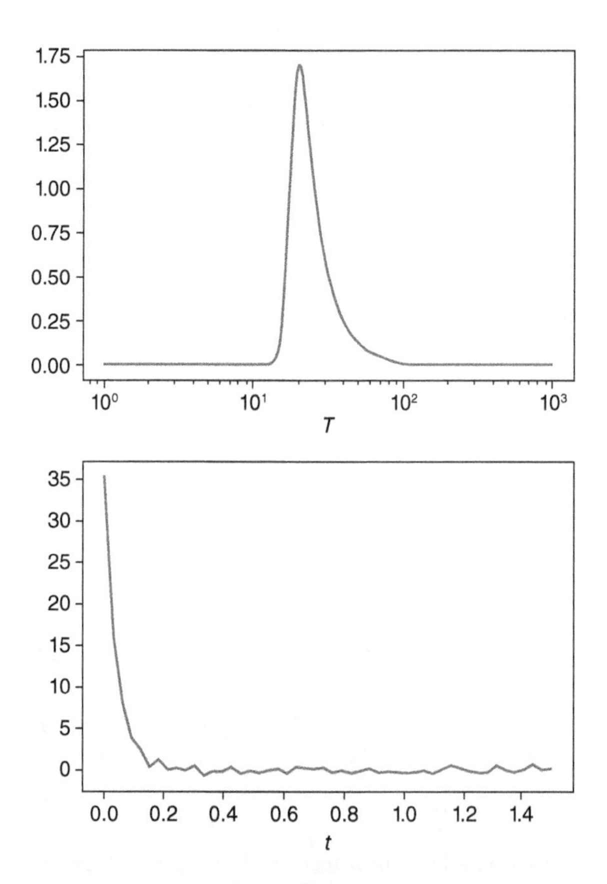

Figure 2.14 Original signal $\overline{\boldsymbol{u}}$ (a) and noisy acquired measure $\boldsymbol{y}$ (b).

means of the normalized mean square error (NMSE):

$$\mathrm{NMSE} = \frac{\|\overline{\boldsymbol{u}} - \hat{\boldsymbol{u}}\|^2}{\|\overline{\boldsymbol{u}}\|^2}, \tag{2.101}$$

with $\hat{\boldsymbol{u}}$ the result of each method.

2.4.1 Quadratic Penalization

Let us start with the following regularization term serving to promote the reconstruction of smooth signals:

$$(\forall \boldsymbol{u} \in \mathbb{R}^N) \quad g(\boldsymbol{u}) = \frac{1}{2}\|\boldsymbol{D}\boldsymbol{u}\|^2 \tag{2.102}$$

where $\boldsymbol{D} \in \mathbb{R}^{N \times N}$ is the discrete gradient operator such that,

$$(\forall n \in \{1, \ldots, N\}) \quad [\boldsymbol{D}\boldsymbol{u}]^{(n)} = u^{(n)} - u^{(n-1)} \tag{2.103}$$

with the circular convention $u^{(0)} = u^{(N)}$. In this case, the cost function involved in Problem (2.100) is convex and quadratic. Assuming that $\boldsymbol{K}^\top \boldsymbol{K} + \beta \boldsymbol{D}^\top \boldsymbol{D}$ is invertible, the solution to Problem (2.100) is unique and reads:

$$\hat{\boldsymbol{u}} = \left(\boldsymbol{K}^\top \boldsymbol{K} + \beta \boldsymbol{D}^\top \boldsymbol{D}\right)^{-1} \boldsymbol{K}^\top \boldsymbol{y}. \tag{2.104}$$

An example of such restored signal is displayed in Figure 2.15.

Now, assume that we want to impose some value range constraints on the sought signal. Then a possible regularization function is:

$$(\forall \boldsymbol{u} \in \mathbb{R}^N) \quad g(\boldsymbol{u}) = \frac{1}{2}\|\boldsymbol{D}\boldsymbol{u}\|^2 + \iota_{[u_{\min}, u_{\max}]^N}(\boldsymbol{u}) \tag{2.105}$$

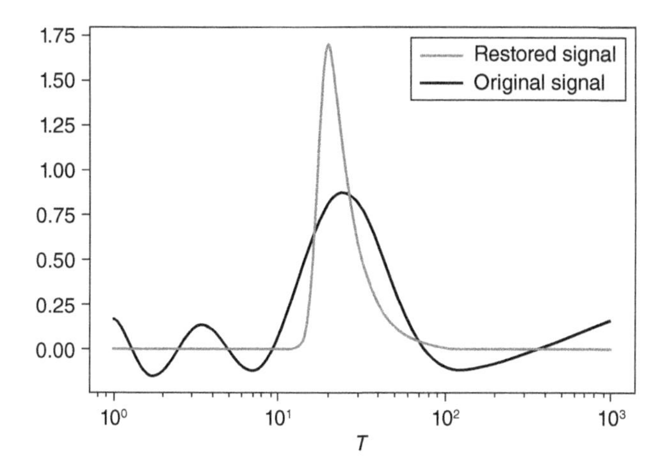

Figure 2.15 Restored signal (2.104) for $\beta = 1$, NMSE = 0.29.

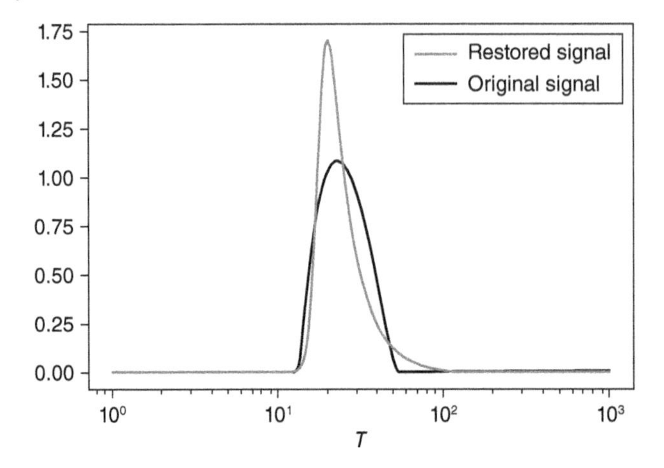

Figure 2.16 Restored signal using Algorithm (2.106) for $\beta = 0.0625$, NMSE = 0.14.

with $0 \leq u_{\min} < u_{\max}$ the minimum and maximum values of the original signal $(\overline{u}^{(n)})_{1 \leq n \leq N}$. The quadratic function $\boldsymbol{u} \mapsto 1/2\|\boldsymbol{K}\boldsymbol{u} - \boldsymbol{y}\|^2 + \beta/2\|\boldsymbol{D}\boldsymbol{u}\|^2$ is Lipschitz differentiable, with constant $|||\boldsymbol{K}^\top \boldsymbol{K} + \beta \boldsymbol{D}^\top \boldsymbol{D}|||$. Problem (2.100) can thus be solved using the projected gradient algorithm introduced in Example 2.15, which reads in this case:

$$
\begin{array}{l}
\boldsymbol{u}_0 \in \mathbb{R}^N \\
\text{For } k = 0, 1, \ldots \\
\left\lfloor
\begin{array}{l}
\alpha_k \in]0, 2/(|||\boldsymbol{K}^\top \boldsymbol{K} + \beta \boldsymbol{D}^\top \boldsymbol{D}|||)[, \\
\tilde{\boldsymbol{u}}_k = \boldsymbol{u}_k - \alpha_k \left(\boldsymbol{K}^\top (\boldsymbol{K}\boldsymbol{u}_k - \boldsymbol{y}) + \beta \boldsymbol{D}^\top \boldsymbol{D}\boldsymbol{u}_k \right), \\
\boldsymbol{u}_{k+1} = \mathrm{P}_{[u_{\min}, u_{\max}]^N}(\tilde{\mathrm{u}}_k) = \min \left(\max \left(\tilde{\mathrm{u}}_k, u_{\min} \right), u_{\max} \right),
\end{array}
\right.
\end{array}
\tag{2.106}
$$

where the min and max operations are performed componentwise. An example of solution obtained using the above algorithm is displayed in Figure 2.16. One can see that the solution is over-smoothed, which shows the considered prior might not be well suited in this particular example.

2.4.2 Entropic Penalization

In the context of NMR relaxometry, a standard strategy for restoring the target signal is to define

$$
(\forall \boldsymbol{u} \in \mathbb{R}^N) \quad g(\boldsymbol{u}) = \sum_{n=1}^{N} \varphi(x^{(n)}), \tag{2.107}
$$

with

$$(\forall u \in \mathbb{R}) \quad \varphi(u) = \begin{cases} u \log u & \text{if } u > 0, \\ 0 & \text{if } u = 0 \\ +\infty & \text{elsewhere.} \end{cases} \tag{2.108}$$

2.4.3 Sparsity Prior in the Signal Domain

Another strategy for regularization is to enforce the sparsity of the sought signal, in addition to the positivity constraint. In that respect, one can use

$$(\forall \boldsymbol{u} = (u_n)_{1\leq n\leq N} \in \mathbb{R}^N) \quad g(\boldsymbol{u}) = \|\boldsymbol{u}\|_1 + \iota_{[0,+\infty[^N}(\boldsymbol{u}). \tag{2.109}$$

Function g is convex, but it is not differentiable. Function $\boldsymbol{u} \mapsto 1/2\|\boldsymbol{K}\boldsymbol{u} - \boldsymbol{y}\|^2$ is Lipschitz differentiable with constant $|||\boldsymbol{K}|||^2$. Problem (2.100) can be solved using the forward–backward algorithm presented in Remark 2.3:

$$\begin{array}{l} \boldsymbol{u}_0 \in \mathbb{R}^N \\ \text{For } k = 0, 1, \ldots \\ \left\lfloor \begin{array}{l} \alpha_k \in]0, 2/|||\boldsymbol{K}|||^2[, \\ \tilde{\boldsymbol{u}}_k = \boldsymbol{u}_k - \theta_k \boldsymbol{K}^\top(\boldsymbol{K}\boldsymbol{u}_k - \boldsymbol{y}), \\ \boldsymbol{u}_{k+1} = \mathrm{prox}_{\alpha_k\beta(\|\cdot\|_1 + \iota_{[0,+\infty[^N})}(\tilde{\boldsymbol{u}}_k) = \max\left(|\tilde{\boldsymbol{u}}_k| - \alpha_k\beta, 0\right). \end{array} \right. \end{array} \tag{2.110}$$

The convergence rate of the above method can be improved, using a variable metric approach ([58], Section 2.2.4) or a Nesterov-based scheme (see for instance, [124]). An example of solution obtained using Algorithm (2.110) is displayed in Figure 2.17.

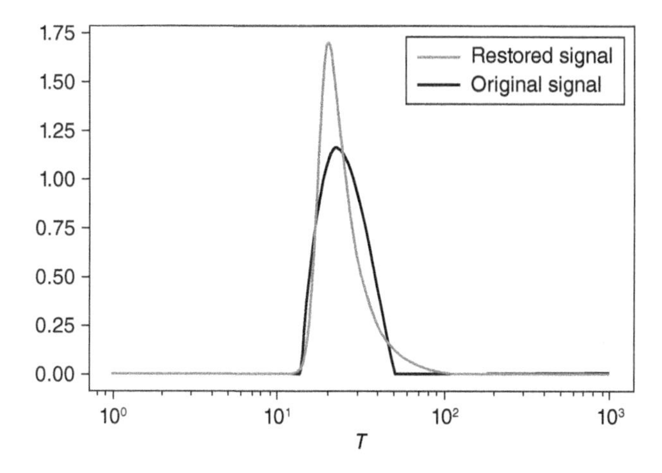

Figure 2.17 Restored signal using Algorithm (2.110) for $\beta = 1$, NMSE = 0.11.

It is also possible to approximate the ℓ_1 norm by a smooth function, in order to go back to a smooth minimization problem. For instance, one can choose:

$$(\forall \boldsymbol{u} = (u_n)_{1\leq n\leq N} \in \mathbb{R}^N) \quad g(\boldsymbol{u}) = \sum_{n=1}^{N} \sqrt{1 + (u_n)^2/\delta^2}, \tag{2.111}$$

with $\delta > 0$ a smoothing parameter. According to Property 2.7, for every $\boldsymbol{v} \in \mathbb{R}^N$, a quadratic tangent majorant for g at $\boldsymbol{v}$ is

$$(\forall \boldsymbol{u} \in \mathbb{R}^N) \quad g(\boldsymbol{u}) \leq g(\boldsymbol{v}) + \nabla g(\boldsymbol{v})^\top(\boldsymbol{u} - \boldsymbol{v}) + \frac{1}{2}\sum_{n=1}^{N} \frac{(u^{(n)} - v^{(n)})^2}{\delta^2\sqrt{1 + (v^{(n)})^2/\delta^2}}. \tag{2.112}$$

Problem (2.100) can then be solved using the MM quadratic algorithm (2.26):

$$\begin{array}{l} \boldsymbol{u}_0 \in \mathbb{R}^N \\ \text{For } k = 0, 1, \ldots \\ \left\lfloor \begin{array}{l} \alpha_k \in]0, 2[\\ \boldsymbol{A}(\boldsymbol{u}_k) = \boldsymbol{K}^\top\boldsymbol{K} + \beta \mathrm{Diag}\left((\delta^{-2}(1 + (u_k^{(n)})^2/\delta^2)^{-1/2})_{1\leq n\leq N} \right) \\ \mathbf{g}(\boldsymbol{u}_k) = \boldsymbol{K}^\top(\boldsymbol{K}\boldsymbol{u}_k - \boldsymbol{y}) + \beta(\delta^{-2}u_k^{(n)}(1 + (u_k^{(n)})^2/\delta^2)^{-1/2})_{1\leq n\leq N} \\ \boldsymbol{u}_{k+1} = \boldsymbol{u}_k - \alpha_k \boldsymbol{A}(\boldsymbol{u}_k)^{-1}\mathbf{g}(\boldsymbol{u}_k). \end{array} \right. \end{array} \tag{2.113}$$

It is worth noting that the computational cost of the above algorithm can be reduced, by making use of a subspace acceleration strategy ([47], Section 2.2.3.3), or a block-coordinate approach ([74], Section 2.2.5).

2.4.4 Sparsity Prior in a Transformed Domain

It can finally be useful to impose the sparsity of the signal in a transformed domain, in order to impose some regularity properties. For instance, it can be assumed that $\boldsymbol{W}\overline{\boldsymbol{u}}$ is sparse, where $\boldsymbol{W}$ is a (possibly overcomplete) wavelet analysis operator [125]. In this case, we may use

$$(\forall \boldsymbol{u} \in \mathbb{R}^N) \quad g(\boldsymbol{u}) = \|\boldsymbol{W}\boldsymbol{u}\|_1. \tag{2.114}$$

Because of the presence of matrix $\boldsymbol{W}$, the proximal step in the forward–backward algorithm is not explicit anymore. In such context, an efficient strategy is to resort to a primal-dual optimization approach as described in Section 2.3. For instance, the primal-dual proximal algorithm (2.74) [92]

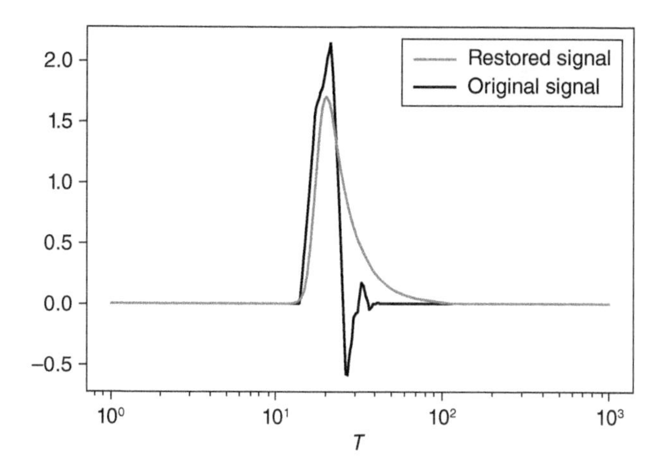

Figure 2.18 Restored signal using Algorithm (2.115) for $\beta = 27$, NMSE = 0.39.

would read in this case:

$$
\begin{array}{l}
\boldsymbol{u}_0 \in \mathbb{R}^N, \boldsymbol{v}_0 \in \mathbb{R}^M \\
(\tau, \gamma) \text{ such that } \tau\gamma |||\boldsymbol{W}|||^2 \leq 1 \\
\text{For } k = 0, 1, \dots \\
\left\lfloor
\begin{array}{l}
\boldsymbol{u}_{k+1} = \mathrm{prox}_{\tau/2\|\boldsymbol{K}\cdot-\boldsymbol{y}\|^2}\left(\boldsymbol{u}_k - \tau \boldsymbol{W}^\top \boldsymbol{v}_k\right) \\
\qquad = (\boldsymbol{I} + \tau \boldsymbol{K}^\top \boldsymbol{K})^{-1}(\boldsymbol{u}_k - \tau \boldsymbol{W}^\top \boldsymbol{v}_k + \tau \boldsymbol{K}^\top \boldsymbol{y}) \\
\boldsymbol{d}_k = \boldsymbol{v}_k + \gamma \boldsymbol{W}(2\boldsymbol{u}_{k+1} - \boldsymbol{u}_k) \\
\boldsymbol{v}_{k+1} = \boldsymbol{d}_k - \gamma \mathrm{prox}_{\beta\gamma^{-1}\|\cdot\|_1}\left(\frac{\boldsymbol{d}_k}{\gamma}\right) \\
\qquad = \boldsymbol{d}_k - \gamma \mathrm{sign}(\boldsymbol{d}_k) \boxdot \max\left(|\frac{\boldsymbol{d}_k}{\gamma}| - \beta\gamma^{-1}, 0\right).
\end{array}
\right.
\end{array}
\tag{2.115}
$$

An example of solution obtained using Algorithm (2.115), where $\boldsymbol{W}$ represents the Symlet orthonormal wavelet analysis operator, with level 3, is displayed in Figure 2.18. While the reconstruction error is rather low, one can observe some negative valued overshoots, typical to the use of orthonormal wavelets. Over-complete dictionaries usually allow to eliminate such behavior, at the price of a higher complexity.

2.4.5 Sparsity Prior and Range Constraints

Let us end this section by focusing on the case when the regularization term includes both sparsity constraints (possibly in a transformed domain), and value range constraints on the sought signal. The problem then becomes:

$$
\underset{\boldsymbol{u} \in \mathbb{R}^N}{\text{minimize}} \ \frac{1}{2}\|\boldsymbol{K}\boldsymbol{u} - \boldsymbol{y}\|^2 + \beta\|\boldsymbol{W}\boldsymbol{u}\|_1 + \iota_{[u_{\min}, u_{\max}]^N}(\boldsymbol{u}). \tag{2.116}
$$

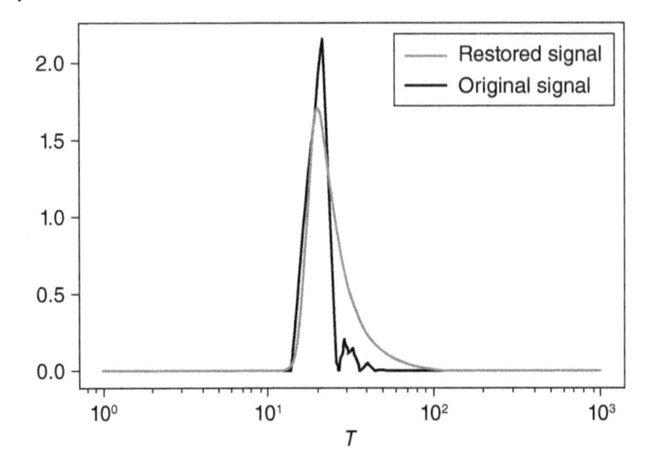

Figure 2.19 Restored signal using Algorithm (2.117) for $\beta = 20$, NMSE = 0.15.

The cost function reads as the sum of three terms, namely a quadratic term, and two non-differentiable terms. An efficient solution to Problem (2.116) can be obtained by using the PPXA+ algorithm (2.69) [87], which reads in this case:

$$
\begin{array}{l}
(\boldsymbol{z}_{0,j})_{1\leq j\leq 3} \in \mathbb{R}^N \\
\gamma \in]0,+\infty[,\ \lambda \in]0,2[\\
\boldsymbol{u}_0 = (\boldsymbol{I}+\boldsymbol{K}^\top\boldsymbol{K}+\boldsymbol{W}^\top\boldsymbol{W})^{-1}\left(\boldsymbol{K}^\top\boldsymbol{z}_{0,1}+\boldsymbol{W}^\top\boldsymbol{z}_{0,2}+\boldsymbol{z}_{0,3}\right) \\
\text{For } k=0,1,\ldots \\
\left\lfloor
\begin{array}{l}
\boldsymbol{v}_{k,1} = \mathrm{prox}_{\frac{\gamma}{2}\|\cdot-\boldsymbol{y}\|^2}(\boldsymbol{z}_{k,1}) = \boldsymbol{y}+(\boldsymbol{z}_{k,1}-\boldsymbol{y})/(\gamma+1) \\
\boldsymbol{v}_{k,2} = \mathrm{prox}_{\gamma\beta\|\cdot\|_1}(\boldsymbol{z}_{k,2}) = \mathrm{sign}(\boldsymbol{z}_{k,2}) \boxdot \max\left(|\boldsymbol{z}_{k,2}|-\beta\gamma,0\right). \\
\boldsymbol{v}_{k,3} = \mathrm{P}_{[u_{\min},u_{\max}]^N}(\mathrm{z}_{k,3}) = \min\left(\max\left(\mathrm{z}_{k,3},u_{\min}\right),u_{\max}\right) \\
\boldsymbol{c}_k = (\boldsymbol{I}+\boldsymbol{K}^\top\boldsymbol{K}+\boldsymbol{W}^\top\boldsymbol{W})^{-1}\left(\boldsymbol{K}^\top\boldsymbol{v}_{k,1}+\boldsymbol{W}^\top\boldsymbol{v}_{k,2}+\boldsymbol{v}_{k,3}\right) \\
\boldsymbol{z}_{k+1,1} = \boldsymbol{z}_{k,1}+\lambda\left(\boldsymbol{K}(2\boldsymbol{c}_k-\boldsymbol{u}_k)-\boldsymbol{v}_{k,1}\right) \\
\boldsymbol{z}_{k+1,2} = \boldsymbol{z}_{k,2}+\lambda\left(\boldsymbol{W}(2\boldsymbol{c}_k-\boldsymbol{u}_k)-\boldsymbol{v}_{k,2}\right) \\
\boldsymbol{z}_{k+1,3} = \boldsymbol{z}_{k,3}+\lambda\left(2\boldsymbol{c}_k-\boldsymbol{u}_k-\boldsymbol{v}_{k,3}\right) \\
\boldsymbol{u}_{k+1} = \boldsymbol{u}_k+\lambda(\boldsymbol{c}_k-\boldsymbol{u}_k).
\end{array}
\right.
\end{array}
\tag{2.117}
$$

An example of solution obtained using Algorithm (2.117) is displayed in Figure 2.19. In this example, the sparse prior combined with range constraints reach the lower NMSE, with an accurate estimation of the maximum peak position which is of interest for chemistry application.

2.4.6 Concluding Remarks

We have illustrated the applicability of the introduced optimization methods, in the context of the restoration of a signal from NMR DOSY measurements. For various choices of the regularization function, we have provided a possible minimization scheme. It is worth noting that there is rarely a single technique available for the resolution of an optimization problem. For a given application, it is always recommended to test and compare different strategies, as we did in the presented example, in order to reach the best possible trade-off in terms of computational complexity and convergence rate.

2.5 Conclusion

Table 2.2 presents a comprehensive list of the algorithms that have been presented in this chapter and the types of problem they can address. Available acceleration strategies to improve their computational efficiency are also

Table 2.2 Summary of the algorithms presented in this chapter.

Algorithm name	Cost function	Particular cases	Acceleration strategy
Generic MM algorithm (2.13)	Any proper lower-semicontinuous function	Expectation-minimization, IRL1, IRLS, Richardson-Lucy, DC programming.	Linesearch strategy [126], block-coordinate approach [13], fixed-point schemes [127].
Quadratic MM algorithm (2.26)	Lipschitz differentiable function.	Gradient descent, SART, half-quadratic (in the smooth case), modified Newton.	Inexact version (2.30), subspace strategy (2.37), block coordinate approach (2.54).
Variable metric forward–backward algorithm (2.39)	Sum of a Lipschitz differentiable and of a convex non-differentiable function.	Forward–backward, ISTA, (scaled) projected gradient, proximal point.	Inexact version [58], block-coordinate approach (2.55).
Primal-dual proximal algorithms [92]	Sum of convex functions composed with linear operators	ADMM (2.67), PPXA+ (2.69), primal-dual algorithm (2.74).	Preconditioning [128], block-coordinate approach [129], distributed strategy [99].
Primal-dual interior point algorithm (2.95)	Twice differentiable function subject to linear constraints		Inexact MM resolution (2.89), linesearch strategy [116].

mentioned. Since optimization constitutes a very rich and continuously evolving scientific area, it was not possible to provide an exhaustive description of all the existing methods and their various, sometimes empirical, extensions. We think however that the presented approaches provide a good basis for sound applicative developments grounded on strong theoretical foundations.

References

1 Rockafellar, R. and Wets, R. (1998) *Variational Analysis, Grundlehren der Mathematischen Wissenschaften*, vol. 317, Springer Berlin, 1st edn.
2 Lee, D.D. and Seung, H.S. (2001) Algorithms for non-negative matrix factorization, in *Advances in Neural and Information Processing Systems*, vol. 13, pp. 556–562.
3 Févotte, C., Bertin, N., and Durrieu, J.L. (2009) Nonnegative matrix factorization with the Itakura-Saito divergence: with application to music analysis. *Neural Computation*, 21 (3), 793–830.
4 Févotte, C. and Idier, J. (2011) Algorithms for nonnegative matrix factorization with the beta-divergence. *Neural Computation*, 23 (9), 2421–2456.
5 Pham, D.T. and Garat, P. (1997) Blind separation of mixtures of independent sources through a quasi maximum likelihood approach. *IEEE Transactions on Signal Processing*, 45 (7), 1712–1725.
6 Cichocki, A., Mandic, D., Lathauwer, L.D., Zhou, G., Zhao, Q., Caiafa, C., and Phan, H.A. (2015) Tensor decompositions for signal processing applications: from two-way to multiway component analysis. *IEEE Signal Processing Magazine*, 32 (2), 145–163.
7 Comon, P. (2014) Tensors: a brief introduction. *IEEE Signal Processing Magazine*, 31 (3), 44–53.
8 Hunter, D.R. and Lange, K. (2004) A tutorial on MM algorithms. *The American Statistician*, 58 (1), 30–37.
9 Lange, K., Hunter, D.R., and Yang, I. (2000) Optimization transfer using surrogate objective functions with discussion. *Journal of Computational and Graphical Statistics*, 9 (1), 1–59.
10 Boyd, S. and Vandenberghe, L. (2004) *Convex Optimization*, Cambridge University Press, New York, 1st edn.
11 Bohning, D. and Lindsay, B.G. (1988) Monotonicity of quadratic-approximation algorithms. *Annals of the Institute of Statistical Mathematics*, 40 (4), 641–663.

12 Zhang, Z., Kwok, J.T., and Yeung, D.Y. (2007) Surrogate maximization/minimization algorithms and extensions. *Machine Learning*, 69, 1–33.

13 Hong, M., Razaviyayn, M., Luo, Z.Q., and Pang, J.S. (2016) A unified algorithmic framework for block-structured optimization involving big data: with applications in machine learning and signal processing. *IEEE Signal Processing Magazine*, 33 (1), 57–77.

14 Horst, R. and Thoai, N.V. (1999) DC programming: overview. *Journal of Optimization Theory and Applications*, 103 (1), 1–43.

15 Nocedal, J. and Wright, S.J. (1999) *Numerical Optimization*, Springer, New York.

16 Combettes, P.L. and Pesquet, J.C. (2010) Proximal splitting methods in signal processing, in *Fixed-Point Algorithms for Inverse Problems in Science and Engineering* (eds H.H. Bauschke, R. Burachik, P.L. Combettes, V. Elser, D.R. Luke, and H. Wolkowicz), Springer, New York, pp. 185–212.

17 Shor, N.Z. (1985) *Minimization Methods for Non-Differentiable Functions, Springer Series in Computational Mathematics*, vol. 3, Springer-Verlag, Berlin Heidelberg.

18 Ahmad, R. and Schniter, P. (2015) Iteratively reweighted ℓ_1 approaches to sparse composite regularization, *Tech. Rep.*. Http://arxiv.org/pdf/1504.05110.pdf.

19 Wipf, D. and Nagarajan, S. (2010) Iterative reweighted ℓ_1 and ℓ_2 methods for finding sparse solutions. *IEEE Journal on Selected Topics in Signal Processing*, 4 (2), 317–329.

20 Carillo, R.E., McEwen, J.D., Van de Ville, D., Thiran, J.P., and Wiaux, Y. (2013) Sparsity averaging for compressive imaging. *IEEE Signal Processing Letters*, 20 (6), 591–594.

21 Candes, E.J., Wakin, M.B., and Boyd, S.P. (2008) Enhancing sparsity by reweighted ℓ_1 minimization. *Journal of Fourier Analysis and Applications*, 14, 877–905.

22 Fuchs, J.J. (2007) Convergence of a sparse representations algorithm applicable to real or complex data. *IEEE Transactions on Signal Processing*, 1 (4), 598–605.

23 Gorodnitsky, I.F. and Rao, B.D. (1997) Sparse signal reconstruction from limited data using FOCUSS: a re-weighted minimum norm algorithm. *IEEE Transactions on Signal Processing*, 45 (3), 600–616.

24 Weiszfeld, E. and Plastria, F. (2009) On the point for which the sum of the distances to n given points is minimum. *Annals of Operations Research*, 167, 7–41.

25 Byrd, R.H. and Payne, D.A. (1979) Convergence of the iteratively reweighted least squares algorithm for robust regression, *Tech. Rep. 131*, The Johns Hopkins University, Baltimore, MD.

26 Rao, B.D., Engan, K., Cotter, S.F., Palmer, J., and Kreutz-Delgao, K. (2003) Subset selection in noise based on diversity measure minimization. *IEEE Transactions on Signal Processing*, 51 (3), 760–770.

27 Allain, M., Idier, J., and Goussard, Y. (2006) On global and local convergence of half-quadratic algorithms. *IEEE Transactions on Image Processing*, 15 (5), 1130–1142.

28 Charbonnier, P., Blanc-Féraud, L., Aubert, G., and Barlaud, M. (1997) Deterministic edge-preserving regularization in computed imaging. *IEEE Transactions on Image Processing*, 6, 298–311.

29 Chan, T.F. and Mulet, P. (1999) On the convergence of the lagged diffusivity fixed point method in total variation image restoration. *SIAM Journal on Numerical Analysis*, 36 (2), 354–367.

30 Geman, D. and Reynolds, G. (1992) Constrained restoration and the recovery of discontinuities. *IEEE Transactions on Pattern Analysis and Machine Intelligence*, 14 (3), 367–383.

31 Geman, D. and Yang, C. (1995) Nonlinear image recovery with half-quadratic regularization. *IEEE Transactions on Image Processing*, 4 (7), 932–946.

32 Idier, J. (2001) Convex half-quadratic criteria and interacting auxiliary variables for image restoration. *IEEE Transactions on Image Processing*, 10 (7), 1001–1009.

33 Nikolova, M. and Ng, M.K. (2005) Analysis of half-quadratic minimization methods for signal and image recovery. *SIAM Journal on Scientific Computing*, 27, 937–966.

34 Beck, A., Teboulle, M., and Chikishev, Z. (2008) Iterative minimization schemes for solving the single source localization problem. *SIAM Journal on Optimization*, 19 (3), 1397–1416.

35 Bissantz, N., Dumbgen, L., Munk, A., and Stratmann, B. (2009) Convergence analysis of generalized iteratively reweighted least squares algorithms on convex function spaces. *SIAM Journal on Optimization*, 19 (4), 1828–1845.

36 Lefkimmiatis, S., Bourquard, A., and Unser, M. (2012) Hessian-based norm regularization for image restoration with biomedical applications. *IEEE Transactions on Image Processing*, 21 (3), 983–995.

37 Fish, D.A., Walker, J.G., Brinicombe, A.M., and Pike, E.R. (1995) Blind deconvolution by means of the Richardson–Lucy algorithm. *Journal of the Optical Society of America A*, 12, 58–65.

38 Vardi, Y., Shepp, L.A., and Kaufman, L. (1985) A statistical model for positron emission tomography (with discussion). *Journal of American Statistical Association*, 80, 8–38.

39 Lanteri, H., Roche, M., and Aime, C. (2002) Penalized maximum likelihood image restoration with positivity constraints: multiplicative algorithms. *Inverse Problems*, 18, 1397–1419.

40 Bertsekas, D.P. (1999) *Nonlinear Programming*, Athena Scientific, Belmont, MA, 2nd edn.

41 Ning, X., Selesnick, I.W., and Duval, L. (2014) Chromatogram baseline estimation and denoising using sparsity (BEADS). *Chemometrics and Intelligent Laboratory Systems*, 139, 156–167.

42 Jiang, M. and Wang, W. (2003) Convergence of the simultaneous algebraic reconstruction technique (SART). *IEEE Transactions on Image Processing*, 12 (8), 957–961.

43 Zibulevsky, M. and Elad, M. (2010) $\ell_2 - \ell_1$ optimization in signal and image processing. *IEEE Signal Processing Magazine*, 27, 76–88.

44 De Pierro, A.R. (1995) A modified expectation maximization algorithm for penalized likelihood estimation in emission tomography. *IEEE Transactions on Medical Imaging*, 14 (1), 132–137.

45 Labat, C. and Idier, J. (2007) Convergence of truncated half-quadratic and Newton algorithms, with application to image restoration, *Tech. Rep.*, IRCCyN, Nantes, France.

46 Chouzenoux, E., Idier, J., and Moussaoui, S. (2011) A majorize–minimize strategy for subspace optimization applied to image restoration. *IEEE Transactions Image Processing*, 20 (18), 1517–1528.

47 Chouzenoux, E., Jezierska, A., Pesquet, J.C., and Talbot, H. (2013) A majorize-minimize subspace approach for ℓ_2-ℓ_0 image regularization. *SIAM Journal Imaging Science*, 6 (1), 563–591.

48 Chouzenoux, E. and Pesquet, J.C. (2017) A stochastic majorize-minimize subspace algorithm for online penalized least squares estimation. *IEEE Transactions on Signal Processing*, 65 (18), 4770–4783.

49 Chouzenoux, E. and Fest, J.B. (2022) SABRINA: a stochastic subspace majorization-minimization algorithm. *Journal of Optimization Theory and Applications*, 195, 919–952.

50 Hager, W.W. and Zhang, H. (2006) A survey of nonlinear conjugate gradient methods. *Pacific Journal on Optimization*, 2 (1), 35–58.

51 Liu, D.C. and Nocedal, J. (1989) On the limited memory BFGS method for large scale optimization. *Mathematical Programming*, 45 (3), 503–528.

52 Chouzenoux, E., Pesquet, J.C., Talbot, H., and Jezierska, A. (2011) A memory gradient algorithm for ℓ_2-ℓ_0 regularization with applications to image restoration, in *18th IEEE International Conference on Image Processing (ICIP 2011)*, Brussels, Belgium, pp. 2717–2720.

53 Florescu, A., Chouzenoux, E., Pesquet, J.C., Ciuciu, P., and Ciochina, S. (2014) A majorize-minimize memory gradient method for complex-valued inverse problem. *Signal Processing*, 103, 285–295. Special issue on Image Restoration and Enhancement: Recent Advances and Applications.

54 Miele, A. and Cantrell, J.W. (1969) Study on a memory gradient method for the minimization of functions. *Journal on Optimization Theory and Applications*, 3 (6), 459–470.

55 Chouzenoux, E. and Pesquet, J.C. (2016) Convergence rate analysis of the majorize–minimize subspace algorithm. *IEEE Signal Processing Letters*, 23 (9), 1284–1288.

56 Chouzenoux, E., Martin, S., and Pesquet, J.C. (2022) A local MM subspace method for solving constrained variational problems in image recovery. *Journal of Mathematical Imaging and Vision*, 65, 253–276.

57 Combettes, P.L. and V u, B.C. (2014) Variable metric forward–backward splitting with applications to monotone inclusions in duality. *Optimization*, 63 (9), 1289–1318.

58 Chouzenoux, E., Pesquet, J.C., and Repetti, A. (2014) Variable metric forward–backward algorithm for minimizing the sum of a differentiable function and a convex function. *Journal of Optimization Theory and Applications*, 162 (1), 107–132.

59 Combettes, P.L. and Wajs, V.R. (2005) Signal recovery by proximal forward-backward splitting. *Multiscale Modeling and Simulation*, 4 (4), 1168–1200.

60 Daubechies, I., Defrise, M., and De Mol, D. (2004) An iterative thresholding algorithm for linear inverse problems with a sparsity constraint. *Communications on Pure and Applied Mathematics*, 57, 1413–1457.

61 Combettes, P., Dung, D., and Vu, B. (2011) Proximity for sums of composite functions. *Journal of Mathematical Analysis and Applications*, 380 (2), 680–688.

62 Abboud, F., Stamm, M., Chouzenoux, E., Pesquet, J.C., and Talbot, H. (2022) Distributed algorithms for scalable proximity operator computation and application to video denoising. *Digital Signal Processing*, 128, 103610.

63 Chambolle, A. and Pock, T. (2015) A remark on accelerated block coordinate descent for computing the proximity operators of a sum

of convex functions. *SMAIJournal of Computational Mathematics*, 1, 29–54.

64 Jaggi, M., Smith, V., Takac, M., Terhorst, J., Krishnan, S., Hofmann, T., and Jordan, M.I. (2014) Communication-efficient distributed dual coordinate ascent, in *Advances in Neural Information Processing Systems 27* (eds Z. Ghahramani, M. Welling, C. Cortes, N. Lawrence, and K. Weinberger), Curran Associates, Inc., pp. 3068–3076.

65 Abboud, F., Chouzenoux, E., Pesquet, J.C., Chenot, J.H., and Laborelli, L. (2015) A distributed strategy for computing proximity operators, in *Proceedings of the 49th Asilomar Conference on Signals, Systems and Computers (ASILOMAR 2015)*, pp. 396–400.

66 Bertsekas, D.P. (1981) Projected Newton methods for optimization problems with simple constraints. *SIAM Journal Control and Optimization*, 20, 762–767.

67 Bonettini, S., Zanella, R., and Zanni, L. (2009) A scaled gradient projection method for constrained image deblurring. *Inverse Problems*, 25 (1), 015 002+.

68 Iusem, A.N. (2003) On the convergence properties of the projected gradient method for convex optimization. *Computational Applied Mathematics*, 22 (1), 37–52.

69 Cherni, A., Chouzenoux, E., Duval, L., and Pesquet, J.C. (2019) A novel smoothed norm ratio for sparse signal restoration application to mass spectrometry, in *Proceedings of Signal Processing with Adaptive Sparse Structured Representations (SPARS 2019)*, Toulouse, France.

70 Cherni, A., Chouzenoux, E., Duval, L., and Pesquet, J.C. (2020) SPOQ lp-Over-lq regularization for sparse signal recovery applied to mass spectrometry. *IEEE Transaction on Signal Processing*, 68, 6070–6084.

71 Jacobson, M.W. and Fessler, J.A. (2007) An expanded theoretical treatment of iteration-dependent majorize-minimize algorithms. *IEEE Transactions on Image Processing*, 16 (10), 2411–2422.

72 Razaviyayn, M., Hong, M., and Luo, Z. (2013) A unified convergence analysis of block successive minimization methods for nonsmooth optimization. *SIAM Journal on Optimization*, 23 (2), 1126–1153.

73 Sotthivirat, S. and Fessler, J.A. (2002) Image recovery using partitioned-separable paraboloidal surrogate coordinate ascent algorithms. *IEEE Transactions on Signal Processing*, 11 (3), 306–317.

74 Chouzenoux, E., Pesquet, J.C., and Repetti, A. (2016) A block coordinate variable metric forward–backward algorithm. *Journal on Global Optimization*, 66 (3), 457–485.

75 Repetti, A., Chouzenoux, E., and Pesquet, J.C. (2014) A preconditioned forward-backward approach with application to large-scale nonconvex spectral unmixing problems, in *Proceedings of the 39th IEEE International Conference on Acoustics, Speech, and Signal Processing (ICASSP 2014)*, Firenze, Italy, pp. 1498–1502.

76 Repetti, A., Pham, M.Q., Duval, L., Chouzenoux, E., and Pesquet, J.C. (2015) Euclid in a taxicab: sparse blind deconvolution with smoothed l1/l2 regularization. *IEEE Signal Processing Letters*, 22 (5), 539–543.

77 Abboud, F., Chouzenoux, E., Pesquet, J., Chenot, J., and Laborelli, L. (2019) An alternating proximal approach for blind video deconvolution. *Signal Processing: Image Communication*, 70, 21–36.

78 Gabay, D. and Mercier, B. (1976) A dual algorithm for the solution of nonlinear variational problems via finite elements approximations. *Computers and Mathematics with Applications*, 2, 17–40.

79 Fortin, M. and Glowinski, R. (eds) (1983) *Augmented Lagrangian Methods: Applications to the Numerical Solution of Boundary-Value Problems*, Elsevier Science Ltd., Amsterdam: North-Holland.

80 Figueiredo, M.A.T. and Bioucas-Dias, J.M. (2010) Restoration of Poissonian images using alternating direction optimization. *IEEE Transactions on Image Processing*, 19 (12), 3133–3145.

81 Boyd, S., Parikh, N., Chu, E., Peleato, B., and Eckstein, J. (2011) Distributed optimization and statistical learning via the alternating direction method of multipliers. *Foundations and Trends in Machine Learning*, 3 (1), 1–222.

82 Lions, P.L. and Mercier, B. (1979) Splitting algorithms for the sum of two nonlinear operators. *SIAM Journal on Numerical Analysis*, 16, 964–979.

83 Combettes, P.L. and Pesquet, J.C. (2007) A Douglas–Rachford splitting approach to nonsmooth convex variational signal recovery. *IEEE Journal of Selected Topics in Signal Processing*, 1 (4), 564–574.

84 Chen, C., He, B., Ye, Y., and Yuan, X. (2014) The direct extension of ADMM for multi-block convex minimization problems is not necessarily convergent. *Mathematical Programming*, 155 (1–2), 1–23.

85 Setzer, S., Steidl, G., and Teuber, T. (2010) Deblurring poissonian images by split bregman techniques. *Journal on Visual Communication and Image Representation*, 21 (3), 193–199.

86 Combettes, P.L. and Pesquet, J.C. (2008) A proximal decomposition method for solving convex variational inverse problems. *Inverse Problems*, 24 (6), 065014.

87 Pesquet, J.C. and Pustelnik, N. (2012) A parallel inertial proximal optimization method. *Pacific Journal of Optimization*, 8 (2), 273–305.

88 Cherni, A., Chouzenoux, E., and Delsuc, M.A. (2016) PALMA, an improved algorithm for DOSY signal processing. *Analyst*, 142 (5), 772–779.

89 Cherni, A., Chouzenoux, E., and Delsuc, M.A. (2016) Proximity operators for a class of hybrid sparsity+entropy priors application to DOSY NMR signal reconstruction, in *Proceedings of the International Symposium on Signal, Image, Video and Communications (ISIVC 2016)*, Tunis, Tunisia.

90 Chambolle, A. and Pock, T. (2011) A first-order primal-dual algorithm for convex problems with applications to imaging. *Journal of Mathematical Imaging and Vision*, 40 (1), 120–145.

91 Esser, E., Zhang, X., and Chan, T. (2010) A general framework for a class of first order primal-dual algorithms for convex optimization in imaging science. *SIAM Journal on Imaging Sciences*, 3 (4), 1015–1046.

92 Komodakis, N. and Pesquet, J.C. (2014) Playing with duality: an overview of recent primal-dual approaches for solving large-scale optimization problems. *IEEE Signal Processing Magazine*, 32 (6), 31–54.

93 Combettes, P. and Pesquet, J.C. (2021) Fixed point strategies in data science. *IEEE Transactions on Signal Processing*, 69, 3878–3905.

94 Combettes, P.L. and Pesquet, J.C. (2012) Primal-dual splitting algorithm for solving inclusions with mixtures of composite, lipschitzian, and parallel-sum type monotone operators. *Set-Valued and Variational Analysis*, 20 (2), 307–330.

95 Condat, L. (2013) A primal-dual splitting method for convex optimization involving Lipschitzian, proximable and linear composite terms. *Journal of Optimization Theory and Applications*, 158 (2), 460–479.

96 V u, B.C. (2013) A splitting algorithm for dual monotone inclusions involving cocoercive operators. *Advances in Computational Mathematics*, 38 (3), 667–681.

97 Raguet, H., Fadili, J., and Peyré, G. (2013) A generalized forward-backward splitting. *SIAM Journal on Imaging Sciences*, 6 (3), 1199–1226.

98 Bricenos-Arias, L., Chierchia, G., Chouzenoux, E., and Pesquet, J.C. (2019) A random block-coordinate douglas-rachford splitting method with low computational complexity for binary logistic regression. *Computational Optimization and Applications*, 72 (3), 707–726.

99 Pesquet, J.C. and Repetti, A. (2015) A class of randomized primal-dual algorithms for distributed optimization. *Journal of nonlinear and convex analysis*, 16 (12), 2453–2490.

100 Cherni, A., Chouzenoux, E., and Delsuc, M.A. (2018) Fast dictionnary-based approach for mass spectrometry data analysis, in *Proceedings of the IEEE International Conference on Acoustics, Speech, and Signal Processing (ICASSP 2018)*, Calgary, Canada.

101 Fiacco, A.V. and McCormick, G.P. (1967) The sequential unconstrained minimization technique (SUMT) without parameters. *Operations Research*, 15 (5), 820–827.

102 Wright, S.J. (1997) *Primal-Dual Interior-Point Methods*, SIAM, Philadelphia, PA, 1st edn.

103 Forsgren, A., Gill, P.E., and Wright, M.H. (2002) Interior methods for nonlinear optimization. *SIAM Review*, 44 (4), 525–597.

104 Moussaoui, S., Chouzenoux, E., and Idier, J. (2012) Primal-dual interior point optimization for penalized least squares estimation of abundance maps in hyperspectral imaging, in *Proceedings of the 4th Workshop on Hyperspectral Image and Signal Processing: Evolution in Remote Sensing (WHISPERS 2012)*, Shangai, China.

105 Chouzenoux, E., Legendre, M., Moussaoui, S., and Idier, J. (2012) Fast constrained least squares spectral unmixing using primal-dual interior-point optimization. *IEEE Journal of Selected Topics in Applied Earth Observations and Remote Sensing*, 7 (1), 59–69.

106 Armand, P., Gilbert, J.C., and Jan-Jégou, S. (2000) A feasible BFGS interior point algorithm for solving strongly convex minimization problems. *SIAM Journal on Optimization*, 11, 199–222.

107 Wright, M.H. (1994) Some properties of the Hessian of the logarithmic barrier function. *Mathematical Programming*, 67 (2), 265–295.

108 Wright, M.H. (1998) Ill-conditioning and computational error in interior methods for nonlinear programming. *SIAM Journal on Optimization*, 9 (1), 84–111.

109 Friedlander, M.P. and Orban, D. (2012) A primal–dual regularized interior-point method for convex quadratic programs. *Mathematical Programming Computation*, 4 (1), 71–107.

110 Conn, A., Gould, N., and Toint, P.L. (1996) A primal-dual algorithm for minimizing a nonconvex function subject to bounds and nonlinear constraints, in *Nonlinear Optimization and Applications* (eds G. Di Pillo and F. Giannessi), Kluwer Academic Publishers, 2nd edn.

111 Armand, P., Benoist, J., and Dussault, J. (2012) Local path-following property of inexact interior methods in nonlinear programming. *Computational Optimization and Applications*, 52 (1), 209–238.

112 Bonettini, S., Galligani, E., and Ruggiero, V. (2007) Inner solvers for interior point methods for large scale nonlinear programming. *Computational Optimization and Applications*, 37 (1), 1–34.

113 Legendre, M., Moussaoui, S., Chouzenoux, E., and Idier, J. (2014) Primal-dual interior-point optimization based on majorization-minimization for edge preserving spectral unmixing, in *Proceedings of the 21st IEEE International Conference on Image Processing (ICIP 2014)*, Shangai, China, pp. 4161–4165.

114 Legendre, M., Moussaoui, S., Schmidt, F., and Idier, J. (2013) Parallel implementation of a primal-dual interior-point optimization method for fast abundance maps estimation, in *Proceedings of the 5th Workshop on Hyperspectral Image and Signal Processing: Evolution in Remote Sensing (WHISPERS 2013)*, Gainesville, FL, USA.

115 Murray, W. and Wright, M.H. (1994) Line search procedures for the logarithmic barrier function. *SIAM Journal on Optimization*, 4 (2), 229–246.

116 Chouzenoux, E., Moussaoui, S., and Idier, J. (2012) Majorize-minimize linesearch for inversion methods involving barrier function optimization. *Inverse Problems*, 28 (6), 065 011.

117 Chouzenoux, E., Moussaoui, S., and Idier, J. (2011) Efficiency of line search strategies in interior point methods for linearly constrained optimization, in *Proceedings of the IEEE Workshop on Statistical Signal Processing (SSP 2011)*, Nice, France, pp. 101–104.

118 Chouzenoux, E., Moussaoui, S., Idier, J., and Mariette, F. (2013) Primal-dual interior point optimization for a regularized reconstruction of NMR relaxation time distributions, in *Proceedings of the 38th IEEE International Conference on Acoustics, Speech, and Signal Processing (ICASSP 2013)*, Vancouver, Canada, pp. 8747–8750.

119 El-Bakry, A.S., Tapia, R.A., Tsuchiya, T., and Zhang, Y. (1996) On the formulation and theory of the Newton interior-point method for nonlinear programming. *Journal of Optimization Theory and Applications*, 89, 507–541.

120 Johnson, C.A., Seidel, J., and Sofer, A. (2000) Interior-point methodology for 3-D PET reconstruction. *IEEE Transactions on Medical Imaging*, 19 (4), 271–285.

121 Chouzenoux, E., Moussaoui, S., Idier, I., and Mariette, F. (2010) Efficient maximum entropy reconstruction of nuclear magnetic resonance T1-T2 spectra. *IEEE Transactions on Signal Processing*, 58 (12), 6040–6051.

122 Kaipin, X. and Shanmin, Z. (2013) Trust-region algorithm for the inversion of molecular diffusion NMR data. *Analytical Chemistry*, 86 (1), 592–599.

123 Johnson, C.S. (1999) Diffusion ordered nuclear magnetic resonance spectroscopy: principles and applications. *Progress in Nuclear Magnetic Resonance Spectroscopy*, 34, 203–256.

124 Beck, A. and Teboulle, M. (2009) A fast iterative shrinkage-thresholding algorithm for linear inverse problems. *SIAM Journal on Imaging Sciences*, 2 (1), 183–202.

125 Pustelnik, N., Benazza-Benhayia, A., Zheng, Y., and Pesquet, J.C. (2016) Wavelet-based image deconvolution and reconstruction. *Wiley Encyclopedia of Electrical and Electronics Engineering*, https://onlinelibrary.wiley.com/action/showCitFormats?doi=10.1002%2F047134608X.W8294.

126 Lange, K. (1995) A gradient algorithm locally equivalent to the EM algorithm. *Journal of the Royal Statistical Society: Series B (Methodological)*, 57 (2), 425–437.

127 Varadhan, R. and Roland, C. (2008) Simple and globally convergent methods for accelerating the convergence of any em algorithm. *Scandinavian Journal of Statistics*, 35 (2), 335–353.

128 Pock, T. and Chambolle, A. (2011) Diagonal preconditioning for first order primal-dual algorithms in convex optimization, in *IEEE International Conference on Computer Vision (ICCV)*, Barcelona, Spain, pp. 1762–1769.

129 Repetti, A., Chouzenoux, E., and Pesquet, J.C. (2015) A random block-coordinate primal-dual proximal algorithm with application to 3D mesh denoising, in *Proceedings of the 40th IEEE International Conference on Acoustics, Speech, and Signal Processing (ICASSP 2015)*, Brisbane, Australia, pp. 3561–3565.

3

Non-negative Matrix Factorization

David Brie[1], Nicolas Gillis[2], and Saïd Moussaoui[3]

[1] *CRAN, Lorraine University, Nancy, France*
[2] *Department of Mathematics and Operational Research, University of Mons, Mons, Belgium*
[3] *LS2N, Nantes Université, Ecole Centrale Nantes, Nantes, France*

Solving a source separation problem when the observations can be interpreted as linear instantaneous mixtures of non-negative sources with non-negative mixing weights reduces to performing a non-negative factorization of the data matrix. This problem is referred to as non-negative matrix factorization (NMF). NMF has a long history originating from linear algebra and analytical chemistry, and extensive developments have been achieved in the signal and image processing fields. The popularity of NMF is guided either by the mathematically challenging question of factorizing a matrix under non-negativity constraint and also by the need to explain observations as purely additive combination of non-negative factors or physically meaning quantities. This chapter addresses the concept of NMF, presents its foundations in terms of model setting and indeterminacy in addition to the main guidelines of existing factorization algorithms. The application of NMF to real situations of chemical data processing is illustrated with two examples of Raman spectroscopy measurements.

3.1 Introduction

The linear instantaneous mixing model assumes that P observations $x_p(t)$, for $t = 1, ..., T$ and $p = 1, \dots, P$, are linear instantaneous combinations of R unknown source signals $s_r(t)$, with $t = 1, ..., T$ and $r = 1, \dots, R$. That is:

$$x_p(t) = \sum_{r=1}^{R} A_{pr}\, s_r(t) + e_p(t), \quad (\forall p = 1, \dots, P)\ (\forall t = 1, \dots, T), \qquad (3.1)$$

Source Separation in Physical-Chemical Sensing, First Edition.
Edited by Christian Jutten, Leonardo Tomazeli Duarte, and Saïd Moussaoui.

where A_{pr} are the mixing coefficients and the additive noise terms $e_p(t)$, for $t = 1, ..., T$ and $p = 1, \dots, P$ represent measurement errors and model uncertainties.

By gathering the P observations in a vector $\boldsymbol{x}(t)$, the model can be written as

$$\boldsymbol{x}(t) = \boldsymbol{A}\ \boldsymbol{s}(t) + \boldsymbol{e}(t), \quad (\forall t = 1, \dots, T), \tag{3.2}$$

where $\boldsymbol{A} \in \mathbb{R}^{P \times R}$ is the matrix and t index refers to the observation variability parameter depending on the considered application. It can correspond for instance to time, frequency, wavelength, and pixel index. A matrix notation is usually used by merging the observations $\boldsymbol{x}(t)$ into a data matrix $\boldsymbol{X} \in \mathbb{R}^{P \times T}$, the source signals $\boldsymbol{s}(t)$ into the source matrix $\boldsymbol{S} \in \mathbb{R}^{R \times T}$, and the noise terms into $\boldsymbol{E} \in \mathbb{R}^{P \times T}$.

Therefore, given the mixture data $\boldsymbol{X}$, the source separation aims at recovering the source matrix $\boldsymbol{S}$ and the mixing matrix $\boldsymbol{A} \in \mathbb{R}^{P \times R}$ such that

$$\boldsymbol{X} = \boldsymbol{A}\boldsymbol{S} + \boldsymbol{E}, \tag{3.3}$$

In this chapter, we consider the specific case where the matrices containing the source signals and the mixing coefficients are component-wise non-negative, that is, $\boldsymbol{A} \geqslant 0$ and $\boldsymbol{S} \geqslant 0$.

Finding non-negative matrices $\boldsymbol{S}$ and $\boldsymbol{A}$ allowing to reproduce the data matrix $\boldsymbol{X}$ according to (3.3) is known as *non-negative matrix factorization* (NMF).

3.1.1 Brief Historical Overview

It is difficult to trace back the first time the model (3.1) with non-negativity constraints was introduced, as it is a rather natural model in many situations; see Section 3.5 for the description of several applications. For example, Imbrie and Van Andel [1] used this model in 1960s for the analysis of mineral data. In the field of linear algebra, the first publications related to the mathematical formulation of the NMF problem concentrated their effort on the conditions for the existence and the uniqueness of such factorization [2–6]. The problem was named as *non-negative rank factorization* and defined as the factorization of a non-negative matrix into the product of two non-negative matrices. However, NMF in its current form (3.2) was introduced by Paatero and Tapper [7] and referred to as positive matrix factorization (PMF). In 1999, Lee and Seung [8] popularized NMF with a paper in Nature “Learning the parts of objects by non-negative matrix factorization” where they applied it to the extraction of facial features in a set of facial images and to identify topics in a set of documents. Regarding the decomposition algorithms, pioneering contributions of Tauler *et al.* [9], Paatero and Tapper [7, 10] proposed original algorithms to find an approximate factorization of a matrix in the case of spectroscopic data and noisy observations, by alternating non-negative least squares estimation in the former and penalized least squares estimation in the latter. They proposed a simple alternating optimization scheme (optimizing over $\boldsymbol{A}$ and $\boldsymbol{S}$ alternatively; see Section 3.4) and applied it to air emission control. More recently, Lee and Seung [8] presented two algorithms based on multiplicative updates dedicated to non-negative matrix factorization (NMF); one for the Frobenius norm and one for the Kullback–Leibler divergence (see Section 3.4). NMF and source separation with non-negativity constraint have since remained an active research topic using several factorization approaches and applications [11–15]. Since then, NMF has become an increasingly popular method and has been used successfully in many different applications. See for instance Section 3.5 and Chapter 5 where some examples are given.

3.2 Geometrical Interpretation of NMF and the Non-negative Rank

Note that, even in noisy settings, the geometrical interpretation that we will describe in this section for NMF is useful because the noisy observations are approximated using the exact linear mixing model.

3.2.1 Non-negative Rank Formulation

Let us assume that there is no noise and that the linear mixture model in (3.3) is exact, that is, each observation can be written as a non-negative linear combination of the R sources. Given $\boldsymbol{X}$, finding $\boldsymbol{A} \geqslant 0$ and $\boldsymbol{S} \geqslant 0$ such that $\boldsymbol{X} = \boldsymbol{A}\boldsymbol{S}$ is referred to as *exact NMF*. The smallest R such that such a decomposition of $\boldsymbol{X}$ exists is called the non-negative rank of $\boldsymbol{X}$ and is denoted $\text{rank}_+\{\boldsymbol{X}\}$. Clearly, we have

$$\text{rank}\{\boldsymbol{X}\} \leq \text{rank}_+\{\boldsymbol{X}\} \leq \min(P, T).$$

3.2.2 Convex Cone Formulation

Let assume the noiseless case and we get rid of the index t in (3.1) by forming vector $\boldsymbol{x}_p$ from the p-th row of $\boldsymbol{X}$ and $\boldsymbol{s}_r$ from the r-th row of $\boldsymbol{S}$. We have:

$$\boldsymbol{x}_p = \sum_{r=1}^{R} A_{pr}\, \boldsymbol{s}_r \quad (\forall p = 1, \dots, P). \tag{3.4}$$

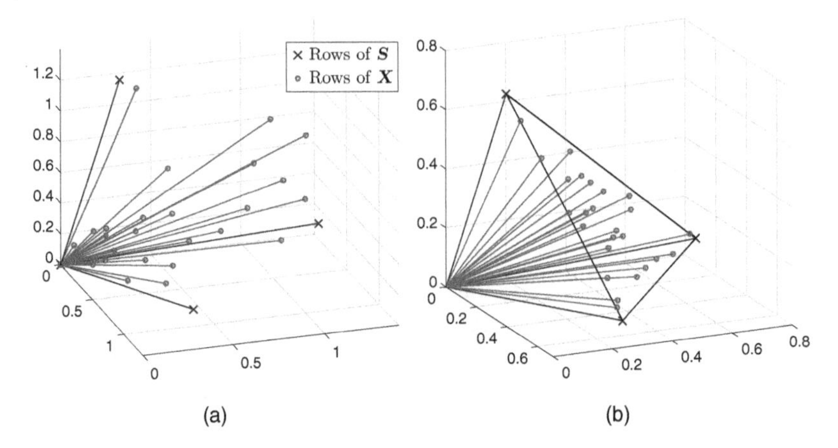

Figure 3.1 Geometric illustration of exact NMF for $T = R = 3$ and $P = 25$. Both figures represent the same data set. The figure (b) is the normalization to unit ℓ_1 norm of the rows of $\boldsymbol{X}$ and $\boldsymbol{S}$ from the Figure (a).

Moreover, let us introduce the vector $\boldsymbol{a}_p$ containing the mixing coefficients of the R sources in the p-th mixture (p-th row of matrix $\boldsymbol{A}$).

Since the mixing coefficients A_{pr} are non-negative, the rows of $\boldsymbol{X}$, that is observations $\boldsymbol{x}_p$, belong to the convex cone generated by the rows of $\boldsymbol{S}$, that is, by the set of sources $\{\boldsymbol{s}_1, \boldsymbol{s}_2, \dots \boldsymbol{s}_R\}$; see Figure 3.1a for an illustration in the case $R = 3$.

Equivalently, we can scale the observations and the sources as follows:

$$\underbrace{\frac{\boldsymbol{x}_p}{\|\boldsymbol{x}_p\|_1}}_{\boldsymbol{x}'_p} = \sum_{r=1}^{R} \underbrace{\left(A_{pr} \frac{\|\boldsymbol{s}_r\|_1}{\|\boldsymbol{x}_p\|_1} \right)}_{A'_{pr}} \underbrace{\frac{\boldsymbol{s}_r}{\|\boldsymbol{s}_r\|_1}}_{\boldsymbol{s}'_r}, \tag{3.5}$$

where $\|\boldsymbol{y}\|_1 = \sum_i |y_i|$. In this way, vectors $\boldsymbol{x}'_p$ and $\boldsymbol{s}'_r$ expressed in (3.5) belong to the probability simplex $\mathcal{S}$, defined as

$$\mathcal{S} = \left\{ \boldsymbol{y} \in \mathbb{R}^T \middle| y_t \geqslant 0 \text{ for } 1 \leq t \leq T, \text{ and } \sum_{t=1}^{T} y_t = 1 \right\}.$$

Consequently, we get $\|\boldsymbol{a}_p\|_1 = \sum_{r=1}^{R} A'_{pr} = 1$, for all $p = 1, \dots, P$, since $\boldsymbol{x}'_p = \sum_{r=1}^{R} A'_{pr}\boldsymbol{s}'_r \in \mathcal{S}$ and $\boldsymbol{s}'_r \in \mathcal{S}$. In other words, the vectors $\boldsymbol{x}'_p$ $(1 \leq p \leq P)$ belong to the convex hull of the set $\{\boldsymbol{s}'_1, \boldsymbol{s}'_2, \dots \boldsymbol{s}'_R\}$, defined as

$$\begin{aligned} \text{conv}\{\boldsymbol{s}'_1, \boldsymbol{s}'_2, \dots \boldsymbol{s}'_r\} &= \left\{ \sum_{r=1}^{R} A'_{pr}\boldsymbol{s}'_r \;\middle|\; \boldsymbol{a} \in \mathcal{S} \right\} \\ &= \text{conv}\{\boldsymbol{S}'\} = \{ \boldsymbol{a}'\boldsymbol{S}' \mid \boldsymbol{a} \in \mathcal{S} \}, \end{aligned}$$

where $\boldsymbol{a} \in \mathbb{R}_+^R$ represents the vector of mixing coefficients. See Figure 3.1b for an illustration.

For simplicity, we assume in the remainder of this section that $\boldsymbol{x}_p$ and $\boldsymbol{s}_r$ are normalized to have unit ℓ_1 norm for all $1 \le p \le P$ and $1 \le r \le R$. Therefore, the entries of $\boldsymbol{x}_p$ and $\boldsymbol{s}_r$ sum-to-one (note that we also assume without loss of generality that the observations $\boldsymbol{x}_p$, the sources $\boldsymbol{s}_r$ and the mixing vectors $\boldsymbol{a}_p$ are different from zero, otherwise we discard them).

3.2.3 Nested Polytope Formulation

Given a normalized $\boldsymbol{X} \geqslant 0$, finding $\boldsymbol{A} \geqslant 0$ and $\boldsymbol{S} \geqslant 0$ such that $\boldsymbol{X} = \boldsymbol{AS}$ is equivalent to finding a set of sources $\{\boldsymbol{s}_1, \boldsymbol{s}_2, \dots \boldsymbol{s}_r\}$ such that

$$\text{conv}\{\boldsymbol{x}_1, \boldsymbol{x}_2, \dots \boldsymbol{x}_P\} \subseteq \text{conv}\{\boldsymbol{s}_1, \boldsymbol{s}_2, \dots \boldsymbol{s}_R\} \subseteq \mathcal{S}.$$

This is an instance of the so-called *nested polytope problem* (NPP) in computational geometry; see, e.g. [16] and the references therein. NPP is defined as follows: given two nested polytopes, $\mathcal{P} \subset \mathcal{Q}$, the goal is to find, if possible, a set of R points $\{\boldsymbol{s}_1, \boldsymbol{s}_2, \dots \boldsymbol{s}_R\}$ such that its convex hull is nested between the two given polytopes $\mathcal{P}$ and $\mathcal{Q}$, that is, such that

$$\mathcal{P} \subset \text{conv}\{\boldsymbol{s}_1, \boldsymbol{s}_2, \dots \boldsymbol{s}_R\} \subset \mathcal{Q}.$$

See Figure 3.2 for an illustration in two dimensions where $\mathcal{P}$ and $\mathcal{Q}$ are squares, and $\text{conv}\{\boldsymbol{s}_1, \boldsymbol{s}_2, \boldsymbol{s}_3\}$ is a nested triangle. For the exact NMF problem above, we have $\mathcal{P} = \text{conv}\{\boldsymbol{X}\}$ and $\mathcal{Q} = \mathcal{S}$. Note that, the outer polytope $\mathcal{Q}$ can have a higher dimension than the inner polytope $\mathcal{P}$. NPP is a very difficult geometric problem, being NP-hard already in dimension three [16], although a polynomial-time algorithm exists when the inner and outer polytopes have dimension two, that is, when they are polygons [17, 18].

The one-to-one equivalence between the exact NMF and NPP was established in [19–22]. In fact, it can also be shown that any NPP instance can be written as an exact NMF problem: Given the outer polytope described with its T facets $\{\boldsymbol{y} | \boldsymbol{a}_p^\top \boldsymbol{y} \le b_t; 1 \le t \le T\}$ and the P vertices $\boldsymbol{v}_p$ with $1 \le p \le P$ of the inner polytope, solving NPP is equivalent to solving exact NMF for the input matrix whose components are

$$x_p(t) = b_t - \boldsymbol{a}_p^\top \boldsymbol{v}_p \geqslant 0 \quad (\forall t = 1,2,\dots,T), (\forall p = 1,2,\dots,P).$$

It is often assumed in practice that the rank of the input matrix $\boldsymbol{X}$ is equal to the number of sources R, that is, $R = \text{rank}\{\boldsymbol{X}\}$. In that case, the NPP problem corresponding to exact NMF can be simplified because the space spanned by the rows of $\boldsymbol{X}$ and $\boldsymbol{S}$ must coincide, since $\boldsymbol{X} = \boldsymbol{AS}$ and $\boldsymbol{S}$ has R rows. Therefore, the rows of $\boldsymbol{S}$ must be contained in the polytope $\mathcal{S} \cap \text{row}(\boldsymbol{X})$, where row($\boldsymbol{X}$) denotes the row space of $\boldsymbol{X}$. Hence we can restrict the outer polytope $\mathcal{S}$ to be $\mathcal{S} \cap \text{row}(\boldsymbol{X})$, so that the dimensions of the inner and outer polytopes coincide and are equal to $\text{rank}\{\boldsymbol{X}\} - 1$. Therefore, we are looking for a matrix $\boldsymbol{S}$ such that

$$\text{conv}\{\boldsymbol{X}\} \subseteq \text{conv}\{\boldsymbol{S}\} \subseteq \mathcal{S} \cap \text{row}(\boldsymbol{X}). \tag{3.6}$$

We will refer to this variant of NPP as the restricted NPP, where the inner and outer polytopes have the same dimension (hence the nested polytope also has the same dimension). The one-to-one correspondence between exact NMF with $R = \text{rank}\{\boldsymbol{X}\}$ and restricted NPP was established in [19, 20]:

Theorem 3.1 *Given a normalized non-negative matrix* $\boldsymbol{X} \in \mathbb{R}_+^{P\times T}$, *deciding whether* $\text{rank}\{\boldsymbol{X}\} = \text{rank}_+\{\boldsymbol{X}\}$ *and computing a corresponding factorization* $\boldsymbol{X} = \boldsymbol{A}\boldsymbol{S}$ *with* $\boldsymbol{A} \in \mathbb{R}_+^{P\times R}$ *and* $\boldsymbol{S} \in \mathbb{R}_+^{R\times T}$ *is equivalent to deciding whether there exists a polytope with* R *vertices nested between* $\text{conv}\{\boldsymbol{X}\}$ *and* $\mathcal{S} \cap \text{row}(\boldsymbol{X})$.

The theorem above can be generalized to check whether the $\text{rank}_+\{\boldsymbol{X}\} = \text{rank}\{\boldsymbol{X}\} + 1$. In fact, when $R = \text{rank}\{\boldsymbol{X}\} + 1$, it can be shown that the nested polytope can be assumed, without loss of generality, to have the same dimension as the inner polytope. In other words, it can be assumed without loss of generality that $\text{rank}\{\boldsymbol{S}\} = \text{rank}\{\boldsymbol{X}\}$. Therefore, checking whether $\text{rank}_+\{\boldsymbol{X}\} = \text{rank}\{\boldsymbol{X}\} + 1$ is equivalent to checking whether there exists a polytope with $R + 1$ vertices nested between $\text{conv}\{\boldsymbol{X}\}$ and $\mathcal{S} \cap \text{row}(\boldsymbol{X})$ [20, Corollary 2]. However, if $\text{rank}_+\{\boldsymbol{X}\} > \text{rank}\{\boldsymbol{X}\} + 1$, then in general $\text{rank}\{\boldsymbol{S}\} > \text{rank}\{\boldsymbol{X}\}$ and it can no longer be assumed that the nested and inner polytopes have the same dimension, that is, we will have that $\text{conv}\{\boldsymbol{S}\} \not\subset \mathcal{S} \cap \text{row}(\boldsymbol{X})$ [21] (see example 2 below).

3.2.4 Non-negative Rank Computation

We can use the equivalence between NPP and the computation of the non-negative rank described above to derive possible values of the non-negative rank of a non-negative matrix $\boldsymbol{X} \in \mathbb{R}^{P\times T}$ in simple cases:

- For $\text{rank}\{\boldsymbol{X}\} = 1$, it is clear that $\text{rank}_+\{\boldsymbol{X}\} = 1$ since all rows of $\boldsymbol{X}$ are multiple of the same vector. Geometrically, the inner polytope corresponding to the NPP instance is a single point and the NPP instance is trivial.
- For $\text{rank}\{\boldsymbol{X}\} = 2$, the inner polytope is a line segment. Therefore, the problem can easily be solved by identifying the two extreme points of that line segment hence $\text{rank}_+\{\boldsymbol{X}\} = 2$ [3].
- For $\min(P, T) = \text{rank}\{\boldsymbol{X}\}$, we have $\text{rank}_+\{\boldsymbol{X}\} = \text{rank}\{\boldsymbol{X}\}$ using the trivial decompositions $\boldsymbol{X} = \boldsymbol{X}\boldsymbol{I}_T = \boldsymbol{I}_P\boldsymbol{X}$ where $\boldsymbol{I}_N$ is the identity matrix of

dimension N. Geometrically, this means that we either take the nested polytope conv$\{\boldsymbol{S}\}$ as the inner polytope conv$\{\boldsymbol{X}\}$ (for $\boldsymbol{X} = \boldsymbol{X}\boldsymbol{I}_T$ where $\boldsymbol{S} = \boldsymbol{X}$) or as the outer polytope $\mathcal{S}$ (for $\boldsymbol{X} = \boldsymbol{I}_P\boldsymbol{X}$ where $\boldsymbol{S} = \boldsymbol{I}_P$).

From the three results above, we have the following theorem from [23].

Theorem 3.2 *Let* $\boldsymbol{X} \in \mathbb{R}_+^{P\times T}$. *If* $\text{rank}\{\boldsymbol{X}\} \leq 2$ *or* $\min(P, T) = \text{rank}\{\boldsymbol{X}\}$, *then* $\text{rank}_+\{\boldsymbol{X}\} = \text{rank}\{\boldsymbol{X}\}$.

When $\min(P, T) \geqslant 4$ and $\text{rank}\{\boldsymbol{X}\} \geqslant 3$, determining the non-negative rank becomes more difficult.

Let us analyze higher dimensional cases.

- For $\text{rank}\{\boldsymbol{X}\} = 3$ and $\min(P, T) \geqslant 4$, the restricted NPP instance is the problem to finding a polygon nested between two given polygons. As mentioned above, this problem can be solved in polynomial time, more precisely in $O((P + T)\log(\min(P, T)))$ operations [17]. Therefore, when $\text{rank}\{\boldsymbol{X}\} = 3$, one can decide in polynomial time whether $\text{rank}_+\{\boldsymbol{X}\} = 3$. Moreover, because it does not help to try to find a higher dimensional nested polytope with only four vertices (this is the case since $R = \text{rank}\{\boldsymbol{X}\} + 1$; see the discussion after Theorem 3.1), this algorithm can also be used to decide in polynomial time whether $\text{rank}_+\{\boldsymbol{X}\} = 4$.
  Deciding whether $\text{rank}_+\{\boldsymbol{X}\} = 5$ becomes more difficult (unless min $(P, T) = 5$) because the nested polytope might live in a higher dimension than conv$\{\boldsymbol{X}\}$; see the second example below. In fact, it is important to realize that, when $\text{rank}\{\boldsymbol{X}\} \geqslant 3$, the non-negative rank of $\boldsymbol{X}$ can be arbitrarily large. For example, for the matrix corresponding to the restricted NPP instances of the regular P-gon nested with itself (for which $T = P$ and conv$\{\boldsymbol{X}\} = \mathcal{S} \cap \text{row}(\boldsymbol{X})$), we have [24]

$$\text{rank}_+\{\boldsymbol{X}\} \geqslant \lceil \log_2(2P + 2) \rceil,$$

  while $\text{rank}\{\boldsymbol{X}\} = 3$; see also [25] the references therein for more details.
- For $R = \text{rank}\{\boldsymbol{X}\} = 4$ and $\min(P, T) \geqslant 5$, it is difficult to compute the non-negative rank [20] since three-dimensional restricted NPP instances are NP-hard [16] (note that this result does not fix a priori the number of vertices of the nested polytope).
  However, checking whether $\text{rank}\{\boldsymbol{X}\} = R$ for fixed R (that is, R is not part of the input) can be done in polynomial time in P and T requiring $\mathcal{O}\left((PT)^{R^2}\right)$ operations [26]. This approach, although theoretically appealing, is not very useful in practice as it requires to solving systems of equations via quantifier elimination theory, and we were not able to identify a software able to solve problems already for $P = T = 4$ and $R = 3$, the first non-trivial case (see the discussion before Theorem 3.2).

3.2.5 Illustrative Examples

Example 1: Nested squares [18]. We consider the smallest possible case where $\text{rank}_+\{\boldsymbol{X}\} > \text{rank}\{\boldsymbol{X}\}$: this requires that P and T are at least 4 (see the discussion in Section 3.2.4). Consider the matrix

$$\boldsymbol{X} = \frac{1}{4}\begin{pmatrix} 1+a & 1-a & 1+a & 1-a \\ 1-a & 1+a & 1+a & 1-a \\ 1-a & 1+a & 1-a & 1+a \\ 1+a & 1-a & 1-a & 1+a \end{pmatrix}, \tag{3.7}$$

where $0 < a \leq 1$ and $\text{rank}\{\boldsymbol{X}\} = 3$. The restricted NPP instances corresponding to the exact NMF problem for $\boldsymbol{X}$ (Theorem 3.3) are two nested squares; see Figure 3.2.

The following can be shown [27]:

- For $0 < a < \sqrt{2} - 1$, the inner square is small enough so that there exists infinitely many triangles in between the two nested squares. This implies that $\text{rank}_+\{\boldsymbol{X}\} = 3$. This also implies that the exact NMF of $\boldsymbol{X}$ is (highly) non-unique; see Section 3.3 for more details.
  (Note that for $a = 0$, the inner "square" is a single point and $\text{rank}\{\boldsymbol{X}\} = 1$.)

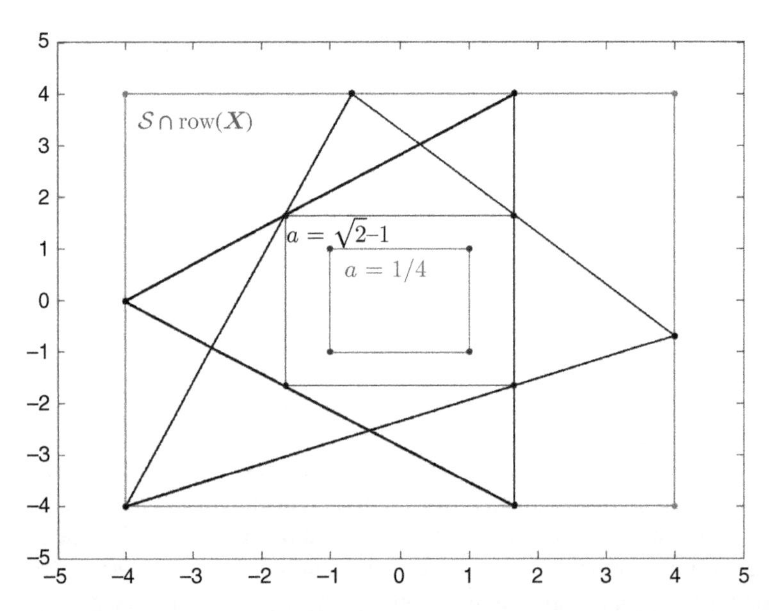

Figure 3.2 Restricted NPP instances of two nested squares corresponding to the exact NMF problem for $\boldsymbol{X}$ from (3.7) for $a = \sqrt{2} - 1$ and $a = 1/4$ [27]. The two triangles correspond to two exact NMFs for $0 < a \leq \sqrt{2} - 1$.

- For $a = \sqrt{2} - 1$, there exists eight different triangles nested between the two squares hence $\text{rank}_+\{\boldsymbol{X}\} = 3$ (leading to eight different exact factorizations, up to permutation and scaling, not infinitely many as above); see Figure 3.2 where two such triangles are represented (the other six solutions are rotations of these two).
- For $a > \sqrt{2} - 1$, there does not exist any triangle between the two nested squares (note that, for $a = 1$, the two squares coincide) hence $\text{rank}_+\{\boldsymbol{X}\} = 4$.

Example 2: Regular hexagon. A popular example is the non-negative matrix corresponding to the restricted NPP instance of the regular hexagon nested with itself. It is the smallest nontrivial case for which $\text{rank}\{\boldsymbol{S}\} > \text{rank}\{\boldsymbol{X}\}$ is necessary to find the exact NMF of minimum rank when $\text{rank}\{\boldsymbol{X}\} = 3$. In fact, for $\text{rank}\{\boldsymbol{S}\} > \text{rank}\{\boldsymbol{X}\}$ to be necessary when $\text{rank}\{\boldsymbol{X}\} = 3$, we need that $\text{rank}_+\{\boldsymbol{X}\}$ is at least five (see the discussion in Section 3.2.4) hence we need $\min(P, T) \geqslant 6$ to have a nontrivial factorization.

Let us consider the following non-negative matrix

$$\boldsymbol{X} = \begin{pmatrix} 0 & 1 & 4 & 9 & 16 & 25 \\ 1 & 0 & 1 & 4 & 9 & 16 \\ 4 & 1 & 0 & 1 & 4 & 9 \\ 9 & 4 & 1 & 0 & 1 & 4 \\ 16 & 9 & 4 & 1 & 0 & 1 \\ 25 & 16 & 9 & 4 & 1 & 0 \end{pmatrix}$$

$$= \begin{pmatrix} 0 & 0 & 4 & 5 & 1 \\ 1 & 0 & 1 & 3 & 0 \\ 4 & 0 & 0 & 1 & 1 \\ 4 & 1 & 0 & 0 & 1 \\ 1 & 3 & 1 & 0 & 0 \\ 0 & 5 & 4 & 0 & 1 \end{pmatrix} \begin{pmatrix} 1 & 0 & 0 & 0 & 0 & 1 \\ 5 & 3 & 1 & 0 & 0 & 0 \\ 0 & 0 & 1 & 1 & 0 & 0 \\ 0 & 0 & 0 & 1 & 3 & 5 \\ 0 & 1 & 0 & 0 & 1 & 0 \end{pmatrix} = \boldsymbol{AS}, \tag{3.8}$$

for which

$$\text{rank}\{\boldsymbol{X}\} = 3 < \text{rank}\{\boldsymbol{S}\} = 4 < \text{rank}_+\{\boldsymbol{X}\} = 5 < \min(P, T) = 6.$$

The restricted NPP instance (3.6) corresponding to $\boldsymbol{X}$ are two hexagons that coincide, with $\text{conv}\{\boldsymbol{X}\} = \mathcal{S} \cap \text{row}(\boldsymbol{X})$. Clearly, it is not possible to find a polygon in between the hexagon and itself with less than six vertices hence (i) $\text{rank}_+\{\boldsymbol{X}\} > 4$ and (ii) it is not possible to find an exact NMF with $R = 5$ and $\text{rank}\{\boldsymbol{S}\} = 3$. However, there exists a three-dimensional polytope $\text{conv}\{\boldsymbol{S}\}$ in $\mathcal{S}$ with five vertices that contains $\text{conv}\{\boldsymbol{X}\}$ hence $\text{rank}_+\{\boldsymbol{X}\} = 5$ and, in any decomposition of rank 5, it is required that $\text{rank}\{\boldsymbol{S}\} > \text{rank}\{\boldsymbol{X}\}$; see Figure 3.3 for an illustration.

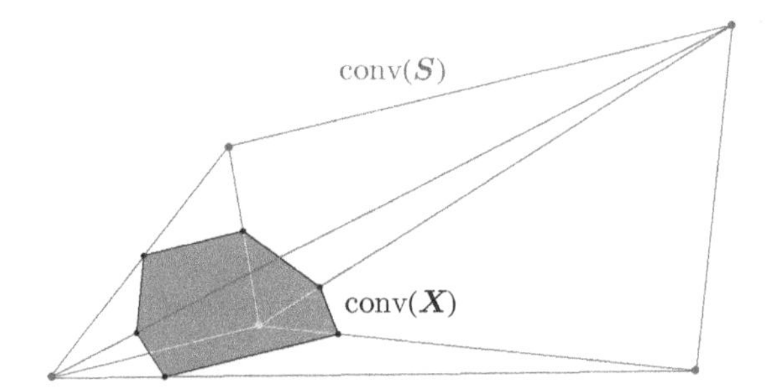

Figure 3.3 Illustration of the polytope conv$\{\boldsymbol{S}\}$ with five vertices containing the hexagon conv$\{\boldsymbol{X}\} = \mathcal{S} \cap$ row($\boldsymbol{X}$) in the unit simplex; see Equation (3.8). Note that the outer polytope $\mathcal{S}$ is not shown here since it has dimension 5.

3.3 Uniqueness and Admissible Solutions of NMF

Before trying to effectively solve any NMF problem, a key point is to answer some questions related to the model indeterminacies and to the uniqueness of the solution. Assume the existence of a non-negative factorization of the data matrix $\boldsymbol{X}$ into matrices $\boldsymbol{A}$ and $\boldsymbol{S}$. Let us start with any pair $(\boldsymbol{A}, \boldsymbol{S})$ that fulfill the mixing model (3.9) and then introduce a non-singular $(p \times p)$ matrix $\boldsymbol{T}$. A new pair $(\tilde{\boldsymbol{A}}, \tilde{\boldsymbol{S}})$ can be defined by

$$\tilde{\boldsymbol{A}} = \boldsymbol{A}\boldsymbol{T}^{-1} \text{and } \tilde{\boldsymbol{S}} = \boldsymbol{T}\boldsymbol{S}, \tag{3.9}$$

with no modification of the recovered data matrix, i.e. $\boldsymbol{X} = \tilde{\boldsymbol{A}}\,\tilde{\boldsymbol{S}}$. In the unconstrained case, this well-known result shows the existence of an infinite number of exact factorizations of the matrix $\boldsymbol{X}$. Matrix $\boldsymbol{T}$ is sometimes called *rotational ambiguity* matrix which includes classical scaling and ordering indeterminacies (see Chapter 1 of this book). In the case of NMF, a possible linear transformation should lead to transformed matrices $\tilde{\boldsymbol{A}}$ and $\tilde{\boldsymbol{S}}$ satisfying the non-negativity constraints

$$\tilde{\boldsymbol{A}} \geqslant 0 \text{ and } \tilde{\boldsymbol{S}} \geqslant 0. \tag{3.10}$$

In that respect, three questions arise:

(1) what are the conditions on the actual source signals and mixing coefficients (factors $\boldsymbol{S}$ and $\boldsymbol{A}$) ensuring the uniqueness of the factorization of $\boldsymbol{X}$ according to (3.2), and satisfying the non-negativity constraints (3.10)?
(2) if the decomposition is not unique, what are the admissible (feasible) solutions?
(3) among all the admissible solutions, can we define a more plausible one?

These questions highlight three aspects of the NMF problem that will be detailed in the sequel, with a special focus on a practical procedure for obtaining a set of admissible NMF solutions.

3.3.1 Uniqueness Conditions

Many necessary and/or sufficient conditions for establishing the NMF uniqueness have been formulated in the literature. In what follows, we will suppose that:

$$\mathrm{rank}\{\boldsymbol{X}\} = \mathrm{rank}_{+}\{\boldsymbol{X}\} = R.$$

The first theorem on uniqueness was proposed by Chen [5]. It gives a necessary and sufficient condition to have a unique NMF. But, this condition does not give any numerical mean to check if a given non-negative matrix admits a unique non-negative factorization. In addition, Park *et al.* [28] and Smilde *et al.* [29] developed some sufficient uniqueness conditions well adapted to some specific applications, but they seem not to be applicable for a general purpose.

First, we give a necessary condition [30] to have the NMF uniqueness. It states that both $\boldsymbol{S}$ and $\boldsymbol{A}$ should have a minimum number of zero entries.

Theorem 3.3 *If the NMF of $\boldsymbol{X}$ into matrices $\boldsymbol{A}$ and $\boldsymbol{S}$ such $\boldsymbol{X} = \boldsymbol{AS}$ is unique, then the following condition are fulfilled:*

(A1) $\forall(r \neq r'), \exists k$ such as : $s_r(k) = 0$ and $s_{r'}(k) > 0$.
(A2) $\forall(r \neq r'), \exists \ell$ such as : $A_{\ell r} = 0$ and $A_{\ell r'} > 0$.

Chen's uniqueness results is the starting point of Donoho and Stodden [31] which gives a sufficient uniqueness condition:

Theorem 3.4 *If the NMF of $\boldsymbol{X}$ into matrices $\boldsymbol{A}$ and $\boldsymbol{S}$ such $\boldsymbol{X} = \boldsymbol{AS}$ is unique if the following conditions are satisfied:*

- *Separability: $\forall r, \exists\ k$ such that: $s_r(k) \neq 0$ and $s_r(\ell) = 0,\ \forall \ell \neq k$*
- *Generative model: the set $\{1, \dots, P\}$ is partitioned into L groups $\mathcal{P}_1, \dots, \mathcal{P}_L$, each containing exactly R elements. $\forall p,\ \forall \ell$, there exists an element a_{pr} such that: $a_{pr} \neq 0,\ r \in \mathcal{P}_\ell$ and $a_{p\ell} = 0,\ \forall \ell \in \mathcal{P}_\ell, \ell \neq r$*
- *Complete Factorial Sampling: $\forall r_1 \in \mathcal{P}_1, \dots, r_L \in \mathcal{P}_L,\ \exists\ k$ such that: $a_{kr_1} \neq 0, \dots, A_{kr_\ell} \neq 0$*

The work of Laurberg *et al.* [32] starts from a different point of view which was initially proposed by [3] and proves the following sufficient uniqueness condition

Theorem 3.5 *The NMF of* $\boldsymbol{X} = \boldsymbol{AS}$ *is unique if the following conditions are satisfied:*

- *Sufficiently spread:* $\forall r, \exists\, k$ *such that:* $s_r(k) \neq 0$ *and* $s_r(\ell) = 0,\ \forall \ell \neq k$
- *Strongly Boundary Close: the matrix* $\boldsymbol{A}$ *satisfies the following conditions*
  1) $\forall r,\ \exists \ell$ *such that:* $a_{\ell r} = 0$ *and* $a_{\ell r'} \neq 0,\ \forall r' \neq r$
  2) There exists a permutation matrix $\boldsymbol{P}$ *such that* $\forall r$*, there exists a set* $k_1, \ldots, k_{R-k}$ *satisfying* $[\boldsymbol{AP}]_{r,k_j} = 0,\ \forall j \leq R-k$*; and the matrix* $[\boldsymbol{AP}]_{r+1:R,k_1:k_{R-r}}$ *is invertible.*

A recent and complete survey on the analysis of NMF uniqueness can be found in [33] where uniqueness conditions are also formulated for the case of symmetric NMF. In fact, although sparsity of the input matrix is neither a necessary nor a sufficient condition for uniqueness, there is a link between uniqueness of NMF and sparsity of the latent variable; see for example Theorem 3.3. It is observed in practice that if the true latent factors are sparse, NMF usually tends to recover the correct solution, the geometric interpretation of NMF shows that sparser matrices lead to more well-posed NMF problems. However, in many applications including multivariate curve resolution and spectral data unmixing, there is at least one factor that is non-sparse. This motivates the development of approaches allowing to assess the extent of the possible results termed as admissible NMF solutions.

3.3.2 Finding the Admissible Solutions

The retained approach for finding the admissible solutions is based on the same idea as the one previously proposed in [30, 34]. It consists in finding a set of parametric transformation matrices $\boldsymbol{T}(\boldsymbol{\theta})$ minimizing a criterion $C_{\text{nneg}}(\boldsymbol{\theta})$ based on a non-negativity measure. This criterion is defined as

$$C_{\text{nneg}}(\boldsymbol{\theta}) = \|\psi(\boldsymbol{T}(\boldsymbol{\theta})\boldsymbol{S})\|_F^2 + \|\psi(\boldsymbol{A}\,\boldsymbol{T}^{-1}(\boldsymbol{\theta}))\|_F^2, \tag{3.11}$$

where $\psi(x) = \min(x, 0)$. Parameters $\boldsymbol{\theta}$ are defined so as to implicitly handle the scaling ambiguity. Such objective function is generally used in constrained optimization as an exterior penalty function since it assigns a high cost to solutions that do not fulfill the non-negativity constraint. The minimization of this criterion with respect to the parameter vector $\boldsymbol{\theta}$ is performed using an unconstrained optimization method such as the Neldar–Mead simplex algorithm with several different random initializations of the transformation matrix parameters $\boldsymbol{\theta}$. The retained solutions are those that cancel C_{nneg}, i.e. that exactly solve the constrained factorization problem. The experiment is repeated with different starting points to get several admissible values of the transformation matrix parameters.

3.3.2.1 Illustration in the Case of Two Sources

For a decomposition rank $R = 2$, it was shown in [30] that the transformation matrix $\boldsymbol{T}(\boldsymbol{\theta})$ reduces to the following parametric form:

$$\boldsymbol{T}(\boldsymbol{\theta}) = \begin{bmatrix} 1-\theta_1 & \theta_1 \\ \theta_2 & 1-\theta_2 \end{bmatrix}, \tag{3.12}$$

with parameter vector $\boldsymbol{\theta} = [\theta_1, \theta_2]^\mathrm{T} \in \mathbb{R}^2$ such that $(\theta_1 + \theta_2) < 1$, to get rid of the ordering indeterminacy and ensuring invertibility of $\boldsymbol{T}$.

Analytical calculations detailed in [30] can be performed in the case of two sources and lead to bounds on the values of the parameters

$$\begin{cases} \theta_1 \in \left[-\min\limits_{t\in\mathbb{T}_1} \left\{ \frac{s_1(t)}{s_2(t)-s_1(t)} \right\}, \quad \min\limits_{\ell} \left\{ \frac{A_{\ell 2}}{A_{\ell 1}+A_{\ell 2}} \right\} \right], \\ \theta_2 \in \left[-\min\limits_{t\in\mathbb{T}_2} \left\{ \frac{s_2(t)}{s_1(t)-s_2(t)} \right\}, \quad \min\limits_{\ell} \left\{ \frac{A_{\ell 1}}{A_{\ell 1}+A_{\ell 2}} \right\} \right] \end{cases} \tag{3.13}$$

with subsets $\mathbb{T}_1$ and $\mathbb{T}_2$ are defined as $\mathbb{T}_1 = \left\{ t \in \{1, \ldots, T\}; s_2(t) > s_1(t) \right\}$ and $\mathbb{T}_2 = \left\{ t \in \{1, \ldots, T\}; s_1(t) > s_2(t) \right\}$. Figure 3.4 gives an illustration of a set of admissible solutions in the case of two spectral sources. This example is inspired from measurement data that can be obtained from the monitoring of kinetic chemical reactions using spectroscopy [35]. One can see here that acceptable NMF solutions deviate from the original sources and present additional peaks, which in practice may lead to data interpretation errors when the separation is performed under the non-negativity constraint alone.

3.3.2.2 Illustration in the Case of More Than Two Sources

In the case of more than two sources, analytical computations cannot be performed to get the feasible values of the transformation matrix parameters. For instance, in the case of three sources, a possible transformation matrix $\boldsymbol{T}$ bypassing permutation and scaling ambiguities [35] takes the form

$$\boldsymbol{T}(\boldsymbol{\theta}) = \begin{bmatrix} 1-\theta_1-\theta_2 & \theta_1 & \theta_2 \\ \theta_3 & 1-\theta_3-\theta_4 & \theta_4 \\ \theta_5 & \theta_6 & 1-\theta_5-\theta_6 \end{bmatrix}, \tag{3.14}$$

with $\boldsymbol{\theta} \in \mathbb{R}^6$. A numerical optimization with several random starting points allows to get several admissible transformation matrices. Figure 3.5 gives an example of the parameters values in the case of three spectral sources. Once again, one can see the existence of several values of the transformation parameters leading to several admissible solutions having different shapes. Actually, the feasible NMF solution is a mixture of the original sources.

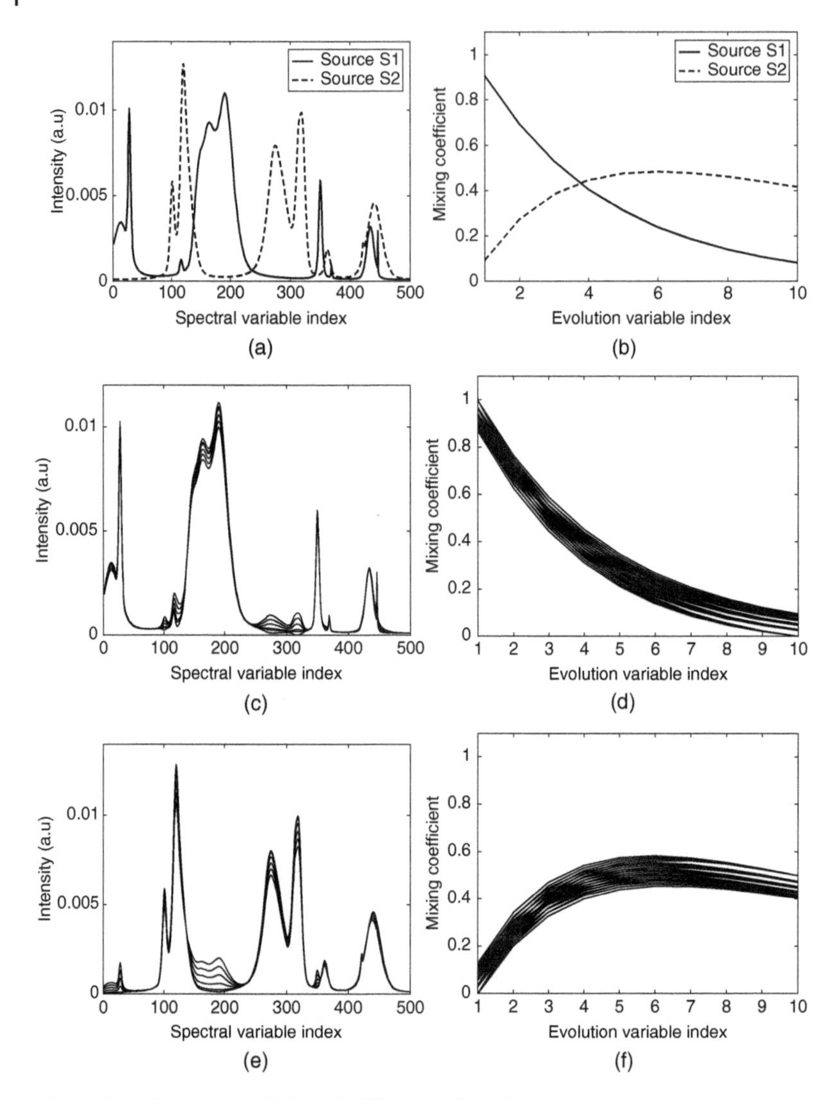

Figure 3.4 Illustration of feasible NMF solutions in the case of a spectral mixture of two components S1 and S2. (a) Actual sources, (b) actual mixing coefficients, (c) NMF solutions for source S1, (d) NMF solutions for S1 mixing coefficients, (e) NMF solutions for source S2, and (f) NMF solutions for S2 Mixing coefficients. The NMF solutions shown in (c), (e) are mixtures of the actual sources plotted in (a).

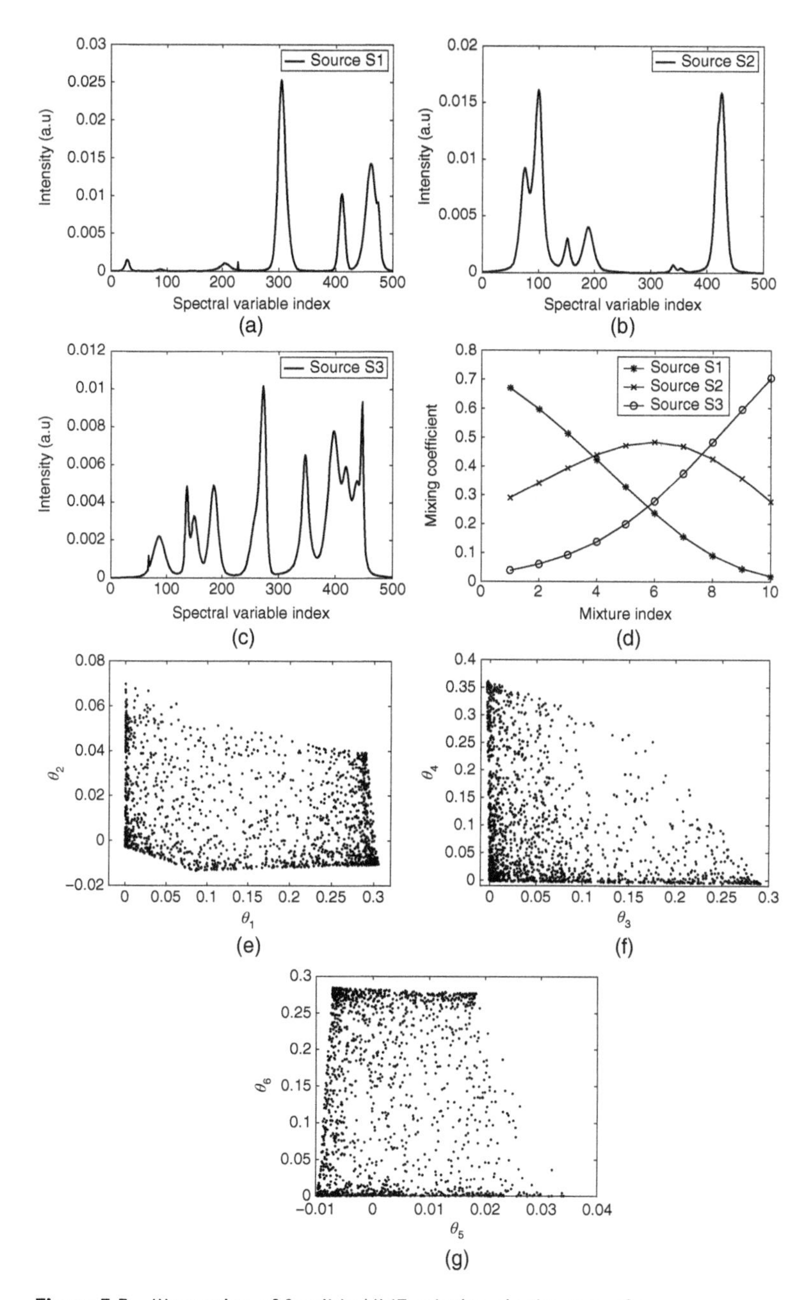

Figure 3.5 Illustration of feasible NMF solutions in the case of a three-component mixture: (a)–(d) simulated sources and mixing coefficients, (e)–(g) joint distribution of the transformation parameters of the actual sources leading to feasible NMF solutions. (a) Source S1, (b) Source S2, (c) Source S3, (d) Mixing coefficients, (e) Feasible transformations for source S1, (f) Feasible transformations for source S2, and (g) Feasible transformations for source S3.

3.4 Non-negative Matrix Factorization Algorithms

3.4.1 Statistical Formulation of Optimization Criteria

Let us consider the NMF problem in the case of noisy observation data and derive the main criteria used in NMF algorithms.

3.4.1.1 Case of a Gaussian Noise

The statistical distribution of each noise term $e_p(t)$ in (3.1) is assumed to be Gaussian with a zero mean and variance σ_p^2. More generally, the distribution of the noise vector $\boldsymbol{e}(t)$ is represented by a multivariate Gaussian distribution with zero mean vector and a covariance matrix $\boldsymbol{\Sigma}_t$. This matrix will be diagonal in the case of mutually independent noise components and its diagonal terms are equal to $[\sigma_1^2, \dots, \sigma_P^2]$. Alternatively, one can consider the same noise statistics for all the observations but with a sample-dependent (or time-dependent) variance value

$$f(\boldsymbol{e}(t)|\sigma_t^2) = \mathcal{N}(\boldsymbol{0}, \sigma_t^2 \boldsymbol{I}_P),$$

where $\boldsymbol{I}_P$ denotes the identity matrix and $\mathcal{N}(\boldsymbol{\mu}, \boldsymbol{\Sigma})$ stands for the Gaussian distribution with mean $\boldsymbol{\mu}$ and covariance $\boldsymbol{\Sigma}$. In addition, by assuming that samples of each noise component are independent and identically distributed, one can write

$$f(\boldsymbol{E}|\boldsymbol{\Sigma}) = \prod_{t=1}^{T} \mathcal{N}(\boldsymbol{0}, \boldsymbol{\Sigma}).$$

Consequently the likelihood can be constructed as

$$f(\boldsymbol{X}|\boldsymbol{S}, \boldsymbol{A}, \boldsymbol{\Sigma}_t) = \prod_{t=1}^{T} \mathcal{N}(\boldsymbol{x}(t) - \boldsymbol{A}\boldsymbol{s}(t), \boldsymbol{\Sigma}_t). \tag{3.15}$$

By taking its negative logarithm, $\mathcal{L}(\boldsymbol{S}, \boldsymbol{A}) = -\log f(\boldsymbol{X}|\boldsymbol{S}, \boldsymbol{A})$, this likelihood leads to a data fitting term

$$\mathcal{L}(\boldsymbol{S}, \boldsymbol{A}) = \sum_{t=1}^{T} (\boldsymbol{x}(t) - \boldsymbol{A}\boldsymbol{s}(t))^{\mathrm{T}} \boldsymbol{\Sigma}_t^{-1} (\boldsymbol{x}(t) - \boldsymbol{A}\boldsymbol{s}(t)). \tag{3.16}$$

This objective function corresponds to a weighted least-squares criterion. However, when $\boldsymbol{\Sigma}_t = \sigma^2 \boldsymbol{I}_R$, this criterion simplifies to a quadratic data fitting objective function, based on the Frobenius norm, used in most NMF algorithms [7, 8]; see Section 3.4.

3.4.1.2 Case of Poissonian Distribution

Poissonian data model is more adequate in the case where measurements in $x_p(t)$ are obtained through a counting process. In this model, each observed data sample $x_p(t) = X_{p,t}$ is assumed to be a realization of a Poisson process of mean $[\mathbf{AS}]_{p,t}$. The likelihood can therefore be expressed as:

$$f(x_p(t)|\mathbf{A},\mathbf{S}) = \frac{([\mathbf{AS}]_{p,t})^{x_p(t)}}{x_p(t)!} \exp(-[\mathbf{AS}]_{p,t}). \tag{3.17}$$

Under the assumption of mutually independent observations and identically distributed samples, the likelihood $f(\mathbf{X}|\mathbf{S},\mathbf{A})$ can be deduced:

$$f(\mathbf{X}|\mathbf{S},\mathbf{A}) = \prod_{p=1}^{P}\prod_{t=1}^{T} f(x_p(t)|\mathbf{S},\mathbf{A}).$$

It can be noted that the data fitting term, $\mathcal{L}(\mathbf{S},\mathbf{A}) = -\log f(\mathbf{X}|\mathbf{S},\mathbf{A})$, resulting from this likelihood is

$$\mathcal{L}(\mathbf{S},\mathbf{A}) = \sum_{p=1}^{P}\sum_{t=1}^{T} \left([\mathbf{AS}]_{p,t} - x_p(t)\log[\mathbf{AS}]_{p,t}\right). \tag{3.18}$$

This criterion is an instance of those used in non-negative matrix factorization algorithms based on divergence measures such as Kullback–Leibler [8, 36]. Actually, the source separation approach using the maximum likelihood approach, whose principle is the maximization of $f(\mathbf{X}|\mathbf{S},\mathbf{A})$ or equivalently the minimization of $-\log f(\mathbf{X}|\mathbf{S},\mathbf{A})$ allows to give a statistical formulation of NMF algorithms.

More generally, adding statistical priors on matrices $\mathbf{X}$ and $\mathbf{S}$ can also be formalized using Bayesian source separation methods [37]. See Chapter 5 of this book for more details on this approach.

In this section, we present several widely used algorithms to solve NMF problems. We will mostly focus on the least squares formulation

$$\min_{\mathbf{A},\mathbf{S}} \|\mathbf{X} - \mathbf{AS}\|_F^2 = \sum_{p,t}(\mathbf{X} - \mathbf{AS})_{p,t}^2 \quad \text{subject to} \quad \mathbf{A} \geqslant \text{and } \mathbf{S} \geqslant 0. \tag{3.19}$$

The Frobenius norm is suitable when assuming Gaussian noise; for example, if $\mathbf{X}$ is a dense matrix representing images [8]. However, this is not always the case, in particular for sparse input matrices $\mathbf{X}$ which are often encountered in the literature (such as document data sets or matrices arising from large networks). In these cases, other objective functions should be used; we refer the reader to [38, 39] and the references therein.

The problem (3.19) is a difficult non-convex optimization problem and, in fact, is NP-hard [20]. In practice, it is in general solved via standard iterative nonlinear optimization approaches; some of them are described in Section 3.4.2. Although these approaches do not guarantee obtaining an optimal solution, they usually generate satisfactory results that are useful for applications. More recently, a new class of NMF methods were introduced that are guaranteed to recover an optimal solution, up to the noise level, given that the input matrix $\boldsymbol{X}$ has a particular structure; this is briefly discussed in Section 3.2, and in much more detail in Chapter 5.

To add to the complexity, as discussed in Sections 3.3.1 and 3.3.2, the optimal solution of (3.19) is in general non-unique (even by considering equivalent solutions due to permutation and scaling); see Section 3.3. To alleviate this problem, a standard approach is to incorporate priors, usually via regularization or additional constraints, into (3.19); see Section 3.4.3 and Chapter 4.

Another problem that we do not discuss in this chapter and which is inherent to most blind source separation problems is the choice of the number of sources (non-negative rank) in the mixture. In the sequel we discuss the NMF algorithm in the case for a fixed value of R.

3.4.2 Iterative Factorization Methods

Although (3.19) is a difficult nonlinear optimization problem, it has several nice properties. In particular, if we assume a known mixing matrix $\boldsymbol{A}$, then the problem becomes a convex optimization problem with respect to $\boldsymbol{S}$,

$$\min_{\boldsymbol{S}\geqslant 0} \|\boldsymbol{X} - \boldsymbol{A}\boldsymbol{S}\|_F^2. \tag{3.20}$$

This is a particular convex quadratic optimization problem with linear inequality constraints referred to as *non-negative least squares* (NNLS) [40, 41]. Clearly, the same property holds for $\boldsymbol{A}$ when $\boldsymbol{S}$ is fixed (this structure is sometimes referred to as bi-convex). Most iterative methods take advantage of this fact: denoting $(\boldsymbol{A}^{(k)}, \boldsymbol{S}^{(k)})$ the solution obtained after k iterations (that is, the kth iterate), most methods obey the following framework:

1. Initialize $(\boldsymbol{A}^{(0)}, \boldsymbol{S}^{(0)})$.
2. For $k = 1,2,\ldots,$
   2.a Compute $\boldsymbol{S}^{(k)} \geqslant 0$ such that $\|\boldsymbol{X} - \boldsymbol{A}^{(k-1)}\boldsymbol{S}^{(k)}\|_F^2 \leq \|\boldsymbol{X} - \boldsymbol{A}^{(k-1)}\boldsymbol{S}^{(k-1)}\|_F^2$.
   2.b Compute $\boldsymbol{A}^{(k)} \geqslant 0$ such that $\|\boldsymbol{X} - \boldsymbol{A}^{(k)}\boldsymbol{S}^{(k)}\|_F^2 \leq \|\boldsymbol{X} - \boldsymbol{A}^{(k-1)}\boldsymbol{S}^{(k)}\|_F^2$.

This practical convergence of this update scheme should be carefully checked by ensuring a sufficient decrease of the objective function by using alternate descent algorithms. The iterative process is usually stopped according to standard convergence rules, e.g. stabilization of the objective function and/or the iterates. See Chapter 2 for more details on iterative optimization schemes.

Before presenting some numerical algorithms, it is interesting to highlight the following issues:

- Because of the symmetry of the problem, since $\boldsymbol{X} \approx \boldsymbol{AS} \iff \boldsymbol{X}^{\mathsf{T}} \approx (\boldsymbol{AS})^{\mathsf{T}} = \boldsymbol{S}^{\mathsf{T}}\boldsymbol{A}^{\mathsf{T}}$, $\boldsymbol{A}$ and $\boldsymbol{S}$ are updated in the same way for most NMF algorithms (that is, steps 2.a and 2.b above use the same strategy).
- The product $\boldsymbol{AS}$ is the sum of R rank-one factors $\boldsymbol{a}_r\boldsymbol{s}_r$, where $\boldsymbol{a}_r$ is the r-th column of $\boldsymbol{A}$ and $\boldsymbol{s}_r$ the r-th row of $\boldsymbol{S}$. Therefore, there is always a scaling and permutation degree of freedom (see also Section 3.3) since we can permute indistinguishably the rank-one factors and since

$$\boldsymbol{a}_r\boldsymbol{s}_r = \left(\lambda_r\boldsymbol{a}_r\right)\left(\frac{1}{\lambda_r}\boldsymbol{s}_r\right) \quad (\forall r = 1, \ldots, R)$$

  for any $\lambda_r > 0$. In practice, to bypass this ambiguity, the columns of $\boldsymbol{A}$ (or rows of $\boldsymbol{S}$) are usually normalized to have unit ℓ_2 or ℓ_1 norm.

3.4.2.1 Initializing NMF Algorithms

The initial matrices $(\boldsymbol{A}^{(0)}, \boldsymbol{S}^{(0)})$ can be selected in many different ways. The most naive approach is to initialize them randomly, using for example the uniform distribution in the interval [0,1] for each entry of $\boldsymbol{A}^{(0)}$ and $\boldsymbol{S}^{(0)}$. This approach is of course very simple and easy to implement, but has the drawback to ignore completely the input matrix $\boldsymbol{X}$. In particular, using such a procedure usually leads to a low-rank approximation $\boldsymbol{A}^{(0)}\boldsymbol{S}^{(0)}$ which can be rather far from $\boldsymbol{X}$. To improve the initial iterate, it is recommended to scale it, that is, to multiply it by a constant $\hat{\alpha}$ such that [42]

$$\hat{\alpha} = \arg\min_{\alpha>0} \|\boldsymbol{X} - \alpha\boldsymbol{A}^{(0)}\boldsymbol{S}^{(0)}\|_F^2 = \frac{\langle \boldsymbol{X}, \boldsymbol{A}^{(0)}\boldsymbol{S}^{(0)}\rangle}{\langle \boldsymbol{A}^{(0)}\boldsymbol{S}^{(0)}, \boldsymbol{A}^{(0)}\boldsymbol{S}^{(0)}\rangle} = \frac{\langle \boldsymbol{X}\boldsymbol{S}^{(0)\mathsf{T}}, \boldsymbol{A}^{(0)}\rangle}{\langle \boldsymbol{A}^{(0)\mathsf{T}}\boldsymbol{A}^{(0)}, \boldsymbol{S}^{(0)}\boldsymbol{S}^{(0)\mathsf{T}}\rangle}.$$

There exist many more sophisticated initialization approaches for NMF. Quite naturally, the goal of this initialization is to locate a good initial point close to a reasonable factorization in order to: (i) avoid bad local minima and (ii) allow the NMF algorithms to converge faster. We list in the following a few standard approaches:

- The most commonly encountered idea is to use clustering algorithms, such as k-means or spherical k-means [43]. They are used to initialize the rows of $\boldsymbol{S}^{(0)}$ (using the cluster centroids), while $\boldsymbol{A}^{(0)}$ is obtained either by using the cluster assignment matrix or by solving the corresponding NNLS subproblem; see, e.g. [44, 45] and the references therein.

- A computationally more expensive but, in general, more effective method is to use the best unconstrained rank-R approximation of matrix $\boldsymbol{X}$ (that can be computed efficiently via the singular value decomposition). Of course, this approximation does not generate non-negative factors and the trick is to (somehow) project them back onto the non-negative orthant [46].
- A procedure that is cheap and effective is to initialize $\boldsymbol{S}^{(0)}$ by selecting a representative subset of the observations, that is, rows of the input matrix $\boldsymbol{X}$. Selecting this subset can be done in many different ways and many algorithms exist for doing so. This is closely related to the column subset selection problem and to separable NMF which is discussed in Section 3.2.

3.4.2.2 Alternating Non-negative Least Squares, An Exact Coordinate Descent Method with 2 Blocks of Variables

The first algorithm for NMF proposed by Paatero and Tapper in their original paper [7] is based on alternating optimization, which is now in general referred to as alternating non-negative least squares. It is a class of methods that solve the NNLS subproblems exactly, alternatively for $\boldsymbol{A}$ and $\boldsymbol{S}$:

1. Initialize $(\boldsymbol{A}^{(0)}, \boldsymbol{S}^{(0)})$.
2. For $k = 1,2,\ldots,$
   2.a Compute $\boldsymbol{S}^{(k)} = \arg\min_{\boldsymbol{Y}\geqslant 0} ||\boldsymbol{X} - \boldsymbol{A}^{(k-1)}\boldsymbol{Y}||_F^2$.
   2.b Compute $\boldsymbol{A}^{(k)} = \arg\min_{\boldsymbol{Z}\geqslant 0} ||\boldsymbol{X} - \boldsymbol{Z}\boldsymbol{S}^{(k)}||_F^2$.

Alternating non-negative least squares (ANLS) is a so-called exact two-block coordinate descent method: there are two blocks of variables ($\boldsymbol{A}$ and $\boldsymbol{S}$) that are alternatively optimized exactly. This method is guaranteed to converge to a stationary point of (3.19) [47]. ANLS methods differ in the way the NNLS subproblems are solved. In fact, any method from (convex) optimization can potentially be used. Here is a possible classification of methods that can be used to solve NNLS:

- *First-order methods* such as standard projected gradient [48], (optimal) fast gradient methods [49], and coordinate descent methods (see Section 3.4.2.5). These methods only use the first-order information, that is, the value and the gradient of the objective function at each iterate.
- *Higher-order methods* such as interior-point methods, Quasi-Newton, or Newton methods [50]. These methods have a faster local convergence rate, but each iteration is more expensive.
- *Active-set methods* that take advantage of the fact that, if one would know the position of the zero entries in $\boldsymbol{A}$ and $\boldsymbol{S}$, then the NNLS subproblems reduce to unconstrained least-squares problems. The set of zero entries (the active set) is updated in a clever way to guarantee the

objective function to decrease at each step and the algorithm to converge [40, 51–53]. This class of algorithms performs well in practice, but the worst-case complexity is exponential; namely, proportional to the number of active sets which is proportional to 2^R (as the simplex method for linear programming).

We refer the reader to [54] for a survey on NNLS methods and to Chapter 2 on the related optimization techniques.

3.4.2.3 Multiplicative Updates

The most popular approach to deal with (3.19) is the multiplicative updates (MU) introduced along with the paper of Lee and Seung [8, 55] that really launched the research on NMF. It updates alternatively over $\boldsymbol{A}$ and $\boldsymbol{S}$ using the following scheme (dropping the iteration index k for convenience):

$$\boldsymbol{A} \leftarrow \boldsymbol{A} \boxdot \frac{[\boldsymbol{X}\boldsymbol{S}^{\top}]}{[\boldsymbol{A}\boldsymbol{S}\boldsymbol{S}^{\top}]} \quad \text{and} \quad \boldsymbol{S} \leftarrow \boldsymbol{S} \boxdot \frac{[\boldsymbol{A}^{\top}\boldsymbol{X}]}{[\boldsymbol{A}^{\top}\boldsymbol{A}\boldsymbol{S}]}, \tag{3.21}$$

where $\boxdot$ denotes the component-wise multiplication. Note that similar updates exists for many other objective functions [38, 55]. Note also that the MU were originally introduced in [56] for updating only one factor in order to solve NNLS problems. The MU (3.21) are guaranteed to decrease the objective function of (3.19) while clearly preserving non-negativity of the iterates.

It is interesting to note that the MU can be interpreted as a scaled gradient descent method, that is, a gradient descent method where each entry of the gradient is multiplied by a non-negative constant (this is equivalent to a quasi-Newton method where one would only use a diagonal matrix as an approximation of the Hessian); see [56, Section 1.3.2]. In fact, (3.21) can be equivalently written as

$$\boldsymbol{A} \leftarrow \boldsymbol{A} - \frac{[\boldsymbol{A}]}{[\boldsymbol{A}\boldsymbol{S}\boldsymbol{S}^{\top}]} \boxdot \left(\boldsymbol{A}\boldsymbol{S}\boldsymbol{S}^{\top} - \boldsymbol{X}\boldsymbol{S}^{\top}\right) \tag{3.22}$$

and similarly for $\boldsymbol{S}$, by symmetry. The term $\boldsymbol{A}\boldsymbol{S}\boldsymbol{S}^{\top} - \boldsymbol{X}\boldsymbol{S}^{\top}$ is the gradient of $\frac{1}{2}||\boldsymbol{X} - \boldsymbol{A}\boldsymbol{S}||_F^2$ with respect to $\boldsymbol{A}$.

It is important to note that, when implementing the MU, the term $\boldsymbol{A}\boldsymbol{S}\boldsymbol{S}^{\top}$ is computed by first performing the matrix-matrix product $\boldsymbol{S}\boldsymbol{S}^{\top}$. In fact, in that case, the computational cost will be in $\mathcal{O}(TR^2 + PR^2)$ operations while computing $\boldsymbol{A}\boldsymbol{S}$ first requires $\mathcal{O}(PTR)$ operations and $\mathcal{O}(PT)$ space in memory. This is particularly crucial for large and sparse matrices since $\boldsymbol{A}\boldsymbol{S}$ could be dense and could be too large to store in memory.

The main advantage of the MU is the ease of its implementation. However, it has several drawbacks. The main one being that it usually converges rather slowly compared to most other approaches. Another drawback is that the MU are not guaranteed to converge to a stationary point. The main reason being that once an entry is fixed to zero, it can no longer be modified (this is the so-called locking phenomenon). However, this can be overcome by unlocking variables at zero using a proper procedure; see the discussion in [39]. Recent work of [58] proposed a regularized update scheme allowing to ensure convergence of the multiplicative update rule.

It is interesting to note that updating $\boldsymbol{A}$ several times before updating $\boldsymbol{S}$ (and similarly for $\boldsymbol{S}$) allows a much faster convergence, because the matrix–matrix products $\boldsymbol{X}\boldsymbol{S}^{\mathsf{T}}$ and $\boldsymbol{S}\boldsymbol{S}^{\mathsf{T}}$ do not need to be recomputed between updates of $\boldsymbol{A}$ when $\boldsymbol{S}$ is fixed [59]. Note also that the multiplicative updates can be used in an ANLS framework, where $\boldsymbol{A}$ and $\boldsymbol{S}$ would be updated until convergence for the NNLS subproblems (see [60] for a convergence analysis when using the MU for NNLS).

3.4.2.4 Alternating Least Squares

A naive approach to solve NMF, referred to as alternating least squares (ALS), takes advantage of the fact that unconstrained least-squares problems can be solved very efficiently. To update $\boldsymbol{A}$ (and similarly for $\boldsymbol{S}$), it first solves the unconstrained least squares problem

$$\boldsymbol{A} \quad \leftarrow \quad \text{argmin}_{\boldsymbol{Y}\in\mathbb{R}^{P\times R}} ||\boldsymbol{X} - \boldsymbol{Y}\boldsymbol{S}||_F^2,$$

and then projects the solution back onto the non-negative orthant

$$\boldsymbol{A} \quad \leftarrow \quad \max(\boldsymbol{0}, \boldsymbol{A}).$$

ALS is very easy to implement (e.g. in Matlab, it requires one line of code to update $\boldsymbol{A}$, namely `A = max(0, (X*S')./(S*S'))`). However, it is not guaranteed to converge and the objective function of (3.19) usually oscillates under the ALS updates, sometimes drastically. However, it is usually efficient to use ALS as an initialization step, before a convergent algorithm is used, especially for sparse input matrices; see, e.g. [38, 61].

3.4.2.5 Exact Coordinate Descent Method with 2*R* Blocks of Variables

A method that works very well in many situations is the so-called method. It is a block coordinate descent method such as ANLS but has more block of variables. Hierarchical alternating least squares (HALS) optimizes alternatively over the columns of $\boldsymbol{A}$ and the rows of $\boldsymbol{S}$; hence, there are R blocks of P variables (the columns of $\boldsymbol{A}$) and R blocks of T variables (the rows of $\boldsymbol{S}$). HALS, although first suggested in [41], was first implemented and analyzed in [62] (and later in [63]) and independently in [42, 64, 65].

The benefit of considering smaller blocks of variables is that the optimization subproblem for each block is much easier to solve. In fact, the optimal solution for each r-th column of $\boldsymbol{A}$ and each r-th row of $\boldsymbol{S}$ can be written in closed form. Let us derive the formula here for the r-th column of $\boldsymbol{A}$ (again, by symmetry, the same holds for the r-th row of $\boldsymbol{S}$). Fixing all variables but the R-th row of $\boldsymbol{A}$, we need to solve the following optimization problem

$$\begin{aligned}\min_{\boldsymbol{a}_r \geqslant 0} \|\boldsymbol{X} - \boldsymbol{A}\boldsymbol{S}\|_F^2 &= \min_{\boldsymbol{a}_r \geqslant 0} \left\| \boldsymbol{X} - \sum_{k \neq r} \boldsymbol{a}_k \boldsymbol{s}_k - \boldsymbol{a}_r \boldsymbol{s}_r \right\|_F^2 \\ &= \min_{\boldsymbol{a}_r \geqslant 0} \|\boldsymbol{Z}^{(r)} - \boldsymbol{a}_r \boldsymbol{s}_r\|_F^2, \end{aligned} \tag{3.23}$$

where $\boldsymbol{a}_r$ is the r-th column of $\boldsymbol{A}$, $\boldsymbol{s}_r$ is the r-th row of $\boldsymbol{S}$, and $\boldsymbol{Z}^{(r)} = \boldsymbol{X} - \sum_{k \neq r} \boldsymbol{a}_k \boldsymbol{s}_k$ is the residual matrix with respect to the r-th rank-one factor $\boldsymbol{a}_r \boldsymbol{s}_r$. Interestingly, (3.23) can be decoupled into P independent NNLS problems in one variable:

$$\min_{\boldsymbol{a}_r \geqslant 0} \|\boldsymbol{Z}^{(r)} - \boldsymbol{a}_r \boldsymbol{s}_r\|_F^2 = \sum_{p=1}^{P} \min_{A_{pr} \geqslant 0} \|\boldsymbol{z}_p^{(r)} - A_{pr} \boldsymbol{s}_r\|_2^2, \tag{3.24}$$

where vector $\boldsymbol{z}_p^{(r)}$ is composed by the p-th row of $\boldsymbol{Z}^{(r)}$. A NNLS problem in one variable is equivalent to a problem of the form

$$\min_{y \geqslant 0} \alpha y^2 - 2\beta y, \tag{3.25}$$

for some $\alpha > 0$ and $\beta \in \mathbb{R}$. Clearly, if the solution of the unconstrained problem $\min_{y \in \mathbb{R}} \alpha y^2 - 2\beta y$ is non-negative, then it is also the solution of (3.25); otherwise, the minimizer is zero. Therefore,

$$\arg\min_{y \geqslant 0} \alpha y^2 - 2\beta y = \max\left(0, \frac{\beta}{\alpha}\right). \tag{3.26}$$

For the problem $\min_{y \geqslant 0} \|\boldsymbol{z}_p^{(r)} - y \boldsymbol{s}_r\|_2^2$, we have

$$\alpha = \boldsymbol{s}_r^{\mathsf{T}} \boldsymbol{s}_r = \|\boldsymbol{s}_r\|_2^2 \quad \text{and} \quad \beta = \boldsymbol{s}_r^{\mathsf{T}} \boldsymbol{z}_p^{(r)},$$

hence

$$\arg\min_{a_{pr} \geqslant 0} \|\boldsymbol{z}_p^{(r)} - A_{pr} \boldsymbol{s}_r\|_2^2 = \max\left(0, \frac{\boldsymbol{s}_r^{\mathsf{T}} \boldsymbol{z}_p^{(r)}}{\|\boldsymbol{s}_r\|_2^2}\right).$$

Finally, in vector form, we have

$$\begin{aligned}\arg\min_{\boldsymbol{a}_r \geqslant 0} \|\boldsymbol{Z}^{(r)} - \boldsymbol{a}_r \boldsymbol{s}_r\|_F^2 &= \max\left(0, \frac{\boldsymbol{Z}^{(r)} \boldsymbol{s}_r^{\mathsf{T}}}{\|\boldsymbol{s}_r\|_2^2}\right) \\ &= \max\left(0, \frac{\boldsymbol{X} \boldsymbol{s}_r^{\mathsf{T}} - \sum_{k \neq r} \boldsymbol{a}_k \boldsymbol{s}_k \boldsymbol{s}_r^{\mathsf{T}}}{\|\boldsymbol{s}_r\|_2^2}\right). \end{aligned} \tag{3.27}$$

HALS updates successively the columns of $\boldsymbol{A}$ and the rows of $\boldsymbol{S}$ using the above closed-form solution:

1. Initialize $(\boldsymbol{A}, \boldsymbol{S})$.
2. For $\ell = 1,2,\ldots,$
   2.a For $r = 1,2,\ldots,R$: Update $\boldsymbol{s}_r \leftarrow \max\left(0, \frac{\boldsymbol{a}_r^{\mathrm{T}}\boldsymbol{X} - \sum_{k \neq r} \boldsymbol{a}_r^{\mathrm{T}}\boldsymbol{a}_k \boldsymbol{s}_k}{\|\boldsymbol{a}_r\|_2^2}\right)$.
   2.b For $r = 1,2,\ldots,R$: Update $\boldsymbol{a}_r \leftarrow \max\left(0, \frac{\boldsymbol{X}\boldsymbol{s}_r^{\mathrm{T}} - \sum_{k \neq r} \boldsymbol{a}_k \boldsymbol{s}_k \boldsymbol{s}_r^{\mathrm{T}}}{\|\boldsymbol{s}_r\|_2^2}\right)$.

Note that the residuals $\boldsymbol{Z}^{(r)}$ are not computed (as they could be dense while $\boldsymbol{X}$ could be sparse). As for ANLS, HALS is guaranteed to converge to a stationary point of (3.19) [66]. As for the MU, it is possible to accelerate HALS by updating several times the columns of $\boldsymbol{A}$ before updating the rows of $\boldsymbol{S}$ since the terms $\boldsymbol{X}\boldsymbol{s}_r^{\mathrm{T}}$ and $\boldsymbol{s}_k\boldsymbol{s}_r^{\mathrm{T}}$ do not need to be recomputed [59, 66]. The computational cost of HALS is, up to some negligible factor, the same as for the MU [59]. However, in practice, HALS converges significantly faster. In fact, in most situations, HALS performs the best among ANLS, the MU, and ALS; see the references above and also, e.g. [38, 67].

To conclude this section, we refer the reader to the survey [68] on the classification of NMF methods as coordinate descent schemes (exact or approximate), where a more detailed analysis can be found along with some numerical comparisons.

3.4.3 Constrained and Penalized Factorization Methods

As explained in detail in Sections 3.2 and 3.3, a crucial aspect for practical applications when designing NMF algorithm is to take into account the non-uniqueness issue. The usual way to tackle this is to take into account, in the model, additional prior information depending on the application at hand. Here is a (non-exhaustive) list of additional constraints that can be added to the NMF model:

- *Minimum volume*. Looking back at the geometric interpretation of NMF from Section 3.2, it often makes sense in practice to look for a source matrix $\boldsymbol{S}$ such that $\mathcal{H}(\boldsymbol{S})$ has minimum volume. This enforces the sources to be as close as possible to the data points $\mathcal{H}(\boldsymbol{X})$, while allowing to approximate them well. This is discussed in much detail in Chapter 5 of this book.
- *Sparsity*. In many cases, the activation matrix $\boldsymbol{A}$ should be sparse because, for most observations, only a few sources are active. For example, in hyperspectral unmixing, most pixels only contain a few constitutive materials (also called endmembers); see Chapter 5. In other cases, the source matrix $\boldsymbol{S}$ should be sparse. For example, in document classification, the sources

are topics (represented by a set of words) which contain only a few words from the dictionary.

- *Orthogonality.* NMF can be used as a clustering model, using an additional orthogonality constraints $\boldsymbol{A}^\top \boldsymbol{A} = \boldsymbol{I}$. In fact, this constraint imposes that each observation is approximated only by a single source since, for any $1 \leq r \leq R$ and $1 \leq p \leq P$,

$$A_{pr} > 0 \quad \Rightarrow \quad A_{p\ell} = 0 \quad (\forall \ell \neq r).$$

where A_{pr} is the activation coefficient of the r-th source for the p-th observation. This follows from the fact that $\sum_{p=1}^{P} A_{pr} A_{p\ell} = 0$ for all $\ell \neq r$ and $\boldsymbol{A} \geqslant 0$.
In practice, this constraint is often too restrictive because most observations result from a combination of several sources. However, adding a penalty term in the objective function of the form $||\boldsymbol{A}^\top \boldsymbol{A} - \boldsymbol{I}_R||_F^2$ to enforce the mixing matrix $\boldsymbol{A}$ to be closer to an orthogonal matrix allows to enforce sparsity.

- *Spatial information.* In imaging applications (see, e.g. Section 3.5), it is possible to take into account the spatial information as most neighboring pixels (the observations) will share similarities and their activation coefficients will be similar. For example, if pixels p and p' for $1 \leq p, p' \leq P$ are neighbors, it is useful to add the following penalty term in the objective function

$$\sum_{r=1}^{R} |A_{pr} - A_{p'r}|.$$

Note that the absolute is usually used because it allows to preserve the edges in the images, instead of the ℓ_2 norm that smooths the edges; see, e.g. [69, 70].

- *Graph regularization.* In several situations, it is possible to embed the observations in a graph using some similarity measure. The vertices of the graph are the observations, and the (weighted) edges indicate whether two observations are similar. The way this graph is constructed depends on the application. For example, in imaging, the graph could be constructed connecting neighboring pixels. In general, this graph is constructed using some similarity measure between observations, and the NMF model takes this information by requiring that similar observations have similar activation coefficients (as shown above for the spatial information). This type of structure can take into account many different priors and is a very flexible way to add prior knowledge in the NMF model [71].
- *Sum-to-one constraints.* In several applications, the mixing coefficients correspond to a proportions (e.g. in hyperspectral imaging) in which case the entries on each row of $\boldsymbol{A}$ should sum-to-one.

The additional constraints added into the NMF model are in general taken into account by adding a penalty term in the objective function, with some penalty parameter balancing the importance between the fitting error and the regularization term.

For example, to obtain sparser $\boldsymbol{A}$ and $\boldsymbol{S}$, it is standard to use the ℓ_1-norm as a proxy for the ℓ_0-"norm" that counts the number of nonzero entries [52], and consider a model of the type

$$\min_{\boldsymbol{A}\geqslant 0,\boldsymbol{S}\geqslant 0} \|\boldsymbol{X} - \boldsymbol{AS}\|_F^2 + \lambda_A\|\boldsymbol{A}\|_1 + \lambda_S\|\boldsymbol{S}\|_1$$

where $\|\boldsymbol{Y}\|_1 = \sum_{i,j}|y_{ij}|$. Doing so, the algorithms described in Section 3.4 can usually be adapted to handle these situations. For example, if the penalty term is convex then the NNLS subproblems remain convex, hence efficiently solvable. For gradient-based methods, it suffices to account for the additional term(s) in the objective function and update the way the gradient is computed. Note that, in general, it will be necessary (and nontrivial) to properly tune the (penalty) parameters of the model to obtain good solutions, although some general strategies can sometimes be designed; see, e.g. [72] for sparse NMF. We refer the reader to the references above, and the references therein, for more details on these methods.

Another way these constraints can be taken care of is by using a projection. If the projection onto the feasible set can be computed efficiently, then gradient-based method can be adapted easily. For example, non-negativity constraints along with the sum-to-one constraint amounts to optimizing over the unit simplex for which the projection can be computed efficiently [73]. For sparsity, Hoyer [12] introduced a projection onto the set of matrices of a given sparsity level, based on the following measure of sparsity: for a nonzero vector $\boldsymbol{y} \in \mathbb{R}^N$,

$$spar(\boldsymbol{y}) = \frac{\sqrt{N} - \|\boldsymbol{y}\|_1/\|\boldsymbol{y}\|_2}{\sqrt{N} - 1} \in [0,1],$$

where $spar(\boldsymbol{y}) = 1$ when $\boldsymbol{y}$ has a single nonzero entry (hence $\|\boldsymbol{y}\|_1 = \|\boldsymbol{y}\|_2$), and $spar(\boldsymbol{y}) = 0$ when all entries of $\boldsymbol{y}$ are positive with similar values (hence $\|\boldsymbol{y}\|_1 = \sqrt{N}\|\boldsymbol{y}\|_2$).

To summarize, when using NMF for a particular application, it is crucial to first think carefully about the constraints that the sources and the mixing coefficients should satisfy. This will allow to design a dedicated NMF algorithm that will be able to recover the sought solution.

3.4.4 Geometrical Approaches and Separability

Another class of NMF problems that has gathered much attention lately is based on intuitions coming from the geometric interpretation of NMF.

This class is referred to as *separable NMF* and requires that the input matrix $\boldsymbol{X}$ satisfies the following separability condition: there exists an index set $\mathcal{K} \subset \{1,2,\ldots,P\}$ of size R, that is, $\mathrm{Card}(\mathcal{K}) = R$, and a non-negative matrix $\boldsymbol{A} \geqslant 0$ with R columns such that

$$\boldsymbol{X} = \boldsymbol{A}\,\underbrace{\boldsymbol{X}(\mathcal{K},:)}_{=\boldsymbol{S}}\,.$$

This means that the sources can be found among the observations. The problem therefore boils down to identifying the rows of $\boldsymbol{X}$ corresponding to the sources.

Geometrically, this means that, if the observations are scaled, we are looking for the vertices of $\mathrm{conv}\{\boldsymbol{X}\}$. This can be done efficiently, even in the presence of noise [74]; see also [75–77] and the references therein for some recent developments. This assumption makes sense in several applications such as document classification [78] and hyperspectral imaging (this is the so-called pure-pixel assumption). These approaches are described and analyzed in detail in Chapter 5.

3.5 Applications of NMF in Chemical Sensing. Two Examples of Reducing Admissible Solutions

As mentioned several times before, NMF is a difficult problem in general, being NP-hard. However, this does not imply that all NMF instances are difficult to solve. In some applications, the input matrix $\boldsymbol{X}$ can have a specific structure or results from a measurement process that makes the NMF problem less difficult; hence, taking this structure into account allows to design much more effective algorithms. For example, we have already seen in Section 3.2 that the NMF problem can be solved easily if $\mathrm{rank}\{\boldsymbol{X}\} \leq 2$. This observation has been used to solve the NMF problem in a hierarchical way and has been shown to perform well in several situations: analyzing Magnetic Resonance Spectroscopy and Imaging (MRSI) data [79], document classification [80], and hyperspectral unmixing [81].

In real-world applications, the given data matrix is not guaranteed to obey the established uniqueness conditions, thereby limiting the practical success of NMF algorithms. Hence, it motivates the development of algorithms to constrain the solution space of a given NMF problem (as presented in Section 3.4.3). However, in some situations, additional information can be taken into account to reduce the set of admissible solutions. We consider first a data augmentation strategy which consists in coupling multiple data sets. This approach is illustrated on a polarized Raman spectroscopy application.

An alternative approach is to perform a data pre-processing stage [82–84]. To illustrate this later approach, we consider the blind unmixing of spectroscopy imaging data.

3.5.1 Polarized Raman Spectroscopy: A Data Augmentation Approach

This section is adapted from [85]. Raman scattering is a light–matter interaction process which reflects the molecular vibration properties of molecules and materials, thus characterizing the chemical composition of the analyzed sample [86, 87]. For materials presenting a regular atomic or molecular structure, a more accurate characterization of the sample can be achieved by using polarizers [88]. In particular, this is the case for crystals as their response to the polarized light excitation will reflect the crystallographic structure of the sample, motivating the development of polarized Raman spectroscopy. This is the type of problem addressed here since we are considering the polarized Raman spectroscopy of a Rutile (TiO_2) crystal. It is worth mentioning that polarized Raman spectroscopy has other possible uses. For example, it can also be used for determining the molecular orientation distribution of polymeric materials [89].

3.5.1.1 Raman Data Description

The Raman measurements presented in this section were carried out in back scattering geometry with the same objective for excitation and collection of light. The confocal Raman spectrometer was equipped with a cooled charge-coupled device (CCD) camera, and the laser source was an ionized argon laser emitting at a wavelength $\lambda = 514.5$ nm. The analyzed crystal sample is fixed on a rotating stage as shown in Figure 3.6. Two coordinate systems are used, one associated with the laboratory space-fixed coordinates (O, X, Y, Z) and another attached to the analyzed sample (O, x, y, z). The incident light is polarized such that the electric field arriving on the sample is oriented along the Y direction. The scattered light is analyzed by positioning an analyzer in front of the entrance slit of the spectrometer. The analyzer is oriented either along the Y-axis (*parallel polarization*) or the X-axis (*crossed polarization*). Thus, the acquisition in one point of the sample yields a pair of spectra, one for the parallel polarization, indexed by Y and another for the crossed polarization, indexed by X. The *rotational diversity* scheme consists in rotating the sample around the Z-axis (Figure 3.6) with a fixed angular step (typically 10°) and acquiring two polarized spectra for each step of the rotation. For the rotational diversity acquisition scheme, m polarized spectra are acquired for m different

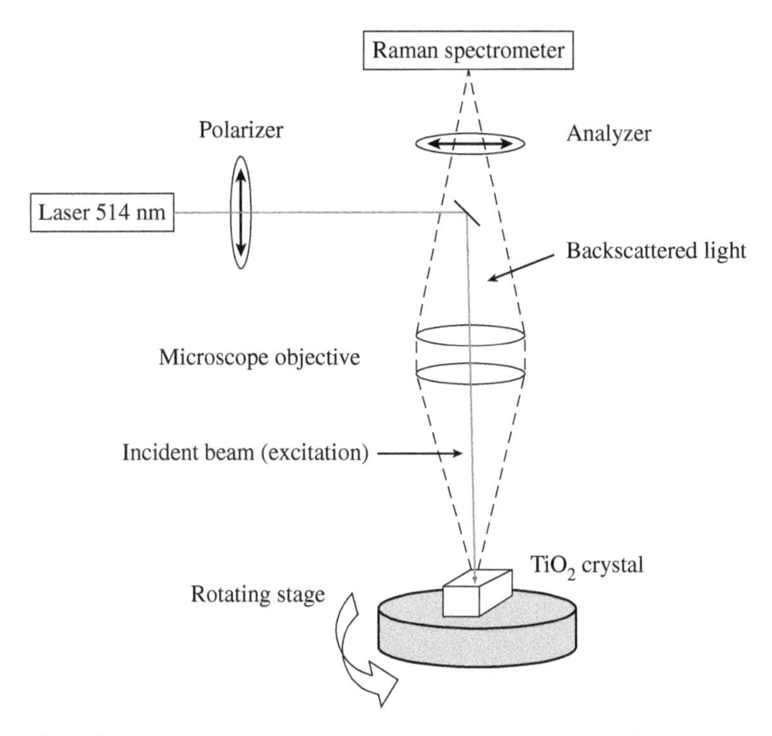

Figure 3.6 Polarized Raman spectroscopy set-up in backscattering geometry.

rotation angles $(\theta_1, \ldots, \theta_m)$ of the analyzed sample. For these angular diversity data, the "sources" are represented by vibrational modes. Indeed, the vibrational modes are characterized by specific displacements of the atoms from their equilibrium position, which dictate the magnitude of the components of the Raman polarizability tensor. The change of polarized Raman intensity versus rotational angle, for a specific vibrational mode, will therefore be different from another one. Each mode in polarized Raman spectra will thus contribute as one source in the full spectrum.

Under the assumption of instantaneous linear mixture, the acquired data can be structured as two $m \times n$ matrices, corresponding to the two polarization orientations:

$$\boldsymbol{X}_1 = \boldsymbol{A}_1 \boldsymbol{S}_1 + \boldsymbol{E}_1. \tag{3.28}$$

and

$$\boldsymbol{X}_2 = \boldsymbol{A}_2 \boldsymbol{S}_1 + \boldsymbol{E}_2. \tag{3.29}$$

In (3.28) and (3.29), matrices $\boldsymbol{E}_1$ and $\boldsymbol{E}_2$ account for the additive noise on the sensors and the model errors.

If we further analyze the underlying physico-chemical phenomenon generating the two data sets, the spectra of pure compounds are the same for the crossed and the parallel polarization [86, 87], since the vibrational modes are imposed by the structure of the crystal. This implies $\boldsymbol{S}_1 = \boldsymbol{S}_2 = \boldsymbol{S}$, which is quite intuitive if we consider a geometrical point of view in which the crossed and parallel polarized spectra are projections of the same signal on two orthogonal axes. By injecting this information into (3.28) and (3.29), we can write:

$$\begin{pmatrix} \boldsymbol{X}_1 \\ \boldsymbol{X}_2 \end{pmatrix} = \begin{pmatrix} \boldsymbol{A}_1 \\ \boldsymbol{A}_2 \end{pmatrix} \boldsymbol{S} + \begin{pmatrix} \boldsymbol{E}_1 \\ \boldsymbol{E}_2 \end{pmatrix}. \tag{3.30}$$

Eq. (3.30) points out a mixing model for the polarized spectra with rotational diversity considering both polarized spectra families jointly. Besides the fact that this is a more natural and compact representation of the data, the sample size is doubled in (3.32) compared to (3.28), (3.29) and the number of unknowns is lower; this should improve the accuracy of the estimated source parameters. In order to simplify the presentation we use the following notations:

$$\boldsymbol{X} = \begin{pmatrix} \boldsymbol{X}_1 \\ \boldsymbol{X}_2 \end{pmatrix}, \quad \boldsymbol{A} = \begin{pmatrix} \boldsymbol{A}_1 \\ \boldsymbol{A}_2 \end{pmatrix}, \quad \boldsymbol{E} = \begin{pmatrix} \boldsymbol{E}_1 \\ \boldsymbol{E}_2 \end{pmatrix}. \tag{3.31}$$

Equation (3.30) can thus be re-written in a more concise manner as:

$$\boldsymbol{X} = \boldsymbol{A}\boldsymbol{S} + \boldsymbol{E}. \tag{3.32}$$

3.5.1.2 Raman Data Processing

Given the physical nature of the data, the sources and the mixing coefficients are non-negative, meaning that (3.32) expresses a NMF model. It should be noticed that stacking the data matrices $\boldsymbol{X}_1$ and $\boldsymbol{X}_2$ into a bigger matrix $\boldsymbol{X}$ corresponds to a data augmentation strategy. This kind of technique was already proposed for diverse problems such as the analysis of multiple runs of gasoline blending processes [90]. Another example is the joint analysis of UV-visible spectra related to the complexation of the aluminum by caffeic acid and the titration of caffeic acid [91]. Actually, the benefit of matrix augmentation strategy is threefold: it allows to decrease estimation error uncertainties, it may remove rank deficiency, and it helps in reducing rotational ambiguities.

The approach was applied to a rutile TiO_2 crystal, as shown in Figure 3.6. The crystallographic face (110) (Hermann–Mauguin international crystallographic symbols) is analyzed. The sample is rotated with respect to Z axis only, meaning $\boldsymbol{\theta} = (0,0,\chi)$. Figure 3.7 presents the acquired polarized data for the parallel and crossed polarizations (matrices $\boldsymbol{X}_1$ and $\boldsymbol{X}_2$).

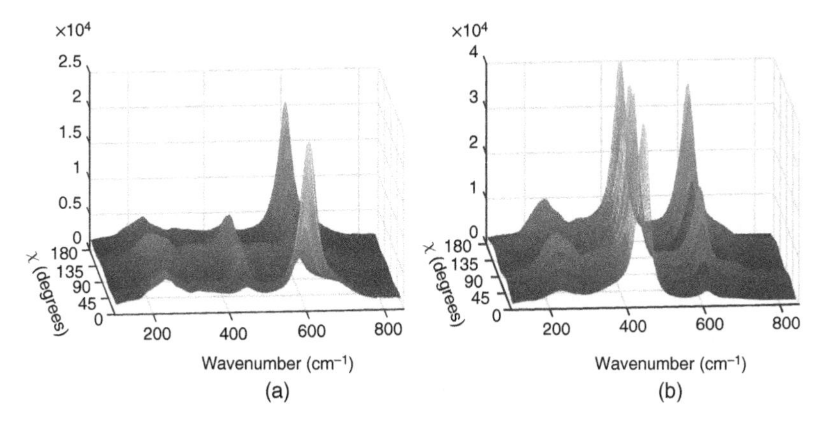

Figure 3.7 Polarized Raman data versus rotational angle χ for rutile $TiO_2(110)$ single crystal. (a) Parallel polarization and (b) crossed polarization.

The data were acquired in a spectral range of 100 cm^{-1} – 800 cm^{-1}, with an angular rotation step of 10 degrees between 0° and 190°.

In the case of TiO_2, four Raman active modes denoted as A_{1g}, E_g, B_{1g}, and B_{2g} (Mulliken symbols for symmetry groups [86]) are expected from theory. However, the B_{2g} mode at 826 cm^{-1} is out of the spectral window used in the present work (and anyway the B_{2g} has a very low Raman crossed section and is often not detected). The B_{1g} mode with the (110) oriented crystal plane is inactive either in parallel or crossed polarizations. Consequently, one can expect two Raman active modes, *i.e.* two sources, in the data collected here. Nevertheless, three sources are necessary to properly describe the data, as indicated by the magnitude analysis of the singular values of data matrices $\boldsymbol{X}_1, \boldsymbol{X}_2$, and $\boldsymbol{X}$. A theoretical explanation for the presence of this third source is provided in [85].

We illustrate the effect of the joint use of the crossed and parallel polarization data sets on the reduction of the NMF admissible solutions set. The NMF algorithm [92] was used to estimate the three source vectors and the corresponding mixing coefficients. The two data sets were processed separately and jointly, and the results are presented in Figure 3.8 for the source spectra and in Figures 3.9 and 3.10 for the mixing coefficients. To evaluate the size of the admissible solutions set, we used 25 independent runs for each plot, with different random initial values for the matrices $\boldsymbol{A}$ and $\boldsymbol{S}$. As one can see, by processing jointly both polarization data sets (Figure 3.8c and 3.10) the admissible solution domain is largely reduced as compared to the case when only one polarization is used (Figures 3.8a,b and 3.9). However, the solution is still not unique, which motivated the use of some regularization techniques. In [85], a penalized NMF algorithm

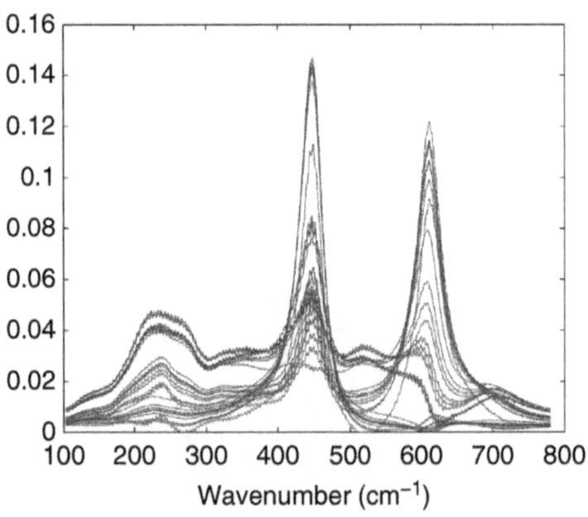

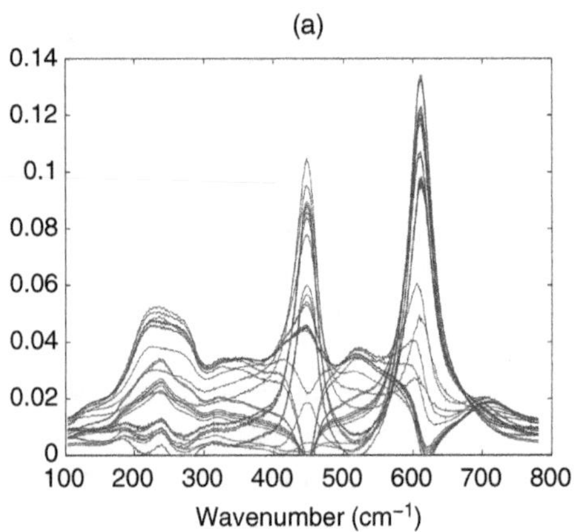

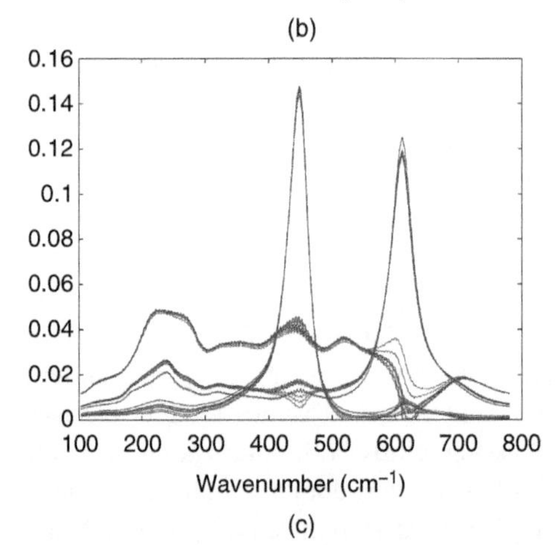

Figure 3.8 Subset of source spectra estimated by NMF (25 runs) on (a) Crossed polarization data processing (b) Parallel polarization data processing and (c) Both datasets processing.

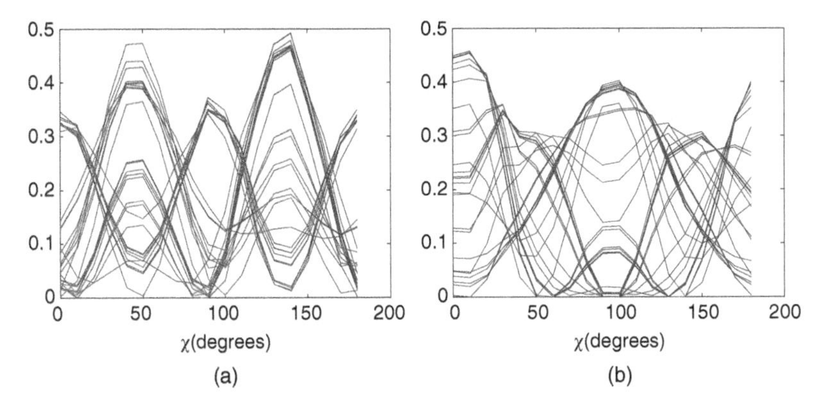

Figure 3.9 Estimated mixing coefficients by NMF (25 runs) for each polarization dataset separately processed. (a) Crossed polarization and (b) parallel polarization.

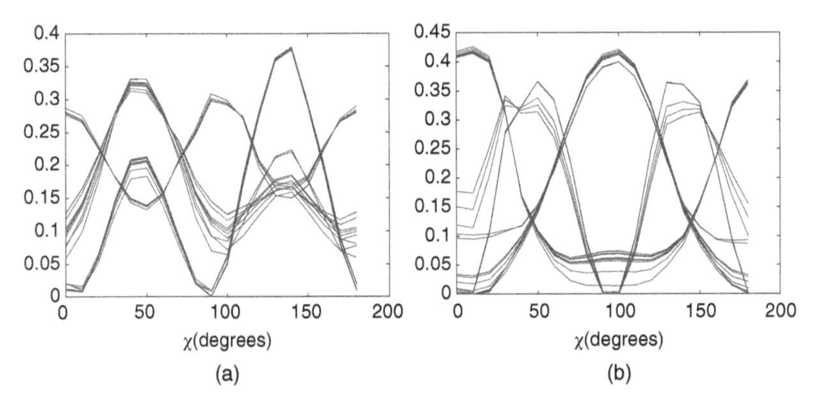

Figure 3.10 Estimated mixing coefficients by NMF (25 runs) for both polarization datasets jointly processed. (a) Crossed polarization and (b) parallel polarization.

derived from a Bayesian source separation approach (called BPSS) [14] was used to further reduce the set of admissible solution. The results and the physical interpretation of the obtained results are not reported here. The interested reader is referred to [85] for the detailed analysis.

3.5.2 Unmixing Blurred Raman Spectroscopy Images

This section is adapted from [84]. Hyperspectral images may be viewed as a collection of highly resolved spectra. In many cases, the image contains a small number of pure materials – termed endmembers – whose spectral signatures are mixed in each pixel because of limited spatial resolution. Blind spectral unmixing usually refers to the estimation of endmembers and their fractional contribution to each pixel, named abundances. A geometrical framework for spectral unmixing has attracted a lot of attention from researchers in the past two decades. In this approach, each pixel spectrum belongs to a simplex whose vertices are the endmembers we seek. In some cases, the data are known to contain at least one pure pixel per endmember, which are subsequently extracted from the hyperspectral scene; this is equivalent to separable NMF; see Section 3.4.4. When the pure pixel hypothesis does not hold, the geometrical approach to unmixing then consists in finding the Minimum Volume Simplex (MVS) enclosing the data using one of many MV algorithms [93–95]; see Chapter 5 for more details. Once endmembers have been extracted from the scene, abundances can be estimated using constrained least squares algorithms [96, 97]. However, highly mixed data are beyond the reach of geometrical algorithms because spectral signatures are located near the center of the true endmember [98].

3.5.2.1 Blurring Effect Modeling

Consider an hyperspectral image measuring radiance on T different spectral bands (*channels*) and P pixels. We gather the data in a $T \times P$ matrix $\boldsymbol{X}$ and use the following notations:

(1) $\boldsymbol{x}^t$ is the t-th row of $\boldsymbol{X}$, that is the 2D image at spectral band t after lexicographical ordering into a row vector of length P;
(2) $\boldsymbol{x}_p$ is the p-th column of $\boldsymbol{X}$, i.e. the $T \times 1$ spectrum of the p-th pixel (also termed spectral vector or pixel vector).

Each spectral vector in the image is a linear combination of an known number R of endmembers $\{\boldsymbol{s}_1, \dots, \boldsymbol{s}_R\}$. When unknown, R can be obtained by some model order estimation method such as *virtual dimensionality* [99]. See also [100, 101] for other approaches. Ignoring noise and modeling errors, the linear mixing model (LMM) writes:

$$\boldsymbol{X} = \boldsymbol{S}\boldsymbol{A}$$

where the r-th column of $L \times R$ *source matrix* $\boldsymbol{S}$ indexes endmember $\boldsymbol{s}_r$ and the p-th column $\boldsymbol{a}_p$ of $P \times R$ *abundance matrix* $\boldsymbol{A}$ contains the fractional abundance coefficients for $\boldsymbol{x}_p$:

$$\boldsymbol{x}_p = \boldsymbol{S}\boldsymbol{a}_p = \sum_{r=1}^{R} A_{pr}\boldsymbol{s}_r. \tag{3.33}$$

One can note here that the data matrix was transposed by interchanging the roles of matrices $\boldsymbol{A}$ and $\boldsymbol{S}$ in order to be coherent with the notations usually employed in spectral imaging. The linear mixing model is generally based on the following assumptions [102]:

(i) The number of endmembers R is much smaller than the number of bands L, that is $R \ll L$;
(ii) Matrix $\boldsymbol{S}$ is of full column rank, *i.e.* endmembers $\{\boldsymbol{s}_1, \ldots, \boldsymbol{s}_R\}$ are linearly independent;
(iii) *Abundance Non-negativity Constraint* (ANC): $A_{pr} \geqslant 0$ for all p and r;
(iv) *Abundance Sum Constraint* (ASC): $\sum_{r=1}^{R} A_{pr} = 1$ for all p.

Assumptions (i) and (ii) seem very reasonable in hyperspectral imaging since many bands are collected and the image is made up of a few distinct materials. Assumptions (iii) and (iv) come from the physical interpretation of abundance coefficient a_{pr} as the fractional spatial area occupied by the r-th endmember in the p-th pixel.

We now account for the fact that the image is degraded during the acquisition process. Under the common linear blur assumption, the 2D image $\boldsymbol{y}^t$ observed at a given channel t is obtained as the 2D convolution product of the true image and the channel point-spread function $\mathcal{H}^t$:

$$\boldsymbol{y}^t = \boldsymbol{x}^t \boldsymbol{H}^t, \tag{3.34}$$

where the $P \times P$ matrix $\boldsymbol{H}^t$ is a convolution matrix corresponding to $\mathcal{H}^t$. For instance, when the blur is space-invariant for different pixels, $(\boldsymbol{H}^t)^{\mathrm{T}}$ is a block-Toeplitz matrix where each block is Toeplitz [103]. Each entry of the observed data matrix $\boldsymbol{Y}$ is given by

$$y_p(t) = \sum_{r=1}^{R} h_{pn}^t x_r(t) \tag{3.35}$$

Using equations (3.33) and (3.35), the overall model combining noise, observation blurring, and linear mixing of endmembers can be written as

$$y_p(t) = \sum_{r=1}^{R} \sum_{p'=1}^{P} h_{p'r}^t A_{pr} s_r(t) + e_p(t), \tag{3.36}$$

where $\boldsymbol{E}$ is the noise term and model (3.36) assumes that the signal-to-noise ratio (SNR) is high enough for the noise to be additive and i.i.d. Gaussian. We observe that the blurred data do not satisfy the linear mixing model since the mixing coefficients $\left(\sum_{r=1}^{R} h_{pr}^t A_{pr}\right)$ depend on the channel index t. However, in the specific case where the point spread function (PSF) is invariant across channels, the model reduces to the following linear mixing model

$$\boldsymbol{Y} = \boldsymbol{SAH} + \boldsymbol{E} \tag{3.37}$$

with fixed matrix $\boldsymbol{H}$.

How does the observation process affect the distribution of pixel vectors inside the simplex? The answer to the question obviously depends on the nature of the PSF. Since the entries of $\boldsymbol{H}$ are known to be non-negative, the blurring process tends to average neighboring pixel intensities. This phenomenon causes observed spectral vectors to cluster toward the center of $\mathcal{S}$. This contraction property has important practical consequence: directly applying a NMF algorithm to the observed data may produce incorrect sources, and thus the subsequent estimation of abundances will also be biased. To improve the unmixing performance, it is necessary to deconvolve the data before applying any NMF algorithm.

3.5.2.2 Application to Raman Spectroscopy Images

In this section, we illustrate the impact of the deblurring process on the performances of the separation of real Raman spectroscopy data. This dataset comprises images of size 98 × 131 pixels, each pixel being 100 nm × 100 nm, acquired on 337 spectral bands ranging from 800 cm^{-1} to 1200 cm^{-1}. The scene of interest consists in a grain of *sodium acetate* (CH_3COONa) covered with *sodium carbonate* (Na_2CO_3) lying on a *silicon* layer (Si). Part of the sodium carbonate reacts with water vapor to yield hydrated sodium carbonate. These four chemical compounds are the endmembers we seek. A thorough inspection of the data reveals that the silicon compound contributes to all pixels of the image. The extraction of these endmembers is a challenging problem, since silicon is the only compound for which the pure pixel assumption is fulfilled. Because of the inherent high mixing of the data, even minimum volume methods are not supposed to produce good results on this data set.

Given the limited spectral range, the point spread function is considered to be invariant across channels and mixels. It is modeled as a 2D Gaussian function [104] with an experimentally measured full-width at half maximum of 300 nm. We apply our deconvolution algorithm to the data by setting regularization parameters through a trial-and-error process to $\mu = 20$ and $\nu = 5$ using the algorithm proposed in [103]. Non-negative Matrix Factorization with volume constraints (*NMF-vol*) [98] was applied for the data processing. The algorithm operates on both the raw data (where negative pixels have been clipped to zero since the algorithm imposes a non-negative data matrix) and restored data.

The resulting abundance maps and endmembers are given in figure 3.11. The first extracted endmember corresponds to the silicon layer, which presents a broad band at 910 – 960 cm^{-1} due to the 2TO harmonic phonon mode of bulk silicon. The deconvolution step allows to denoise its abundance map and, more importantly, uncovers structure that was distributed

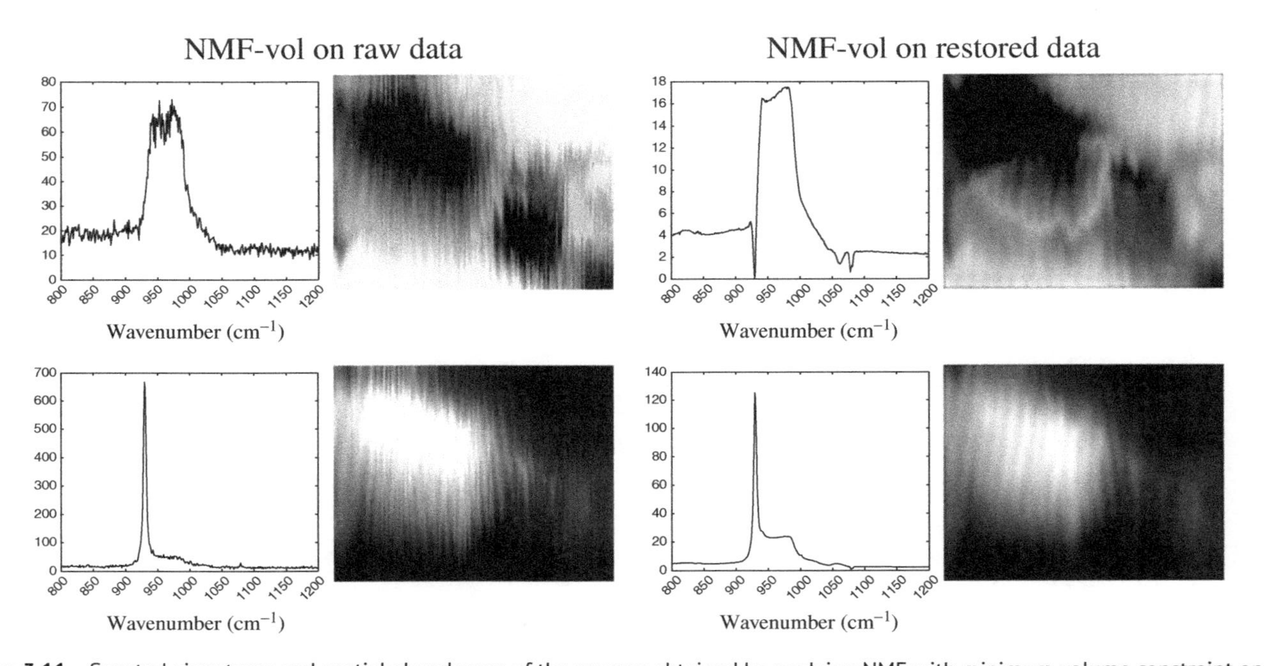

Figure 3.11 Spectral signatures and spatial abundances of the sources obtained by applying NMF with minimum volume constraint on both the raw Raman data and the deblurred data.

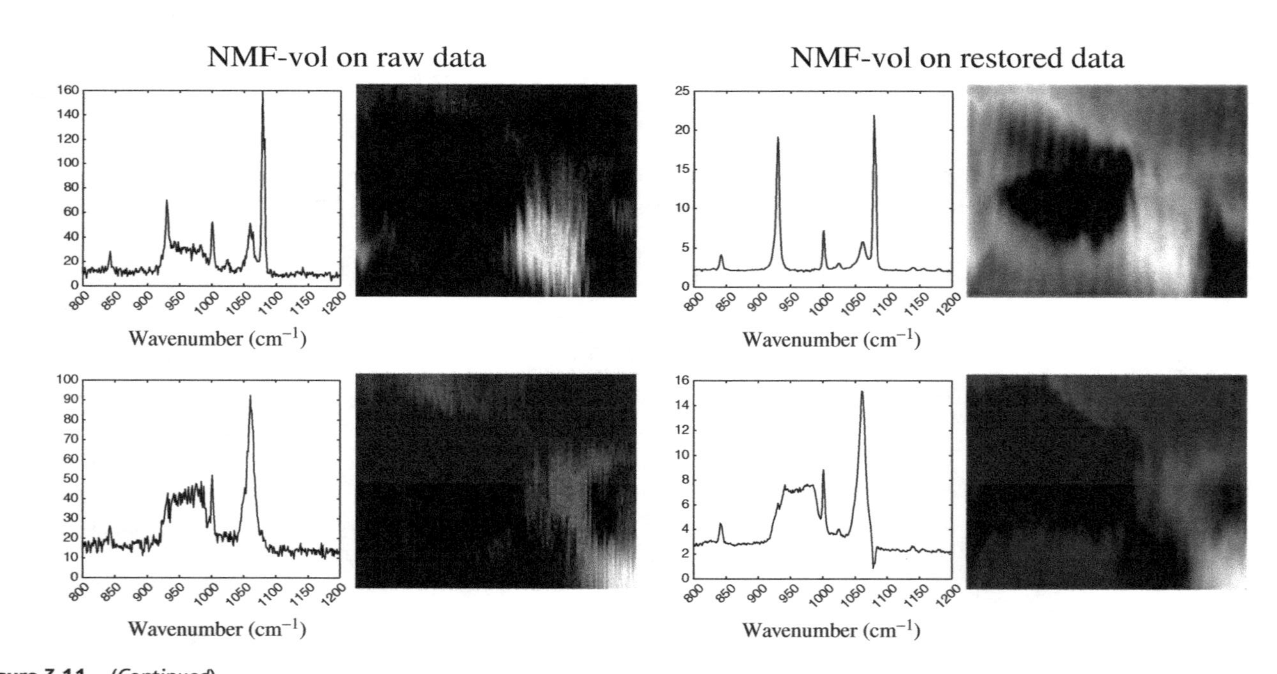

Figure 3.11 (*Continued*)

throughout other abundance maps. The second endmember is sodium acetate, with a peak at 930 cm^{-1} due to the intense C-C stretching mode of the acetate molecule [105]. The sodium acetate endmember displays more undesirable contribution from the silicon compound. The spectral shapes of endmembers extracted from the raw data appear noisier, a problem solved by the deconvolution step. Moreover, the algorithm is able to separate the sodium carbonate compound (third endmember, characterized by the peak at 1080 cm^{-1}) from the hydrated sodium carbonate compound (fourth endmember, peak at 1060 cm^{-1}); both are mixed with the sodium benzoate, as expected. The main gain of deconvolution clearly appears on the third endmember, where the silicon contribution is completely suppressed. Another benefit of the restoration step is to reveal structure hidden in raw abundance maps (first and third endmembers) that was not displayed by applying NMF-vol to the raw data.

3.6 Conclusions

This chapter described the concept of non-negative matrix factorization and its application in the context of source separation in chemical–physical sensing. Although the NMF problem is described in a simple mathematical way, the theoretical formulation of its solution existence and uniqueness remains an active area of study. Moreover, the resolution algorithms are generally application dependent since additional constraints (to non-negativity) should be firstly determined and then accounted for to get an acceptable solution. Actually, these additional constraints are exploited by developing dedicated factorization algorithms calling to some algebraic tools, optimization techniques, geometrical concepts, and statistical inference methods.

Active open problems in the NMF field rely on the reduction of the numerical complexity of the algorithms through the development of sequential factorization methods and data compression techniques.

References

1 Imbrie, J. and Van Andel, T.H. (1964) Vector analysis of heavy-mineral data. *Geological Society of America Bulletin*, 75, 1131–1156.

2 Markham, T.L. (1972) Factorization of non-negative matrices. *Proceedings of American Mathematical Society*, 32, 45–47.

3 Thomas, L.B. (1974) Rank factorizations of nonnegative matrices. *SIAM Review*, 16, 393–394.

4 Campbell, S.I. and Poole, G.D. (1981) Computing non-negative rank factorizations. *Linear Algebra and its Applications*, 35, 175–182.

5 Chen, J.C. (1984) The Nonnegative rank factorizations of nonnegative matrices. *Linear Algebra and its Applications*, 62, 207–217.
6 Cohen, J. and Rothblum, U. (1993) Non-negative ranks decompositions and factorizations of non-negative matrices. *Linear Algebra and its Applications*, 190, 149–168.
7 Paatero, P. and Tapper, U. (1994) Positive matrix factorization: a non-negative factor model with optimal utilization of error estimates of data values. *Environmetrics*, 5, 111–126.
8 Lee, D. and Seung, H. (1999) Learning the parts of objects by non-negative matrix factorization. *Nature*, 401, 788–791.
9 Tauler, R., Kowalski, B., and Fleming, S. (1993) Multivariate curve resolution applied to spectral data from multiple runs of an industrial process. *Analytical Chemistry*, 65, 2040–2047.
10 Paatero, P. (1997) Least squares formulation of robust non-negative factor analysis. *Chemometrics and Intelligent Laboratory Systems*, 37, 23–35.
11 Sajda, P., Du, S., Brown, T.R., Stoyanova, R., Shungu, D.C., Mao, X., and Parra, L.C. (2004) Nonnegative matrix factorization for rapid recovery of constituent spectra in magnetic resonance chemical shift imaging of the brain. *IEEE Transaction on Medical Imaging*, 23 (12), 1453–1465.
12 Hoyer, P. (2004) Nonnegative matrix factorization with sparseness constraints. *Journal of Machine Learning Research*, 5, 1457–1469.
13 Pascual-Montano, A., Carazo, J., Kochi, K., and Pascual-Marqui, R. (2006) Nonsmooth nonnegative matrix factorization. *IEEE Transactions on Pattern Analysis and Machine Intelligence*, 28 (3), 403–415.
14 Moussaoui, S., Brie, D., Mohammad-Djafari, A., and Carteret, C. (2006) Separation of non-negative mixture of non-negative sources using a Bayesian approach and MCMC sampling. *IEEE Transactions on Signal Processing*, 54 (11), 4133–4145.
15 Gillis, N. (2020) *Nonnegative Matrix Factorization*, SIAM, Philadelphia, PA.
16 Das, G. and Joseph, D. (1990) The complexity of minimum convex nested polyhedra, in *Proceedings of the 2nd Canadian Conference on Computational Geometry*, pp. 296–301.
17 Aggarwal, A., Booth, H., O'Rourke, J., and Suri, S. (1989) Finding minimal convex nested polygons. *Information and Computation*, 83 (1), 98–110.
18 Silio, C.B. (1979) An efficient simplex coverability algorithm in E2 with application to stochastic sequential machines. *IEEE Transactions on Computers*, 28 (2), 109–120.

19 Mond, D., Smith, J., and van Straten, D. (2003) Stochastic factorizations, sandwiched simplices and the topology of the space of explanations. *Proceedings of the Royal Society of London A: Mathematical, Physical and Engineering Sciences*, 459 (2039), 2821–2845.

20 Vavasis, S. (2010) On the complexity of nonnegative matrix factorization. *SIAM Journal on Optimization*, 20 (3), 1364–1377.

21 Gillis, N. and Glineur, F. (2012) On the geometric interpretation of the nonnegative rank. *Linear Algebra and its Applications*, 437 (11), 2685–2712.

22 Chistikov, D., Kiefer, S., Marusic, I., Shirmohammadi, M., and Worrell, J. (2016) On restricted nonnegative matrix factorization, in *43rd International Colloquium on Automata, Languages, and Programming (ICALP 2016), Leibniz International Proceedings in Informatics (LIPIcs)*, vol. 55 (eds I. Chatzigiannakis, M. Mitzenmacher, Y. Rabani, and D. Sangiorgi), Schloss Dagstuhl–Leibniz-Zentrum fuer Informatik, Dagstuhl, Germany, pp. 103:1–103:14.

23 Cohen, J. and Rothblum, U. (1993) Nonnegative ranks, decompositions and factorization of nonnegative matrices. *Linear Algebra and its Applications*, 190, 149–168.

24 Goemans, M. (2015) Smallest compact formulation for the permutahedron. *Mathematical Programming*, 153 (1), 5–11.

25 Vandaele, A., Gillis, N., and Glineur, F. (2017) On the linear extension complexity of regular *n*-gons. *Linear Algebra and its Applications*, 521, 217–239.

26 Moitra, A. (2013) An almost optimal algorithm for computing nonnegative rank, in *Proceedings of the 24th Annual ACM-SIAM Symposium on Discrete Algorithms (SODA '13)*, pp. 1454–1464.

27 Gillis, N. (2012) Sparse and unique nonnegative matrix factorization through data preprocessing. *Journal of Machine Learning Research*, 13, 3349–3386.

28 Park, E.S., Spiegelman, C.H., and Henry, R.C. (2002) Bilinear estimation of pollution source profiles and amounts by using multivariate receptor models. *Environmetrics*, 13, 775–798.

29 Smilde, A., Hoefsloot, H., Kiers, H., Bijlsma, S., and Boelens, H. (2001) Sufficient condition for unique solutions within a certain class of curve resolution models. *Journal of Chemometrics*, 15, 405–411.

30 Moussaoui, S., Brie, D., and Idier, J. (2005) Non-negative source separation: range of admissible solutions and conditions for the uniqueness of the solution, in *Proceedings of IEEE International Conference on Acoustics, Speech, and Signal Processing (ICASSP'2005)*, Philadelphia, PA, USA.

31 Donoho, D. and Stodden, V. (2003) When does non-negative matrix factorization give a correct decomposition into parts? in *Advances in Neural Information Processing Systems 16*, MIT Press, Cambridge, United States.

32 Laurberg, H., Christensen, M., Plumbley, M.D., Hansen, L.K., and Jensen, S.H. (2008) Theorems on positive data: on the uniqueness of NMF. *Computational Intelligence and Neuroscience*, 2008, Article ID 764206, 9 pages.

33 Fu, X., Huang, K., Sidiropoulos, N.D., and Ma, W.K. (2019) Nonnegative matrix factorization for signal and data analytics: identifiability, algorithms, and applications. *IEEE Signal Processing Magazine*, 36 (2), 59–80.

34 Sasaki, K., Kawata, S., and Minami, S. (1983) Constrained nonlinear method for estimating component spectra from multicomponent mixtures. *Applied Optics*, 22 (22), 3599–3606.

35 Moussaoui, S., Carteret, C., Brie, D., and Mohammad-Djafari, A. (2006) Bayesian analysis of spectral mixture data using Markov chain Monte Carlo methods. *Chemometrics and Intelligent Laboratory Systems*, 81 (2), 137–148.

36 Févotte, C. and Idier, J. (2011) Algorithms for nonnegative matrix factorization with the beta-divergence. *Neural Computation*, 23 (9), 2421–2456.

37 Mohammad-Djafari, A. (1999) A Bayesian approach to source separation. *American Institute of Physics (AIP) Proceedings*, 567, 221–244.

38 Cichocki, A., Zdunek, R., Phan, A., and Amari, S.I. (2009) *Nonnegative Matrix and Tensor Factorizations: Applications to Exploratory Multi-way Data Analysis and Blind Source Separation*, Wiley-Blackwell.

39 Chi, E. and Kolda, T. (2012) On tensors, sparsity, and nonnegative factorizations. *SIAM Journal on Matrix Analysis and Applications*, 33 (4), 1272–1299.

40 Lawson, C. and Hanson, R. (1974) *Solving Least Squares Problems*, Prentice-Hall.

41 Bro, R. (1998) *Multi-Way Analysis in the Food Industry: Models, Algorithms, and Applications*, Ph.D. thesis, University of Copenhagen.

42 Ho, N.D. (2008) *Nonnegative Matrix Factorization - Algorithms and Applications*, Ph.D. thesis, Université catholique de Louvain.

43 Wild, S., Curry, J., and Dougherty, A. (2004) Improving non-negative matrix factorizations through structured initialization. *Pattern Recognition*, 37 (11), 2217–2232.

44 Langville, A., Meyer, C., Albright, R., Cox, J., and Duling, D. (2006) Initializations for the nonnegative matrix factorization, in *Proceedings of the 12th ACM SIGKDD International Conference on Knowledge Discovery and Data Mining*, pp. 23–26.

45 Casalino, G., Del Buono, N., and Mencar, C. (2013) Subtractive clustering for seeding non-negative matrix factorizations. *Information Sciences*, 257, 369–387.

46 Boutsidis, C. and Gallopoulos, E. (2008) SVD based initialization: a head start for nonnegative matrix factorization. *Pattern Recognition*, 41, 1350–1362.

47 Grippo, L. and Sciandrone, M. (2000) On the convergence of the block nonlinear Gauss–Seidel method under convex constraints. *Operations Research Letters*, 26, 127–136.

48 Lin, C.J. (2007) Projected gradient methods for nonnegative matrix factorization. *Neural Computation*, 19, 2756–2779.

49 Guan, N., Tao, D., Luo, Z., and Yuan, B. (2012) NeNMF: an optimal gradient method for nonnegative matrix factorization. *IEEE Transactions on Signal Processing*, 60 (6), 2882–2898.

50 Cichocki, A., Zdunek, R., and Amari, S. (2006) *Non-negative Matrix Factorization with Quasi-Newton Optimization, Lecture Notes in Artificial Intelligence*, Springer, vol. 4029, pp. 870–879.

51 Bro, R. and De Jong, S. (1997) A fast non-negativity-constrained least squares algorithm. *Journal of Chemometrics*, 11 (5), 393–401.

52 Kim, H. and Park, H. (2007) Sparse non-negative matrix factorizations via alternating non-negativity-constrained least squares for microarray data analysis. *Bioinformatics*, 23 (12), 1495–1502.

53 Kim, J. and Park, H. (2011) Fast nonnegative matrix factorization: an active-set-like method and comparisons. *SIAM Journal on Scientific Computing*, 33 (6), 3261–3281.

54 Chen, D. and Plemmons, R. (2009) Nonnegativity constraints in numerical analysis, in *Symposium on the Birth of Numerical Analysis*, (eds A. Bultheel and R. Cools), World Scientific Press, pp. 109–139.

55 Lee, D. and Seung, H. (2001) Algorithms for non-negative matrix factorization, in *Advances in Neural Information Processing 13 (NIPS '01).*

56 Daube-Witherspoon, M. and Muehllehner, G. (1986) An iterative image space reconstruction algorithm suitable for volume ECT. *IEEE Transactions on Medical Imaging*, 5, 61–66.

57 Bertsekas, D.P. (2014) *Constrained Optimization and Lagrange Multiplier Methods*, Academic Press.

58 Repetti, A., Chouzenoux, E., and Pesquet, J.C. (2014) A preconditioned forward-backward approach with application to large-scale nonconvex spectral unmixing problems, in *Proceedings of the 39th IEEE International Conference on Acoustics, Speech, and Signal Processing (ICASSP 2014)*, Firenze, Italy, pp. 1498–1502.

59 Gillis, N. and Glineur, F. (2012) Accelerated multiplicative updates and hierarchical ALS algorithms for nonnegative matrix factorization. *Neural Computation*, 24 (4), 1085–1105.

60 Sha, F., Lin, Y., Saul, L., and Lee, D. (2007) Multiplicative updates for nonnegative quadratic programming. *Neural Computation*, 19 (8), 2004–2031.

61 Gillis, N. (2014) The why and how of nonnegative matrix factorization, in *Regularization, Optimization, Kernels, and Support Vector Machines, Machine Learning and Pattern Recognition Series* (eds J. Suykens, M. Signoretto, and A. Argyriou), Chapman & Hall/CRC, pp. 257–291.

62 Cichocki, A., Zdunek, R., and Amari, S.I. (2007) *Hierarchical ALS Algorithms for Nonnegative Matrix and 3D Tensor Factorization, Lecture Notes in Computer Science*, vol. 4666, Springer, pp. 169–176.

63 Cichocki, A. and Phan, A. (2009) Fast local algorithms for large scale nonnegative matrix and tensor factorizations. *IEICE Transactions on Fundamentals of Electronics*, E92-A (3), 708–721.

64 Li, L. and Zhang, Y.J. (2009) FastNMF: highly efficient monotonic fixed-point nonnegative matrix factorization algorithm with good applicability. *Journal of Electronic Imaging*, 18 (3), 033004.

65 Liu, J., Liu, J., Wonka, P., and Ye, J. (2012) Sparse non-negative tensor factorization using columnwise coordinate descent. *Pattern Recognition*, 45 (1), 649–656.

66 Hsieh, C.J. and Dhillon, I. (2011) Fast coordinate descent methods with variable selection for non-negative matrix factorization, in *Proceedings of the 17th ACM SIGKDD International Conference on Knowledge Discovery and Data Mining*, pp. 1064–1072.

67 Vandaele, A., Gillis, N., Glineur, F., and Tuyttens, D. (2015) Heuristics for exact nonnegative matrix factorization. *Journal of Global Optimization*, 65, 369–400.

68 Kim, J., He, Y., and Park, H. (2013) Algorithms for nonnegative matrix and tensor factorizations: a unified view based on block coordinate descent framework. *Journal of Global Optimization*, 58, 285–319.

69 Zymnis, A., Kim, S.J., Skaf, J., Parente, M., and Boyd, S. (2007) Hyperspectral image unmixing via alternating projected subgradients, in *Signals, Systems and Computers, 2007*, pp. 1164–1168.

70 Iordache, M.D., Bioucas-Dias, J., and Plaza, A. (2011) Total variation regulatization in sparse hyperspectral unmixing, in *3rd Worskshop on*

Hyperspectral Image and Signal Processing: Evolution in Remote Sensing (WHISPERS), Lisbon.

71 Cai, D., He, X., Han, J., and Huang, T. (2011) Graph regularized nonnegative matrix factorization for data representation. *IEEE Transactions on Pattern Analysis and Machine Intelligence*, 33 (8), 1548–1560.

72 Rapin, J., Bobin, J., Larue, A., and Starck, J.L. (2013) Sparse and non-negative BSS for noisy data. *IEEE Transactions on Signal Processing*, 61 (22), 5620–5632.

73 Condat, L. (2015) Fast projection onto the simplex and the ℓ_1 ball. *Mathematical Programming*, 158, 575–585.

74 Arora, S., Ge, R., Kannan, R., and Moitra, A. (2012) Computing a nonnegative matrix factorization – provably, in *Proceedings of the 44th Symposium on Theory of Computing (STOC '12)*, pp. 145–162.

75 Gillis, N. and Vavasis, S. (2015) Semidefinite programming based preconditioning for more robust near-separable nonnegative matrix factorization. *SIAM Journal on Optimization*, 25 (1), 677–698.

76 Kumar, A. and Sindhwani, V. (2015) Near-separable non-negative matrix factorization with ℓ_1- and Bregman loss functions. in *SIAM International Conference on Data Mining*.

77 Gillis, N. and Ma, W.K. (2015) Enhancing pure-pixel identification performance via preconditioning. *SIAM Journal on Imaging Sciences*, 8 (2), 1161–1186.

78 Arora, S., Ge, R., Halpern, Y., Mimno, D., Moitra, A., Sontag, D., Wu, Y., and Zhu, M. (2013) A practical algorithm for topic modeling with provable guarantees, in *International Conference on Machine Learning (ICML '13)*, vol. 28, pp. 280–288.

79 Li, Y., Sima, D., Van Cauter, S., Croitor Sava, S., Himmelreich, U., Pi, Y., and Van Huffel, S. (2012) Hierarchical non-negative matrix factorization (hNMF): a tissue pattern differentiation method for glioblastoma multiforme diagnosis using mrsi. *NMR in Biomedicine*, 26 (3), 307–319.

80 Kuang, D. and Park, H. (2013) Fast rank-2 nonnegative matrix factorization for hierarchical document clustering, in *19th ACM SIGKDD Conference on Knowledge Discovery and Data Mining (KDD '13)*, pp. 739–747.

81 Gillis, N., Kuang, D., and Park, H. (2015) Hierarchical clustering of hyperspectral images using rank-two nonnegative matrix factorization. *IEEE Transactions on Geoscience and Remote Sensing*, 53 (4), 2066–2078.

82 Gillis, N. (2012) Sparse and unique nonnegative matrix factorization through data preprocessing. *Journal of Machine Learning Research*, 13 (1), 3349–3386.

83 Kumar, N., Moussaoui, S., Idier, J., and Brie, D. (2015) Impact of sparse representation on the admissible solutions of spectral unmixing by non-negative matrix factorization, in *Proceeding of IEEE Workshop on Hyperspectral Image and Signal Processing (WHISPERS)*, Tokyo, Japan.

84 Henrot, S., Soussen, C., Dossot, M., and Brie, D. (2014) Does deblurring improve geometrical hyperspectral unmixing? *IEEE Transactions on Image Processing*, 23 (3), 1169–1180.

85 Miron, S., Dossot, M., Carteret, C., Margueron, S., and Brie, D. (2011) Joint processing of the parallel and crossed polarized Raman spectra and uniqueness in blind nonnegative source separation. *Chemometrics and Intelligent Laboratory Systems*, 105 (1), 7–18.

86 Long, D.A. (2002) *The Raman Effect: A Unified Treatment of the Theory of Raman Scattering by Molecules*, John Wiley & Sons.

87 Turrell, G. (1972) *Infrared and Raman spectra of crystals*, Academic Press, New York.

88 Jiménez, C., Caroff, T., Bartasyte, A., Margueron, S., Abrutis, A., Chaix-Pluchery, O., and Weiss, F. (2009) Raman study of CeO_2 texture as a buffer layer in the $CeO_2/La_2Zr_2O_7/Ni$ architecture for coated conductors. *Applied spectroscopy*, 63 (4), 401–406.

89 Tanaka, M. and Young, R. (2006) Review Polarised Raman spectroscopy for the study of molecular orientation distributions in polymers. *Journal of Materials Science*, 41 (3), 963–991.

90 Jaumot, J., Menezes, J.C., and Tauler, R. (2006) Quality assessment of the results obtained by multivariate curve resolution analysis of multiple runs of gasoline blending processes. *Journal of Chemometrics*, 20 (1–2), 54–67.

91 Ruckebusch, C., De Juan, A., Duponchel, L., and Huvenne, J. (2006) Matrix augmentation for breaking rank-deficiency: a case study. *Chemometrics and Intelligent Laboratory Systems*, 80 (2), 209–214.

92 Lee, D. and Seung, H. (2000) Algorithms for non-negative matrix factorization, in *Advances on Neural Information Processing Systems 13, (NIPS'2000)*, MIT Press, pp. 556–562.

93 Craig, M. (1994) Minimum-volume transforms for remotely sensed data. *IEEE Transactions on Geoscience and Remote Sensing*, 32 (3), 542–552.

94 Winter, M. (1999) N-FINDR: an algorithm for fast autonomous spectral end-member determination in hyperspectral data, in *Proceedings of SPIE Conference on Imaging Spectrometry V*.

95 Chang, C.I. (2007) *Hyperspectral Data Exploitation. Theory and Applications*, Wiley Interscience.
96 Heinz, D.C. and Chang, C.I. (2001) Fully constrained least squares linear spectral mixture analysis method for material quantification in hyperspectral imagery. *IEEE Transactions on Geoscience and Remote Sensing*, 39 (3), 529–545.
97 Chouzenoux, E., Legendre, M., Moussaoui, S., and Idier, J. (2014) Fast constrained least squares spectral unmixing using primal-dual interior-point optimization. *IEEE Journal of Selected Topics in Applied Earth Observations and Remote Sensing*, PP (99), 1–11.
98 Miao, L. and Qi, H. (2007) Endmember extraction from highly mixed data using minimum volume constrained nonnegative matrix factorization. *IEEE Transactions on Geoscience and Remote Sensing*, 45 (3), 765–777.
99 Chang, C.I. and Du, Q. (2004) Estimation of number of spectrally distinct signal sources in hyperspectral imagery. *IEEE Transactions on Geoscience and Remote Sensing*, 42 (3), 608–619.
100 Bioucas-Dias, J. and Nascimento, J. (2008) Hyperspectral subspace identification. *IEEE Transactions on Geoscience and Remote Sensing*, 46 (8), 2435–2445.
101 Abderrahim, H., Honeine, P., Kharouf, M., Richard, C., and Tourneret, J.Y. (2019) Estimating the intrinsic dimension of hyperspectral images using a noise-whitened eigengap approach. *IEEE Transactions on Geoscience and Remote Sensing*, 54 (27), 3811–3821.
102 Bioucas-Dias, J., Plaza, A., Dobigeon, N., Parente, M., Du, Q., Gader, P., and Chanussot, J. (2012) Hyperspectral unmixing overview: geometrical, statistical, and sparse regression-based approaches. *IEEE Journal of Selected Topics in Applied Earth Observations and Remote Sensing*, 5 (2), 354–379.
103 Henrot, S., Soussen, C., and Brie, D. (2013) Fast positive deconvolution of hyperspectral images. *IEEE Transactions on Image Process*, 22 (2), 828–833.
104 De Grauw, C., Sijtsema, N., Otto, C., and Greve, J. (1997) Axial resolution of confocal Raman microscopes: Gaussian beam theory and practice. *Journal of Microscopy*, 188 (3), 273–279.
105 Wang, L.Y., Zhang, Y.H., and Zhao, L.J. (2005) Raman spectroscopic studies on single supersaturated droplets of sodium and magnesium acetate. *The Journal of Physical Chemistry A*, 109 (4), 609–614.

4

Bayesian Source Separation

Saïd Moussaoui[1], Leonardo Tomazeli Duarte[2], Nicolas Dobigeon[3], and Christian Jutten[4]

[1] *LS2N, Nantes Université, Ecole Centrale Nantes, Nantes, France*
[2] *School of Applied Sciences (FCA), University of Campinas, Limeira, Brazil*
[3] *IRIT, INP-ENSEEIHT, IRIT University of Toulouse, Toulouse, France*
[4] *GIPSA-lab, Univ. Grenoble Alpes, CNRS, Institut Univ., de France, Grenoble, France*

4.1 Introduction

Blind Source separation is an ill-posed inverse problem since the solution may not be unique. Actually, without any assumption on the source signals and the mixing coefficient several solutions are possible. Therefore, in order to get a particular solution, one has to use any available knowledge or additional assumptions on the sought source signals and mixing coefficients. It has been shown in Chapter 1 that source separation using a maximum likelihood approach offers the possibility to encode, in addition to the mutual statistical independence of the sources, additional prior information on these signals by specifying their probability distribution functions. Such setting also corresponds to some specific choice of the nonlinear functions in independent component analysis (ICA) algorithms based on nonlinear decorrelation (see also Chapter 1). However, the maximum likelihood approach does not allow any constraint on the mixing matrix to be taken into account. In this context, the Bayesian approach arises as a natural and convenient way to encode available information on both the source signals and the mixing coefficients, thus allowing one to get a physically meaningful solution [1]. In that respect, additional assumption and constraints on the noise statistics, the source signals, and the mixing coefficients are modeled in terms of probability distribution functions.

The goal of this chapter is to present the Bayesian approach for source separation and to illustrate its application in the case of physical and chemical

Source Separation in Physical-Chemical Sensing, First Edition.
Edited by Christian Jutten, Leonardo Tomazeli Duarte, and Saïd Moussaoui.

sensing. The general framework of the Bayesian source separation approach will be presented in the first part of this chapter. It includes the specification of the likelihood resulting from the statistical description of the noise and the formulation of statistical models encoding the available information and constraints on the sought source signals and mixing coefficients. In a second step, separation models and methods for linear and nonlinear mixing models in the context of physical–chemical sensing will be presented. The Bayesian approach will be illustrated through some examples of case studies of the processing spectral data resulting from chemical component analysis and ion-selective electrodes and spectrometry measurements in chemical sensing.

4.2 Overview of Bayesian Source Separation

The first works on the formulation of the source separation problem within the Bayesian estimation theory have been initially proposed in [2–5], in which a more general Bayesian inference framework was stated. Subsequently, several separation algorithms have been proposed by exploiting different assumptions on the source signals and the mixing coefficients. For instance, in [6, 7] the sparsity of the sources has been exploited, while the authors in [8] proposed to perform the separation in the wavelet domain. However, its application to the separation of linear spectral mixture data in the context of chemical–physical sensing has been performed firstly in [9, 10] and later in [11, 12]. For instance, the method proposed in [11] incorporates the non-negativity constraint and has been successfully applied in [13] and [14] for the analysis of hyperspectral/multispectral images. Bayesian separation methods in the case of linear mixing can be directly linked to the constrained non-negative matrix factorization algorithms presented in Chapter 3 of this book. On the other hand, the Bayesian approach offers a very well-stated framework for addressing the case of nonlinear mixing in chemical sensing [15, 16] and remote sensing [17, 18].

4.2.1 General Framework

Without loss of generality, let us consider a source separation problem where the mixing process allows the observed data to be expressed according to

$$\boldsymbol{X} = \mathcal{H}(\boldsymbol{S}; \boldsymbol{\Psi}) + \boldsymbol{E}, \tag{4.1}$$

where each row vector $\boldsymbol{x}_p$ of the $(P \times T)$ data matrix $\boldsymbol{X}$ contains the p-th observation signal $\{x_p(t), t = 1, \ldots, T\}$. The $(R \times T)$ matrix $\boldsymbol{S}$ is composed of R row vectors $\boldsymbol{s}_r$ containing the source signals. $\boldsymbol{\Psi}$ comprises the parameters of the mixing process, here represented by the mapping $\mathcal{H}(\cdot)$. For instance, in the linear mixing case, which will guide the discussion throughout the chapter, $\mathcal{H}(\boldsymbol{S};\boldsymbol{\Psi}) = \boldsymbol{A}\boldsymbol{S}$, that is, the parameters $\boldsymbol{\Psi}$ correspond to a mixing matrix $\boldsymbol{A}$ of size $(P \times R)$; its column vectors $\boldsymbol{a}_r$ contain the mixing coefficients of each source signal in the P observations signals. Finally, in (4.1), the $(P \times T)$ matrix $\boldsymbol{E}$ is an additive residual term that stands for the measurement noise and any modeling error resulting from the choice of the mapping function $\mathcal{H}(\cdot)$.

Since both matrices $\boldsymbol{S}$ and $\boldsymbol{A}$ represent, in the linear case, physical quantities (for instance absorbance spectra and abundance fractions, in vibrational spectroscopy), their estimation using a Bayesian source separation approach is firstly motivated by the opportunity of encoding additional available knowledge through the assignment of prior probability density functions (p.d.f)[1] $f(\boldsymbol{S})$ and $f(\boldsymbol{A})$ associated with the source signals and the mixing matrices, respectively. According to Bayes' theorem, the posterior distribution $f(\boldsymbol{S},\boldsymbol{A}|\boldsymbol{X})$ can be derived from the likelihood function $f(\boldsymbol{X}|\boldsymbol{S},\boldsymbol{A})$ and the prior distributions as follows:

$$f(\boldsymbol{S},\boldsymbol{A}|\boldsymbol{X}) = \frac{f(\boldsymbol{X}|\boldsymbol{S},\boldsymbol{A}) \times f(\boldsymbol{S}) \times f(\boldsymbol{A})}{f(\boldsymbol{X})}, \tag{4.2}$$

where the prior independence between $\boldsymbol{A}$ and $\boldsymbol{S}$ has been assumed. The term $f(\boldsymbol{X})$ is the probability density function of the measured data, whose calculation is intractable and not necessary for the estimation of $\boldsymbol{S}$ and $\boldsymbol{A}$. Therefore, one can write

$$f(\boldsymbol{S},\boldsymbol{A}|\boldsymbol{X}) \propto f(\boldsymbol{X}|\boldsymbol{S},\boldsymbol{A}) \times f(\boldsymbol{S}) \times f(\boldsymbol{A}). \tag{4.3}$$

This posterior distribution combines the prior knowledge and assumptions on the source signals and on the mixing coefficients with the information brought by the observed data (mixing model and measurement noise). Indeed, the likelihood function $f(\boldsymbol{X}|\boldsymbol{S},\boldsymbol{A})$ is formulated by exploiting the mixing model and the noise statistics, while the prior distributions $f(\boldsymbol{S})$ and $f(\boldsymbol{A})$ explicitly model the properties of the sought source signals and mixing coefficients.

4.2.2 Choice of the Prior Distributions

The multivariate prior probability distribution functions $f(\boldsymbol{S})$ and $f(\boldsymbol{A})$ should be defined in such a way to encode the available *a priori* information.

---

1 For the sake of simplification, the *probability density* function of a random variable X is noted $f(x)$ rather than $f_X(x)$. It is also called *probability distribution function* and referred to by *distribution*, and should not be confused with the *cumulative distribution function* defined as $F_X(x) = \int_{-\infty}^{x} f(t)dt$

Firstly, according to the most classical prior model, which assumes prior mutual statistical independence of the sources, one may factorize the joint distribution of the sources as follows:

$$f(\mathbf{S}) = \prod_{r=1}^{R} f(\mathbf{s}_r). \tag{4.4}$$

Then, by assuming that the samples of each source signal are independent and identically distributed (iid), the prior distribution of each source signal, $p(\mathbf{s}_r)$, can be expressed as

$$f(\mathbf{s}_r) = \prod_{t=1}^{T} f(s_r(t)). \tag{4.5}$$

Finally, the prior distribution of the source matrix $p(\mathbf{S})$ can be deduced:

$$f(\mathbf{S}) = \prod_{r=1}^{R}\prod_{t=1}^{T} f(s_r(t)). \tag{4.6}$$

According to Eq. (4.4), it is interesting to note that, as in ICA [19, 20], most of Bayesian source separation methods take the independence hypothesis into account. However, in the Bayesian case, such an assumption means that no particular measure of dependence among the sources is taken into account in the inference process. Conversely, in the ICA case, the independence assumption lies at the core of almost every separation criterion. As a result, compared to ICA, Bayesian methods are more robust in the cases where the sources present some degree of correlation and some knowledge is available on the mixing coefficients [3]. We shall further discuss this point later in this chapter. In the same spirit, the assumption that the each source is iid (see (4.5)) means, in a Bayesian approach, that possible information related to the temporal (or spatial) structure of the given source is not taken into account by the inference algorithm.

Note that a similar reasoning on the mixing coefficients leads to

$$f(\mathbf{A}) = \prod_{p=1}^{P}\prod_{p=1}^{P} f(A_{pr}). \tag{4.7}$$

Actually, the key point for the specification of the prior distribution functions resides in the choice of the univariate statistical distributions $f(A_{pr})$ and $f(s_r(t))$ so that they can encode the available information or assumptions. Among the archetypal examples are:

- non-negativity can be ensured by deriving prior models based on a truncated (non-negative support) Gaussian or a Gamma distribution [11, 12];

- sparsity can be encoded using generalized Gaussian (e.g. Laplacian distribution) or Student-t distributions as in [6, 7] (the interested reader is invited to consult [21] for a comprehensive review of sparsity-inducing prior modeling);
- a sum-to-one (i.e. additivity) constraint, which is typical in physical–chemical applications and arises when, for instance, the sources represent species which are connected via reactions. Consequently the mixing coefficients for each observation p are dependent and the prior distribution expressed in (4.7) becomes

  $$f(\boldsymbol{A}) = \prod_{r=1}^{R} f(\boldsymbol{a}_p).$$

  To encode the sum-to-one constraint, the joint distribution $f(\boldsymbol{a}_p)$ is expressed for instance using either a truncated Gaussian or a Dirichlet distribution [12, 22].

Note that more complex prior models should be designed when more specific prior knowledge can be exploited. For instance, relaxing the independence in (4.5) between the sources components allows joint priors $f(\boldsymbol{s}_r)$ to be designed to promote particular behavior of the sources, e.g. smoothness [23, 24]. Similarly, non-separable joint prior distributions $f(\boldsymbol{S})$ for the source matrix (4.4) can be considered, as in [25], to constrain the intrinsic volume spanned by the source vectors.

4.2.3 Source Signal and Mixing Matrix Estimation from the Posterior Distribution

Having defined the priors and the posterior distribution, various Bayesian estimators can be considered [1] in order to get estimates of the source signals and the mixing coefficients. The most popular ones are the joint maximum *a posteriori* (JMAP), the marginal maximum *a posteriori* (MMAP), and the minimum mean square error (MMSE) estimators.

4.2.3.1 Joint Maximum A Posteriori (JMAP)

The JMAP estimates are obtained by maximizing the joint posterior distribution of the source and mixing matrices

$$\left(\boldsymbol{S}_{\text{JMAP}}, \boldsymbol{A}_{\text{JMAP}}\right) = \arg\max_{\boldsymbol{S},\boldsymbol{A}} \ f(\boldsymbol{S}, \boldsymbol{A}|\boldsymbol{X}). \tag{4.8}$$

This maximization is equivalent to a joint minimization of the negative log-posterior, defined as the objective function by

$$J(\boldsymbol{S}, \boldsymbol{A}) = -\log f(\boldsymbol{S}, \boldsymbol{A}|\boldsymbol{X}), \tag{4.9}$$

and depends not only on the measured data but also on the prior distributions of the sources signals and the mixing coefficients. Therefore, according to Bayes' theorem, given in Eq. (4.6), this objective function is

$$J(\boldsymbol{S},\boldsymbol{A}) = -\log f(\boldsymbol{X}|\boldsymbol{S},\boldsymbol{A}) - \log f(\boldsymbol{S}) - \log f(\boldsymbol{A}). \tag{4.10}$$

which in turn can be written as

$$J(\boldsymbol{S},\boldsymbol{A}) = \mathcal{L}(\boldsymbol{X},\boldsymbol{S},\boldsymbol{A}) + \mathcal{R}_1(\boldsymbol{S}) + \mathcal{R}_2(\boldsymbol{A}). \tag{4.11}$$

where Q is a data fitting criterion, $\mathcal{R}_1$ and $\mathcal{R}_2$ are regularization terms which, by taking (4.4) and (4.5) into account, can be written as:

$$\begin{cases} \mathcal{R}_1(\boldsymbol{S}) = -\log f(\boldsymbol{S}) = -\sum_{r=1}^{R}\sum_{t=1}^{T} \log f(s_r(t)), \\ \mathcal{R}_2(\boldsymbol{A}) = -\log f(\boldsymbol{A}) = -\sum_{p=1}^{P}\sum_{r=1}^{R} \log f(a_{pr}). \end{cases} \tag{4.12}$$

Actually, a given prior distribution function leads to a particular regularization function $\psi(s_r) = -\log f(s_r)$. This aspect corresponds to an implicit regularization function synthesis thanks to the statistical formulation of the prior knowledge. This formulation is known as a Bayesian interpretation of a regularized estimation methods [26]. The numerical calculation of the solution is most often performed using iterative optimization methods [27]. Chapter 2 provides an overview of methods that can be used for solving this minimization problem. One can also notice that this formulation arises in penalized NMF algorithms described in Chapters 3 and 5.

4.2.3.2 Marginal Maximum *A Posteriori* (MMAP)

MMAP estimates are obtained by first integrating (marginalizing) the joint posterior distribution $f(\boldsymbol{S},\boldsymbol{A}|\boldsymbol{X})$ with respect to $\boldsymbol{S}$ or $\boldsymbol{A}$ to get the two corresponding marginal posteriors

$$f(\boldsymbol{S}|\boldsymbol{X}) = \int f(\boldsymbol{S},\boldsymbol{A}|\boldsymbol{X})d\boldsymbol{A}, \tag{4.13}$$

$$f(\boldsymbol{A}|\boldsymbol{X}) = \int f(\boldsymbol{S},\boldsymbol{A}|\boldsymbol{X})d\boldsymbol{S}, \tag{4.14}$$

and then maximizing these marginal posterior distributions $f(\boldsymbol{A}|\boldsymbol{X}), f(\boldsymbol{S}|\boldsymbol{X})$, i.e.

$$\boldsymbol{S}_{\mathrm{MMAP}} = \arg\max_{\boldsymbol{S}} f(\boldsymbol{S}|\boldsymbol{X}) \tag{4.15}$$

$$\boldsymbol{A}_{\mathrm{MMAP}} = \arg\max_{\boldsymbol{A}} f(\boldsymbol{A}|\boldsymbol{X}) \tag{4.16}$$

In such a context, one difficulty may arise because of the potentially computationally intensive numerical integration required by (4.13). To alleviate this issue, methods such as the expectation-maximization (EM) algorithm and its statistical variants (Stochastic EM [28] and Monte Carlo EM [29]) can be used to approximate the MMAP estimates.

4.2.3.3 Posterior Mean (PM) or Minimum Mean Square Error (MMSE)

These estimators are obtained by computing the mean of the marginal posterior distributions $f(\boldsymbol{A}|\boldsymbol{X})$ and $f(\boldsymbol{S}|\boldsymbol{X})$, i.e.

$$\boldsymbol{S}_{\text{MMSE}} = \text{E}\left[\boldsymbol{S}|\boldsymbol{X}\right] = \int \boldsymbol{S}f(\boldsymbol{S}|\boldsymbol{X})d\boldsymbol{S} \tag{4.17}$$

$$\boldsymbol{A}_{\text{MMSE}} = \text{E}\left[\boldsymbol{A}|\boldsymbol{X}\right] = \int \boldsymbol{A}f(\boldsymbol{A}|\boldsymbol{X})d\boldsymbol{A} \tag{4.18}$$

In practice, unless in some particular cases, these integrals cannot be computed analytically. The analytical integration is therefore replaced by an empirical approximation of the estimates thanks to Monte Carlo methods [30]. These techniques generate a sufficiently high set of samples distributed according to the posterior distribution. These samples are then used to evaluate empirically the statistics of this distribution (means, covariance, etc.).

Most popular techniques are Markov chain Monte Carlo (MCMC) algorithms [31, 32] designed to generate iteratively a series of N_{MC} random realizations $\left\{\left(\boldsymbol{A}^{(k)}, \boldsymbol{S}^{(k)}\right)\right\}_{k=1}^{N_{\text{MC}}}$ asymptotically distributed according to the posterior distribution $f(\boldsymbol{S}, \boldsymbol{A}|\boldsymbol{X})$. At each iteration, the simulation from $f\left(\boldsymbol{S}, \boldsymbol{A}|\boldsymbol{X}\right)$ is realized using the Gibbs sampler [31] that alternatively sample according to the two corresponding conditional posteriors, leading to the following two-step procedure

1. the sources signals $\boldsymbol{S}^{(k+1)}$ are generated from[2]

$$f\left(\boldsymbol{S}|\boldsymbol{X}, \boldsymbol{A}^{(k)}\right) \propto f\left(\boldsymbol{X}|\boldsymbol{S}, \boldsymbol{A}^{(k)}\right) \times f\left(\boldsymbol{S}\right),$$

2. the mixing coefficients $\boldsymbol{A}^{(k+1)}$ are generated from

$$f\left(\boldsymbol{A}|\boldsymbol{X}, \boldsymbol{S}^{(k+1)}\right) \propto f\left(\boldsymbol{X}|\boldsymbol{S}^{(k+1)}, \boldsymbol{A}\right) \times f\left(\boldsymbol{A}\right).$$

Specific techniques can be used to perform the sample generation from a simple target distribution [33], and more sophisticated ones may be needed in the general case [31].

According to this algorithmic scheme, a burn-in period is required in order to ensure that the Markov chain converges to its stationary distribution. This means that an initial set of generated samples are discarded in the inference step. Therefore, the two sets of generated samples $\left\{\boldsymbol{A}^{(k)}\right\}_{k=N_b+1}^{N_{\text{MC}}}$

---

2 Note here that $f\left(\boldsymbol{S}|\boldsymbol{X}, \boldsymbol{A}^{(k)}\right)$ refers to a probability distribution with respect to the variable $\boldsymbol{S}$. This conditional distribution has, as parameters, the observations $\boldsymbol{X}$, and the current sample $\boldsymbol{A}^{(k)}$.

and $\left\{\boldsymbol{S}^{(k)}\right\}_{k=N_b+1}^{N_{MC}}$ are asymptotically distributed according to the conditional posteriors $f(\boldsymbol{A}|\boldsymbol{X})$ and $f(\boldsymbol{S}|\boldsymbol{X})$, respectively – N_b corresponds to the burn-in iterations and N_{MC} to the maximal number of iterations. Thus, these realizations can be used to get a comprehensive statistical description of the sought sources and the mixing coefficients. In particular, the MMSE estimates can be approximated according to

$$\hat{\boldsymbol{S}} = \frac{1}{N_{MC} - N_b} \sum_{k=N_b+1}^{N_{MC}} \boldsymbol{S}^{(k)}, \tag{4.19}$$

$$\hat{\boldsymbol{A}} = \frac{1}{N_{MC} - N_b} \sum_{k=N_b+1}^{N_{MC}} \boldsymbol{A}^{(k)}. \tag{4.20}$$

Other statistical quantities of interest (variance, histograms, confidence intervals) can also be similarly approximated.

The critical point here is the choice of efficient methods for sampling the conditional distributions appearing in Steps 1 and 2 of the algorithm. In Section 4.5, we shall provide a simple example illustrating how this sampling step can be accomplished.

4.2.4 Hierarchical Bayesian Modeling and Inference

In practice, the statistical models related to the distributions of the source signals, the mixing coefficients, and the observation noise are given in a parametric form $f(\boldsymbol{S}|\theta_s)$ and $f(\boldsymbol{A}|\theta_a)$ and $f(\boldsymbol{E}|\theta_e)$, respectively. However, the appropriate values of these hyperparameters, gathered in a vector $\theta = \{\theta_s, \theta_a, \theta_e\}$, are data dependent and, generally, they are not known in advance. Therefore, within an unsupervised framework, they must also be considered in the inference scheme. The additional and noteworthy benefit of the Bayesian approach relies on the possibility of naturally including them into the inference procedure, following the hierarchical model

$$f(\boldsymbol{S}, \boldsymbol{A}, \theta|\boldsymbol{X}) \propto f(\boldsymbol{X}|\boldsymbol{S}, \boldsymbol{A}, \theta_e)\, f(\boldsymbol{S}|\theta_s)\, f(\boldsymbol{A}|\theta_a)\, f(\theta). \tag{4.21}$$

The joint estimation of the source signals, the mixing coefficients, and the hyperparameters is then performed from this extended posterior distribution. MCMC techniques can therefore be used to draw samples and to get empirical approximations of posterior statistics.

According to this hierarchical modeling, at each new iteration k, the main steps of the sampling algorithm allowing to get realizations of $f(\boldsymbol{S}, \boldsymbol{A}, \theta|\boldsymbol{X})$ are given by:

1. Simulate the sources signals $\boldsymbol{S}^{(k+1)}$ from

$$f\left(\boldsymbol{S}|\boldsymbol{X}, \boldsymbol{A}^{(k)}, \theta^{(k)}\right) \propto f\left(\boldsymbol{X}|\boldsymbol{S}, \boldsymbol{A}^{(k)}, \theta_e^{(k)}\right) \times f\left(\boldsymbol{S}|\theta_s^{(k)}\right),$$

2. Simulate the mixing coefficients $\boldsymbol{A}^{(k+1)}$ from
$$f\left(\boldsymbol{A}|\boldsymbol{X},\boldsymbol{S}^{(k+1)},\boldsymbol{\theta}^{(k)}\right) \propto f\left(\boldsymbol{X}|\boldsymbol{S}^{(k+1)},\boldsymbol{A},\boldsymbol{\theta}_e^{(k)}\right) \times f\left(\boldsymbol{A}|\boldsymbol{\theta}_a^{(k)}\right).$$
3. Simulate the noise distribution hyperparameters $\boldsymbol{\theta}_e^{(k+1)}$ from
$$f\left(\boldsymbol{\theta}_e|\boldsymbol{X},\boldsymbol{S}^{(k+1)}\boldsymbol{A}^{(k+1)}\right) \propto f\left(\boldsymbol{X}|\boldsymbol{S}^{(k+1)},\boldsymbol{A}^{(k+1)},\boldsymbol{\theta}_e^{(k)}\right) \times f\left(\boldsymbol{\theta}_e\right).$$
4. Simulate the source distribution hyperparameters $\boldsymbol{\theta}_s^{(k+1)}$ from
$$f\left(\boldsymbol{\theta}_s|\boldsymbol{S}^{(k+1)}\right) \propto f\left(\boldsymbol{S}^{(k+1)}|\boldsymbol{\theta}_s\right) \times f\left(\boldsymbol{\theta}_s\right).$$
5. Simulate the mixing coefficient distribution hyperparameters $\boldsymbol{\theta}_a^{(k+1)}$ from
$$f\left(\boldsymbol{\theta}_a|\boldsymbol{A}^{(k+1)}\right) \propto f\left(\boldsymbol{A}^{(k+1)}|\boldsymbol{\theta}_a\right) \times f\left(\boldsymbol{\theta}_a\right).$$

It is worth mentioning that this inference procedure also needs the definition of appropriate priors for the hyperparameters in steps 3, 4, and 5. In practice, this is done by defining a vague choice for the prior distributions. As a result, such a procedure allows one to set the range of admissible values of the hyperparameters [34], which is much easier than assigning a specific value of each parameter.

4.3 Statistical Models for the Separation in the Linear Mixing

In the context of physical–chemical sensing the observation signals are obtained from various measurements techniques, such as for instance spectroscopy, microscopy, and chromatography. The analysis of such spectral mixture data is also called *factor analysis* [35] or *multivariate curve resolution* [36] in chemistry. A review on these methods can be found in [37]. The source signals correspond to the pure component spectra and the mixing coefficients are related to their abundances in the mixtures. The aim of this section is to present some statistical separation models using the Bayesian approach that was successfully applied to physical–chemical data in the case of linear mixing. The linear mixing model is assumed in most of source separation problems in chemical sensing since it can be seen as a first-order approximation of a nonlinear mixing model. Actually, when the physical mixing model cannot be expressed analytically, a first natural attempt is to solve the separation problem under a linear mixing model.

4.3.1 Mixing Model

According to the linear mixing model, the observation data are interpreted as a weighted sum of unknown components, which in the context of physical–chemical applications, can represent spectra. In such a case, the spectral mixing model assumes that P measured spectra $\{x_p(t), t = 1, \dots, T\}_{p=1}^{P}$ are linear combinations of R unknown pure component spectra $\{s_r(t), t = 1, \dots, T\}_{r=1}^{R}$. This mixing model is expressed as

$$x_p(t) = \sum_{r=1}^{R} A_{pr}\, s_r(t) + e_p(t), \tag{4.22}$$

where $p = 1, \dots, P$ and $r = 1, \dots, R$ are indexes related to the measured spectra and unknown pure component spectra, respectively, and the index t corresponds to the spectral variable $\{\lambda_t, t = 1, \dots T\}$. Each mixing coefficient A_{pr} is proportional to the concentration of the rth pure component in the pth mixture. The additive noise terms $\{e_p(t), t = 1, \dots, T\}_{p=1}^{P}$ represent the measurement errors and model imperfections. Using matrix notations, this model can be rewritten as

$$\boldsymbol{X} = \boldsymbol{A}\, \boldsymbol{S} + \boldsymbol{E}, \tag{4.23}$$

where the row vectors of the $(P \times T)$ data matrix $\boldsymbol{X}$ contain the P measured spectra, $\boldsymbol{A}$ is the $(P \times R)$ mixing matrix, whose column vectors represent the mixing coefficient of each pure component, $\boldsymbol{S}$ is the $(R \times T)$ matrix, whose row vectors contain the R pure component spectra and $\boldsymbol{E}$ is the $(P \times R)$ additive noise matrix.

4.3.2 Likelihood Functions in the Linear Mixing Case

4.3.2.1 Gaussian Noise

The noise matrix $\boldsymbol{E}$ is usually assumed to be distributed according to a zero-mean multivariate random normal distribution and by considering mutual independence between noise samples. The resulting distribution in this case is given by

$$f\left(\boldsymbol{E}|\boldsymbol{\Sigma}_1, \dots, \boldsymbol{\Sigma}_T\right) = \prod_{t=1}^{T} f\left(\boldsymbol{e}(t)|\boldsymbol{\Sigma}_t\right)$$

with

$$f(\boldsymbol{e}(t)|\boldsymbol{\Sigma}_t) = \mathcal{N}\left(\boldsymbol{0}, \boldsymbol{\Sigma}_t\right)$$

where $\mathcal{N}(\cdot, \cdot)$ denotes the multivariate normal distribution and $\boldsymbol{\Sigma}_t$ corresponds to its covariance matrix. As a consequence, the likelihood function is given by

$$f(\boldsymbol{X}|\boldsymbol{S}, \boldsymbol{A}, \boldsymbol{\Sigma}_1, \dots, \boldsymbol{\Sigma}_T) = \prod_{t=1}^{T} \mathcal{N}(\boldsymbol{x}(t) - \boldsymbol{A}\boldsymbol{s}(t), \boldsymbol{\Sigma}_t). \tag{4.24}$$

The maximization of this likelihood is equivalent to the minimization of the following data fitting term

$$\mathcal{L}(\boldsymbol{S}, \boldsymbol{A}) = \sum_{t=1}^{T} (\boldsymbol{x}(t) - \boldsymbol{A}\boldsymbol{s}(t))^{\mathrm{T}} \boldsymbol{\Sigma}_t^{-1} (\boldsymbol{x}(t) - \boldsymbol{A}\boldsymbol{s}(t)). \tag{4.25}$$

Additionally, when the noise components are assumed to be mutually independent and the $T \times P$ noise components are identically distributed, the covariance matrix is given by $\boldsymbol{\Sigma}_t = \sigma^2 \boldsymbol{I}$, where σ^2 corresponds to the noise variance at each mixture. It is worth noticing that, in such case, the objective function (4.25) corresponds to the weighted least square criterion used, for instance, in PMF algorithms [27, 38]. See Chapter 3 for more details on penalized NMF.

4.3.2.2 Poissonian Noise

The use of Poissonian noise is adequate in the case of counting data. In this model, each observed data $x_p(t)$ is assumed to be a realization of a Poisson random variable of mean given by the entry (p, t) of the matrix $\boldsymbol{AS}$ – we shall denote this element by $[\boldsymbol{AS}]_{p,t}$. The likelihood function can therefore be expressed as

$$f(x_p(t)|\boldsymbol{S}, \boldsymbol{A}) = \frac{([\boldsymbol{AS}]_{p,t})^{x_p(t)}}{x_p(t)!} \exp(-[\boldsymbol{AS}]_{p,t}), \tag{4.26}$$

where $x_p(t)$ is a discrete number in this model. By using the assumption of mutually independent observations and identically distributed samples, the likelihood function $f(\boldsymbol{X}|\boldsymbol{S}, \boldsymbol{A})$ can be deduced:

$$f(\boldsymbol{X}|\boldsymbol{S}, \boldsymbol{A}) = \prod_{p=1}^{P} \prod_{t=1}^{T} f(x_p(t)|\boldsymbol{S}, \boldsymbol{A}).$$

It can be noted that the data fitting term, $Q(\boldsymbol{S}, \boldsymbol{A}) = -\log p(\boldsymbol{X}|\boldsymbol{S}, \boldsymbol{A})$, associated with this likelihood function is

$$\mathcal{L}(\boldsymbol{S}, \boldsymbol{A}) = \sum_{p=1}^{P} \sum_{t=1}^{T} \left([\boldsymbol{AS}]_{p,t} - x_p(t) \log [\boldsymbol{AS}]_{p,t} + \log x_p(t)! \right). \tag{4.27}$$

This criterion is similar to those of data fitting based on Kullback-Leibler divergence [39, 40]. More generally, the separation using the maximum likelihood approach, whose principle is the maximization of $f(\boldsymbol{X}|\boldsymbol{S}, \boldsymbol{A})$ or equivalently the minimization of $-\log f(\boldsymbol{X}|\boldsymbol{S}, \boldsymbol{A})$, provides a statistical formulation of the NMF algorithms presented in Chapter 3 of this book.

4.3.3 Priors on the Source Signals and Mixing Coefficients in Chemical Sensing

Several constraints should be applied when dealing with chemical or physical data sets. Such constraints are needed in order to preserve the physical link between the estimated source signals, the mixing coefficients, and the quantities they are expected to represent. The purpose of this section is to review some of these constraints and to introduce the statistical distributions that are suited to encode these constraints in a Bayesian framework. The main constraints that will be discussed are non-negativity, bounds, sum-to-one, and smoothness since they are the most encountered in the literature and for several applications.

4.3.3.1 Non-negativity Constraint

One of the most usual constraints when dealing with physical–chemical sensing is the non-negativity of both the source signals and the mixing coefficients:

$$A_{pr} \geqslant 0, \quad (\forall p = 1, \ldots, P, \forall r = 1, \ldots, R) \tag{4.28}$$

$$s_r(t) \geqslant 0, \quad (\forall r = 1, \ldots R, \forall t = 1, \ldots, T) \tag{4.29}$$

These constraints are due to physical considerations when the sources correspond to non-negative quantities such as spectral absorbance, concentrations, reflectance, or diffusion values, and when the mixing coefficients are related to the abundances (proportions, concentrations) of the components in the mixtures. The data processing can be addressed using the non-negative matrix factorization approach, fully described in Chapter 3.

From a statistical point of view, several distributions can be used to encode non-negativity [41]. A first candidate can be the Gaussian distribution restricted to a positive support. Conversely, the Gamma distribution family offers a natural way to model non-negative data since it consists of probability density functions defined on the positive domain. In addition, the Gamma distribution is defined with two parameters that allow a wide range of distributions with various shapes to be described. For example, Exponential, Erlang, and Chi-squared distributions are particular instances of the Gamma distribution family. However, there exist other continuous non-negative support distributions such as the log-normal (taking logs of data produces normality), Weibull, Rayleigh, and the inverse Gaussian that can also be used.

Table 4.1 gives an illustration of some of these distributions. It can be seen that the shape of the probability density function of theses distributions depends on the values of their corresponding hyperparameters.

Table 4.1 Example of (scalar) probability distributions encoding non-negativity.

(a) Half Gaussian	(b) Gamma
$f(x;\sigma) = \frac{\sqrt{2}}{\sigma\sqrt{\pi}} \exp\left(-\frac{x^2}{2\sigma^2}\right)$	$f(x;\alpha,\beta) = \frac{\beta^\alpha}{\Gamma(\alpha)} x^{\alpha-1} \exp -\beta x$
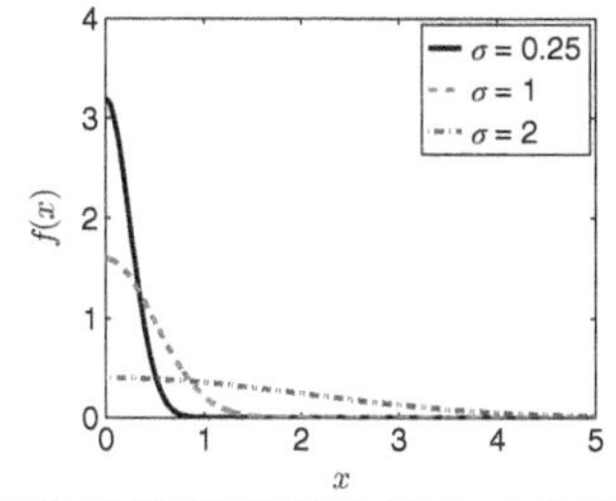 $\sigma = 0.25$ $\sigma = 1$ $\sigma = 2$	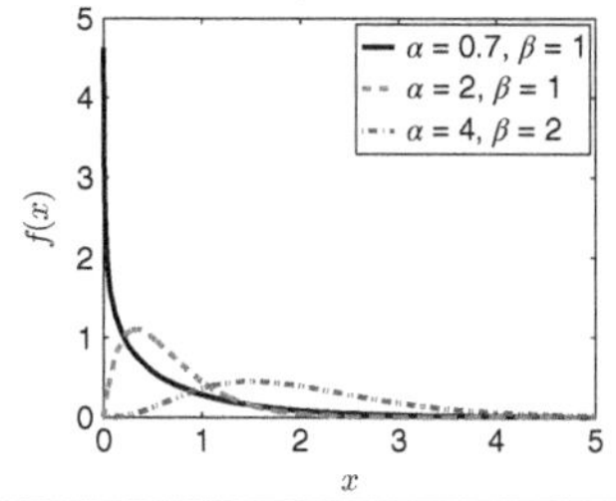 $\alpha = 0.7, \beta = 1$ $\alpha = 2, \beta = 1$ $\alpha = 4, \beta = 2$
(c) Rayleigh	**(d) Inverse Gaussian**
$f(x;\sigma,\mu) = \frac{x}{\sigma^2} e^{-x^2/(2\sigma^2)}$	$f(x;\mu,\beta) \left[\frac{\beta}{2\pi x^3}\right]^{1/2} \exp \frac{-\beta(x-\mu)^2}{2\mu^2 x}$
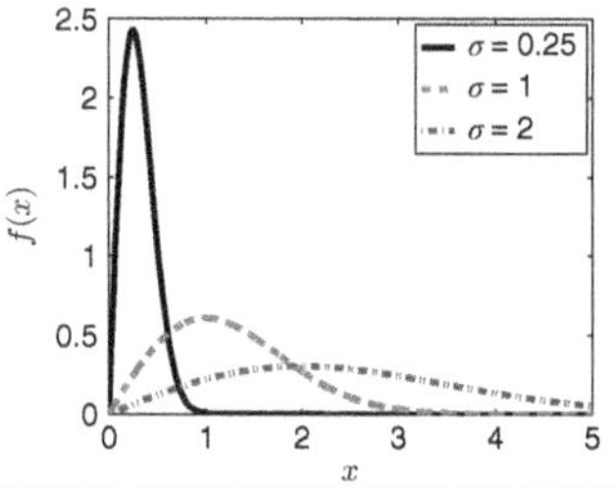 $\sigma = 0.25$ $\sigma = 1$ $\sigma = 2$	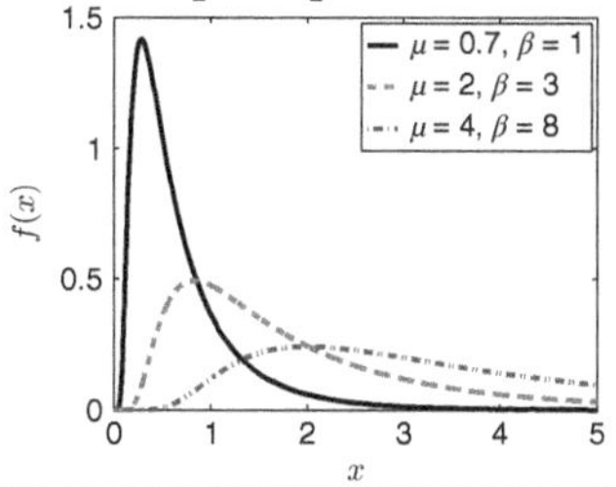 $\mu = 0.7, \beta = 1$ $\mu = 2, \beta = 3$ $\mu = 4, \beta = 8$
(e) Log-normal	**(f) Weibull**
$f(x\sigma) = \frac{1}{\sigma\sqrt{2\pi}}\frac{1}{x} \exp\left[-\frac{(\log x-\mu)^2}{2\sigma^2}\right]$	$f(x;\sigma,k) = \frac{k}{\sigma}\left(\frac{x}{\sigma}\right)^{k-1} e^{-(x/\sigma)^k}$
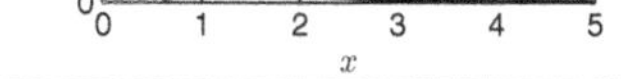 $\mu = -1, \sigma = 1$ $\mu = 0, \sigma = 1$ $\mu = 1, \sigma = .5$	 $k = 1, \sigma = .2$ $k = 1, \sigma = 3$ $k = 3, \sigma = 5$

Their probability density functions $f(x;\theta)$ are defined for $x \geqslant 0$. The shape of each density function depends on the values of the hyperparameters.

These distributions can present similar shapes and their choice in the Bayesian inference are guided by two principles: their ability to encode additional information, such as sparsity, and the complexity of the Bayesian computations that should be performed to get the estimates from the posterior distribution.

A practical example that aims at dealing with the above-mentioned (simplicity of calculation and capacity of encoding real aspects, such a non-negativity) can be found at [11], where the authors proposed a Bayesian source separation algorithm based on Gamma distribution priors for both the source signals and the mixing coefficients. More sophisticated models assume mixtures of distributions, such as mixture of exponential distributions that has been used in [10] to enforce the non-negativity constraint in infrared imaging. Alternatives to deal with the non-negativity constraint are also presented in [42–44].

4.3.3.2 Bound Constraints

This prior consists of limiting the range of the values that can be taken by the variables (sources or mixing coefficients). For example, when the values of the mixing coefficients a_{pr} should not be out of an interval $[a_r^{\min}; a_r^{\max}]$, a (shifted and scaled) Beta distribution can be used [45]. The distribution is expressed by

$$f(x|\alpha,\beta) = \frac{(x - x_{\min})^{(\alpha-1)}(x_{\max} - x)^{(\beta-1)}}{B(\alpha,\beta)\,(x_{\max} - x_{\min})^{(\alpha+\beta-1)}}\,\mathbb{I}_{(\{x_{\min},x_{\max}\})}(x), \tag{4.30}$$

where $\mathbb{I}_E(x)$ is the indicator function (equals 1 when $x \in E$ and 0 otherwise). $B(\alpha,\beta)$ is the Beta function and (α,β) are two hyperparameters governing the shape of this density. In the particular case $\alpha = \beta = 1$, (4.30) leads to a uniform distribution over the bounded set $(x_{\min}, x_{\max})$. The Beta distribution is illustrated in Figure 4.1.

4.3.3.3 Sum-to-One Constraint

This constraint appears in the case of measurement data obtained from the monitoring of chemical reactions or in remote sensing by hyperspectral imaging where the mixing coefficients (called abundances) are related to proportions and therefore should sum-to-one. Note that this property can also be related to the principle of conservation. In chemometrics community, this constraint is called *closure*. In hyperspectral imaging community and chemometrics (see Chapter 6), it is also called *additivity*.

In order to jointly encode the non-negativity and the sum-to-one constraints for a R dimensional random vector $\boldsymbol{x}$, the Dirichlet distribution $\mathcal{D}(\boldsymbol{x};\boldsymbol{\alpha})$ is an appropriate candidate, where $\boldsymbol{\alpha}$ is a vector of hyperparameters

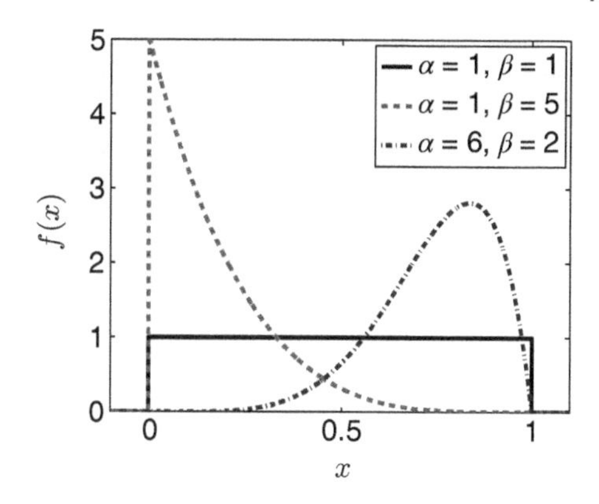

Figure 4.1 Illustration of the standard Beta distribution whose support is constrained within $[x_{\min}, x_{\max}] = [0, 1]$.

governing the shape of the density. Its probability density function is defined as

$$f(\boldsymbol{x}; \boldsymbol{\alpha}) = \frac{\Gamma\left(\sum_{r=1}^{R} \alpha_r\right)}{\prod_{r=1}^{R} \Gamma(\alpha_r)} \prod_{r=1}^{R} x_r^{\alpha_r - 1} \, \mathbb{1}_{\Delta_{R-1}}(\boldsymbol{x})$$

where $\mathbb{1}_{S_{R-1}}(\cdot)$ represents the indicator function over the region S_{R-1}, which is the simplex defined as

$$S_{R-1} = \left\{ \boldsymbol{x} = \left[x_1, \ldots, x_R\right]^{\mathrm{T}}; x_r \geqslant 0, (\forall r = 1, \ldots, R), \sum_{r=1}^{R} x_r = 1 \right\},$$

and $\Gamma(\cdot)$ corresponds to the gamma function which is defined as follows:

$$\Gamma(x) = \int_0^{\infty} y^{x-1} e^{-y} \, dy.$$

The Dirichlet distribution is frequently used in Bayesian statistics for modeling positive quantities summing to one. The most popular example is its use as a prior distribution of mixture weights in statistical mixture models estimation using Bayesian approach [46, 47]. An important property of the Dirichlet distribution is that, when it is applied to model the mixing coefficients $[a_1, \ldots, a_R]$, and when the parameters of the Dirichlet distribution are given by $\boldsymbol{\alpha} = [1, \ldots, 1]$, it means that the mixing coefficients are uniformly distributed on the simplex Δ_{R-1}. The choice of this prior for Bayesian source separation [12] presents the advantage to give the same expectation for all the components and there is no additional hyperparameter to estimate. In Figure 4.2, examples of the Dirichlet distribution are given for different values of $\boldsymbol{\alpha}$.

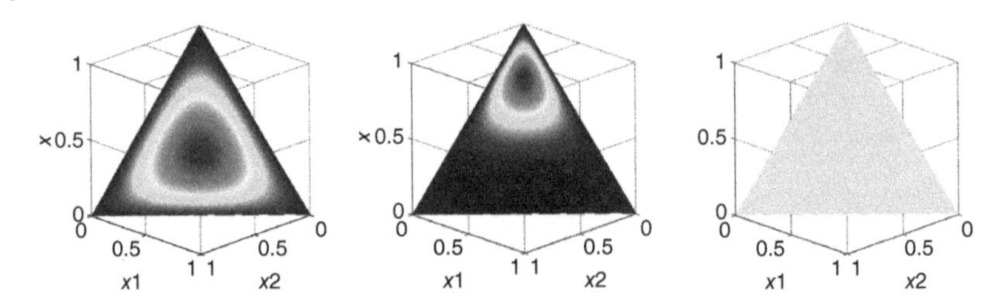

Figure 4.2 Illustration of the Dirichlet distribution shape in $\mathbb{R}^3$ for different combinations of its parameter values. One can particularly note that the distribution is uniform over a simplex when $\alpha = [1, 1, 1]$.

4.3.3.4 Smoothness Constraint

In some applications, the source signal samples are considered as identically distributed, but they present a non-negligible correlation between successive samples. The mostly used prior to encode such information is a first-order Markovian model quite similar to the classical first order autoregressive model driven by Gaussian innovation sequence, with the non-negativity constraint of the sources [16]. The prior density on each source signal is thus expressed by

$$f(s_r(t)|s_r(t-1)) = \frac{1}{K} \exp\left(-\frac{[s_r(t) - s_r(t-1)]^2}{2\sigma_r^2}\right) \mathbb{1}_{\mathbb{R}^+}(s_r(t)), \quad (4.31)$$

where $K = K(\sigma_r, s_r(t-1))$ is a normalization constant. We emphasize that such prior distribution will promote the smoothness of the estimated source signals, since it gives the highest probability densities for successive samples with small squared differences.

4.3.4 Application to the Separation of Synthetic Spectral Mixtures

Figure 4.3a shows an example of $R = 3$ source signals of $T = 1000$ spectral bands. The shape of these simulated spectral sources is similar to those obtained from absorption spectroscopy data. Each source signal is constructed by a superposition of Gaussian and Lorentzian functionals with randomly chosen parameters (location, amplitude, and width) [48]. For this application, a "spectral" band corresponds to a given value of the wavelength λ (expressed in nanometers). The mixing coefficients have been chosen to obtain time evolution profiles similar to component abundance variation in a kinetic reaction, as depicted in Figure 4.3b. The abundance fraction profiles have been simulated for $P = 10$ observation times, which provides

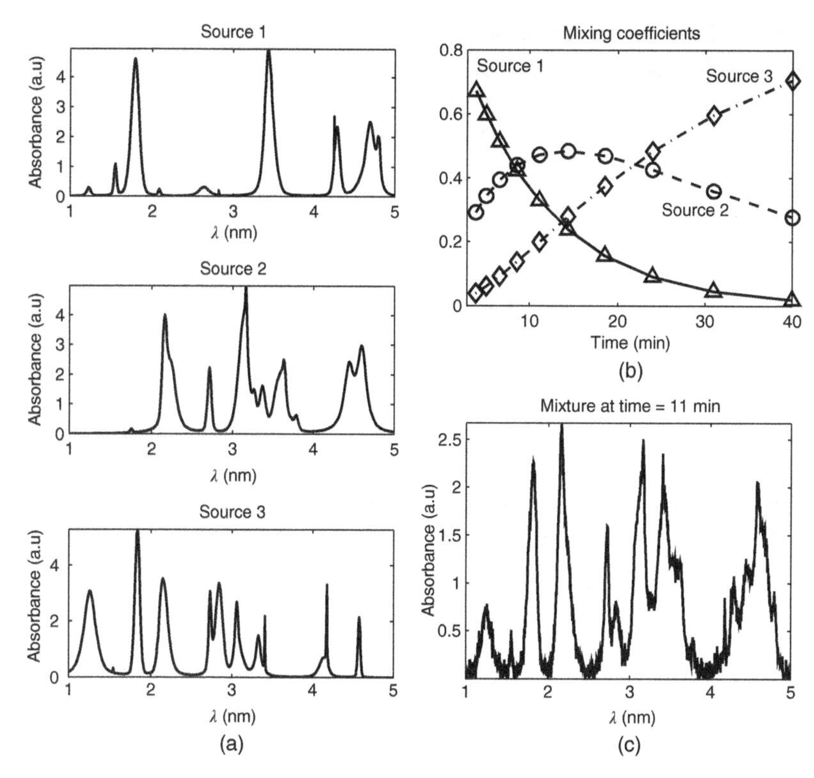

Figure 4.3 (a) Example of $R = 3$ simulated spectral sources where the x-axis corresponds to the wavelength, expressed in *nm*, and the y-axis corresponds to the absorbance of the components. (b) Abundance evolution profiles with respect to a kinetic reaction time. (c) One typical realization of the observed mixture spectra (for $t = 11$ min).

$P = 10$ observation spectra. An independent and identically distributed (i.i.d.) Gaussian sequence has been added as additive noise to each observation with appropriate standard deviation to reach a signal-to-noise ratio (SNR) equal to 20 dB. One typical realization of the observed spectra is shown in Figure 4.3c.

A Bayesian source separation algorithm was applied for the processing of the mixture data illustrated in Figure 4.3. The Bayesian inference is conducted in the linear mixing case with Gaussian errors, by focusing on the non-negativity constraint and the sum-to-one of the mixing coefficients. The joint estimation of the source signals and all the hyperparameters is performed using MCMC applied to a hierarchical Bayesian model. A step-by-step description of the approach is detailed in the sequel.

4.3.4.1 Bayesian Separation Model

Noise Model and Likelihood. The noise components $e_p(t)$, $p = 1, \dots, P$, are assumed to be independent and distributed according to a Gaussian distribution with a zero-mean and a variance σ_p^2,

$$f(e_p(t)|\sigma_p^2) = \mathcal{N}\left(0, \sigma_p^2\right) \quad (\forall p = 1, \dots, P)$$

Consequently, for the noise vector $\boldsymbol{e}(t)$, the distribution is given by:

$$f(\boldsymbol{e}(t)|\sigma_1^2, \dots, \sigma_P^2) = \left(\prod_{p=1}^{P} \frac{1}{2\pi\sigma_p^2}\right)^{1/2} \exp\left[-\frac{1}{2}\sum_{p=1}^{P} \frac{e_p(t)^2}{\sigma_p^2}\right],$$

which, by introducing vector norm notations, can also be expressed as

$$f(\boldsymbol{e}(t)|\boldsymbol{\Sigma}) = \left(\frac{1}{(2\pi)^P \det \boldsymbol{\Sigma}}\right)^{1/2} \exp\left[-\frac{1}{2}(\boldsymbol{e}(t)^T \boldsymbol{\Sigma}^{-1} \boldsymbol{e}(t))\right].$$

with $\boldsymbol{\Sigma} = diag\{\sigma_1^2, \dots \sigma_P^2\}$. Therefore, according to the mixing model $\boldsymbol{x}(t) = \boldsymbol{A}\boldsymbol{s}(t) + \boldsymbol{e}(t)$, one can obtain the following likelihood function

$$f(\boldsymbol{x}(t)|\boldsymbol{s}(t), \boldsymbol{A}, \boldsymbol{\Sigma}) = \left(\frac{1}{(2\pi)^P \det \boldsymbol{\Sigma}}\right)^{1/2} \exp\left[-\frac{1}{2}\|\boldsymbol{x}(t) - \boldsymbol{A}\boldsymbol{s}(t)\|_{\boldsymbol{\Sigma}}^2\right],$$

where $\|\boldsymbol{x}(t) - \boldsymbol{A}\boldsymbol{s}(t)\|_{\boldsymbol{\Sigma}}^2$ is a simplified notation for $(x(t) - \boldsymbol{A}\boldsymbol{s}(t))^T \boldsymbol{\Sigma}^{-1} (x(t) - \boldsymbol{A}\boldsymbol{s}(t))$. Finally, by assuming independent and identically distributed observation samples $\boldsymbol{x}(t)$, for $t = 1, \dots, T$, the likelihood of the data matrix will be expressed as

$$f(\boldsymbol{X}|\boldsymbol{S}, \boldsymbol{A}, \boldsymbol{\Sigma}) = \left(\frac{1}{(2\pi)^P \det \boldsymbol{\Sigma}}\right)^{T/2} \exp\left[-\frac{1}{2}\|\boldsymbol{X} - \boldsymbol{A}\boldsymbol{S}\|_{\boldsymbol{\Sigma}}^2\right].$$

Prior Distributions of the Source Signals. In order to account for the non-negativity constraint, the prior distribution of each source signal is chosen as a Gamma distribution

$$f(s_r(t)|\alpha_r, \beta_r) = \frac{\beta_r^{\alpha_r}}{\Gamma(\alpha_r)} s_r(t)^{\alpha_r - 1} \exp(-\beta_r s_r(t)).$$

As discussed in Section 4.3.3.1, this Gamma distribution allows a wide range of possible shapes to be described. Especially, this distribution is suitable to fit spectral data for absorption spectroscopy since it promotes sparsity when $\alpha_r \leqslant 1$ [11].

Prior Distributions of the Mixing Coefficients. The mixing coefficient vectors $\boldsymbol{a}_p = \left[A_{p1}, \dots, A_{pR}\right]^{\mathrm{T}}$ are assumed to be *a priori* independent and identically distributed according to uniform distributions in order to jointly account for the non-negativity and the sum-to-one constraint. As discussed in

Section 4.3.3.3, this distribution corresponds to the Dirichlet distribution with the hyperparameter vector $\boldsymbol{\alpha} = (1, \dots, 1)$.

Noise Distribution Hyperparameters. A distribution with a non-negative support should be chosen to ensure the non-negativity of the noise variance. In addition, to make the Bayesian computation easier, a Gamma density is used as prior distribution of $1/\sigma_p^2$ with parameters (α_n, β_n) where α_n and β_n are set in such a way to get a non-informative (vague) prior. According to this choice, the prior is also said to be a conjugate prior since the posterior distribution of the inverse of the noise variance will also be a Gamma distribution [1]. In practice, the value of α_n can be set equal to 2, which ensures a non-informative prior [1]. The value of β_n can be inferred from

$$f(\beta_n|\sigma_1^2, \dots, \sigma_P^2) = \prod_{p=1}^{P} f(\sigma_1^2, \dots, \sigma_P^2|\beta_n) f(\beta_n)$$

where $f(\beta_n)$ is a non-informative probability distribution, called Jeffreys' prior, which reflects the lack of knowledge regarding this hyperparameter value

$$f(\beta_n) = \frac{1}{\beta_n} \mathbb{I}_{\mathbb{R}^+}(\beta_n).$$

Source Distribution Hyperparameters. The prior distribution of each source signal s_r being a Gamma distribution, two hyperparameters α_r and β_r should therefore be estimated. The posterior density of each hyperparameter α_r is given as

$$f\left(\alpha_r | \boldsymbol{s}_r, \beta_r\right) \propto \prod_{t=1}^{T} \frac{\beta_r^{\alpha_r}}{\Gamma(\alpha_r)} s_r(t)^{\alpha_r - 1} f(\alpha_r), \tag{4.32}$$

$$\propto \frac{1}{\Gamma(\alpha_r)^T} \exp\left\{ \left(T \log \beta_r + \sum_{t=1}^{T} \log s_r(t) \right) \alpha_r \right\} f(\alpha_r). \tag{4.33}$$

In order to account for the non-negativity of the parameter α_r, we use an exponential density of parameter $\lambda_{\alpha_r}^{prior}$ as prior distribution, so its posterior density takes the form

$$f\left(\alpha_r | \boldsymbol{s}_r, \beta_r\right) \propto \left(\frac{1}{\Gamma(\alpha_r)} \exp\left\{ -\lambda_{\alpha_r}^{post} \alpha_r \right\} \right)^T \mathbb{I}_{[0,+\infty]}\left(\alpha_r\right), \tag{4.34}$$

where $\lambda_{\alpha_r}^{post} = \log \beta_r + \frac{1}{T} \sum_{t=1}^{T} \log s_r(t) - \frac{1}{T} \lambda_{\alpha_r}^{prior}$. This posterior distribution does not belong to a known family, so its simulation needs a Metropolis–Hastings algorithm [31].

Concerning the hyperparameter β_r, the related posterior distribution $f\left(\beta_r | \boldsymbol{s}_r, \alpha_r\right)$ is expressed as

$$f\left(\beta_r \middle| \boldsymbol{s}_r, \alpha_r\right) \propto \beta_r^{T\alpha_r} \exp\left(-\beta_r \sum_{t=1}^{T} s_r(t)\right) f(\beta_r). \tag{4.35}$$

Therefore, one can note that the conjugate prior for the parameter β_r is a Gamma density,

$$\beta_r \sim \mathcal{G}\left(\alpha_{\beta_r}^{prior}, \beta_{\beta_r}^{prior}\right), \tag{4.36}$$

leading to an *a posteriori* Gamma distribution

$$\left(\beta_r \middle| \boldsymbol{s}_r, \alpha_r\right) \sim \mathcal{G}\left(\alpha_{\beta_r}^{post}, \beta_{\beta_r}^{post}\right), \tag{4.37}$$

with parameters

$$\begin{cases} \alpha_{\beta_r}^{post} &= 1 + T\alpha_r + \alpha_{\beta_r}^{prior}, \\ \beta_{\beta_r}^{post} &= \sum_{t=1}^{T} s_r(t) + \beta_{\beta_r}^{prior}. \end{cases} \tag{4.38}$$

4.3.4.2 Bayesian Separation Algorithm

The estimation is performed using the MMSE estimator after sampling from the posterior distribution using MCMC algorithm. More details concerning the sampling strategies are available in [11, 12]. We also present an example in Section 4.5.

4.3.4.3 Bayesian Separation Results

Figures 4.4 and 4.5 give an excerpt of the simulated Markov chains and the empirical distributions of some parameters (the hyperparameters of the first source in Figure 4.4 and one mixing coefficient in Figure 4.5). One can notice here the main advantage of the Bayesian separation approach, which consists in estimating the hyperparameters and getting the full posterior distributions of the parameters of interest.

Figure 4.6 shows a comparison between the actual mixing coefficient (cross) and their MMSE estimates (circles) obtained from a Markov chain of $N_{\text{MC}} = 5000$ iterations including $N_{\text{b}} = 1000$ burn-in iterations. The estimated mixing coefficients are clearly in high agreement with the actual abundances, and the estimates satisfy the positivity and sum-to-one constraints. By comparing Figures 4.3a and 4.6a, it can also be observed that the source signals have been correctly estimated.

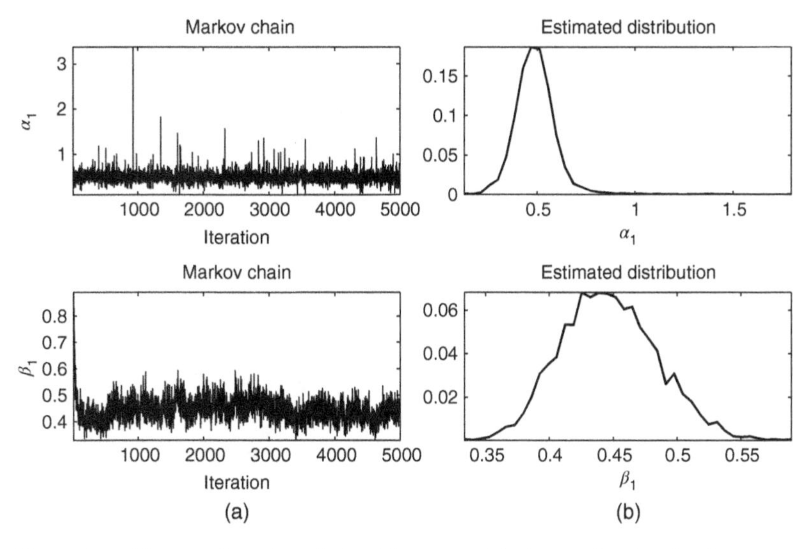

Figure 4.4 Simulated Markov chains (a) and empirical posterior distributions (b) of the two hyperparameters of the first source.

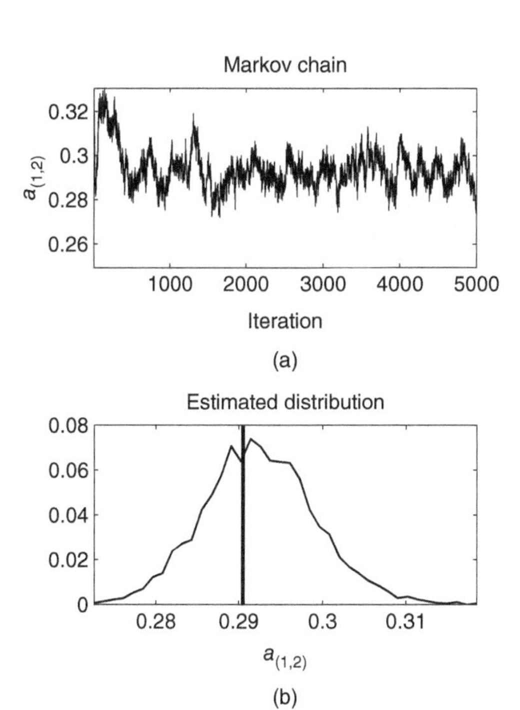

Figure 4.5 Markov chains (a) and empirical posterior distribution (b) of mixing coefficient a_{12}. The vertical bar indicates its actual value.

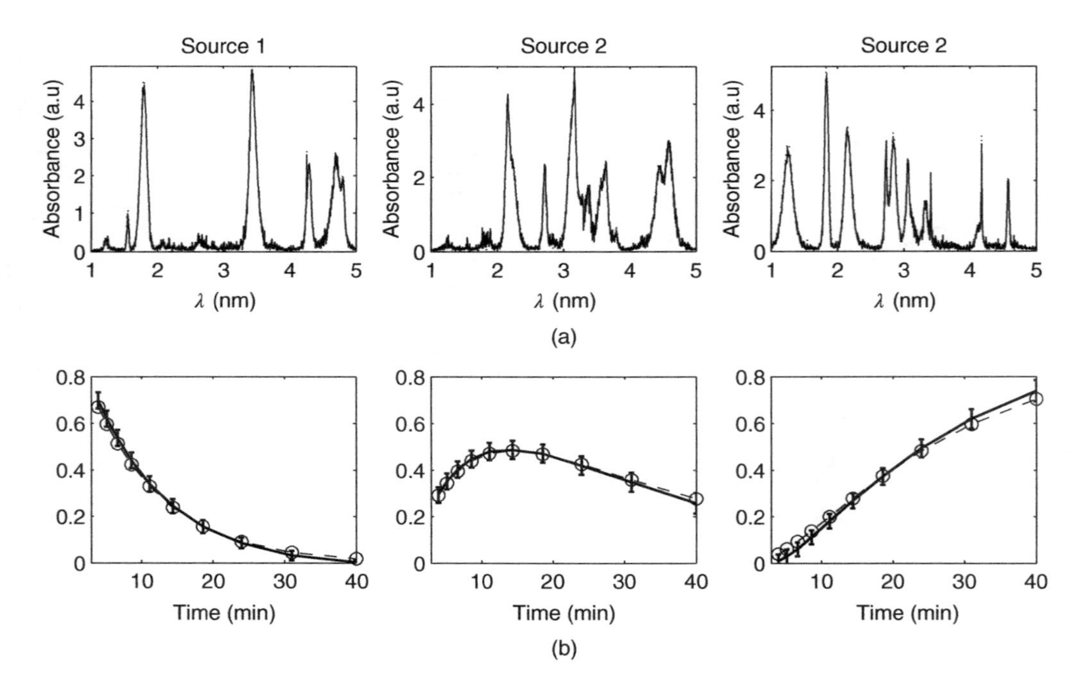

Figure 4.6 (a) Simulated (dotted) and estimated (continuous line) source sources signals.(b) Simulated values (cross) and MMSE estimates (circles) of the mixing coefficients.

4.4 Statistical Models and Separation Algorithms for Nonlinear Mixtures

Although linear mixing models are physically valid or provide a satisfactory first-order approximation of the measurement process in several applications, there are some cases in which a nonlinear model is mandatory. Dedicated studies and algorithms have been proposed to address source separation in the nonlinear mixing case. An overview on source separation methods in the context on nonlinear models can be found in [49, 50]. Indeed, exploiting a nonlinear models allows to increase the amount of information that can be extracted from the data. However, such modeling will lead to an increase of the source separation complexity [51, 52].

Solving the source separation in the context of nonlinear mixing requires a first step for establishing the mixing model and a second step for prior encoding. For a Bayesian separation approach, the form of the nonlinearity will impact the likelihood function and the posterior distribution. The strategy that can be adopted for encoding the prior information on the sources and the mixing coefficients is the same as in the linear case. Therefore, the focus of this section is concentrated on the mixing model derivation and on its impact on the Bayesian separation algorithms in the context of chemical and physical sensing.

There exist two main approaches for establishing the mixing model: the first one is based on the physical theory behind the measurement process and the second one can be seen as a second (or higher)-order empirical approximation of this model.

4.4.1 Nonlinear Mixing Models from Physical–Chemical Sensing Theory

Nonlinear models are derived from physical principles inherent in the sensing modality that relate the measured data with the signatures of the mixture components. In general, these models exhibit nonlinear functions which depend on the sensing modality. A particular instance of nonlinear models is the post-nonlinear (PNL) mixing model introduced in [51] in the context of source separation.

For PNL mixtures, when using only the statistical independence assumption, the invertibility of the nonlinear functions is a crucial (but not sufficient) condition to ensure the separability and the uniqueness of the independent components (up to ordering and scaling indeterminacies) [51, 52]. In this case, the resort to Bayesian approaches is an adequate strategy to get a more plausible solution by incorporating additional assumptions and constraints, in agreement with the properties of the sought sources and mixing coefficients. Let us introduce two examples of two physical–chemical sensing modalities involving two nonlinear mixing models.

4.4.1.1 First Example. Interference in Potentiometric Sensors

As discussed in Chapter 1, potentiometric sensors, such as ion-selective electrodes (ISEs) and ion-sensitive field-effect transistors (ISFETs), provide a practical approach to estimate ionic activities in a solution. These devices find applications in many domains, from food and beverage industry to clinical analysis. On the other hand, they suffer from a major issue, which is the problem of selectivity. In other words, it is difficult to build an electrode that exclusively responds to the target ion.

A potentiometric sensor is a nonlinear device, and, therefore, its output is non-linearly related to the target ion and also to the interfering ones. The classical Nicolsky–Eisenman equation, expressed in (4.39), provides a simple model to describe the nonlinear behavior of potentiometric sensors. The Nicolsky–Eisenman equation can be seen as an empirical extension of the Nernst equation [53]. More recently, alternatives models to the Nicolsky–Eisenman equation were proposed with the aim of achieving a better description of potentiometric sensors [54].

A possible strategy to address the problem of interference in ISEs is the use of arrays of electrodes combined with signal processing methods. If the Nicolsky–Eisenman equation is considered in the array representation, then the response of the p-th sensor (dedicated to measure the p-th ion) within the array is given by

$$x_p(t) = c_p + d_p \log\left(\sum_{r=1}^{R} A_{pr} s_r(t)^{z_p/z_r} \right), \tag{4.39}$$

where d_p and c_p are constants that depend on some physical parameters, z_r and $s_r(t)$ denote the valence and the activity (a measure of effective ionic concentration) of the r-th ion, respectively. The non-negative parameters a_{pr}, known as selectivity coefficients, explain the influence (interference) of the r-th ion on the p-th sensor; t stands for the temporal index. In the context of source separation, the ionic activities correspond to the sources and the outputs of the array to the mixtures.

When the valences of the ions are different, the model (4.39) becomes difficult to deal with because a nonlinearity (power term) arises inside the logarithm term [55]. However, when the valences are equal, which is a common situation, Eq. (4.39) becomes a post-nonlinear (PNL) model [15], with a logarithmic nonlinearity function $g_p(x) = d_p \log(x)$. Moreover, since the parameters d_p are associated with the Nernst slope [53], one can use this theoretical value, which are not necessarily observed in practice, at least to carry out an approximate inversion of the logarithm function. Then, the remaining nonlinear mixing model can be modeled by, for instance, polynomial mixtures, which will be discussed in Section 4.4.2.2.

4.4.1.2 Second Example. Intimate Mixtures in Hyperspectral Imaging

Hyperspectral imaging is a remote sensing technique allowing the composition of a surface to be analyzed by jointly processing a set of images acquired in several spectral bands [56]. Each image pixel intensity is related to the spectral reflectance of the components of its associated surface area. The linear mixing model is adequate for modeling geometrical mixtures (pixels whose components are separated spatially) since it expresses the reflected light from a pixel as a combination of the reflected light from the components individually. In the case of intimate mixtures, the linear mixing occurs in the albedo domain [57], and the observation model corresponds to a nonlinear mixing model

$$x_p(t) = \mathcal{G}\left(\sum_{r=1}^{R} B_{pr}\tilde{s}_r(t)\right) \tag{4.40}$$

where $\tilde{s}_r = \mathcal{G}^{-1}(s_r)$ corresponds to the source signature in the albedo domain and $\mathcal{G}$ to the nonlinear mapping from albedo to reflectance domains. Note that a similar model arises in reflection-diffuse spectroscopy where the measurements correspond to reflectance values while the mixing occurs in the absorbance domain and the mapping $\mathcal{G}$ is derived from Kubelka–Munk theory [58].

Since both intimate and geometrical mixtures can occur in the image pixels [59], the whole mixing model can be written as a weighted combination of the two models

$$x_p(t) = \sum_{r=1}^{R} A_{pr}s_r(t) + \beta_p\, \mathcal{G}\left(\sum_{r=1}^{R} B_{pr}\tilde{s}_r(t)\right), \tag{4.41}$$

where β_p indicates the amount of nonlinear mixing in the p-th pixel.

4.4.2 Empirical Nonlinear Mixing Models Used in Separation Algorithms

Empirical mixing models are very useful when the physical model is not available or when the separation using such model is not possible due to identifiability issues [51, 52]. These empirical models are used to give an approximation of the actual mixing model by adding to the linear mixing additional terms allowing higher order approximation of the nonlinear model. Some of these models are summarized in the sequel. Note that these nonlinear models are more deeply discussed in [60, 61] in the specific context of hyperspectral remote sensing. In addition, an overview on the separation algorithms for nonlinear mixtures is also available in [49, 50].

4.4.2.1 Post-Nonlinear Mixing Models

Such kind of mixing models are intermediate between purely nonlinear models and linear mixing since they assume that each observed mixture signal corresponds to a univariate nonlinear function of a linear mixture of the sources

$$x_p(t) = g_p\left(\sum_{r=1}^{R} A_{pr} s_r(t)\right), \tag{4.42}$$

where functions g_p are often assumed to be strictly monotonics and, thus, admit inverse functions. For instance, such models arise in sensors array processing [15] and in many biological systems [62]. Several works have been directed at analyzing the identifiability of such model in the case of the statistical independence of the sources [51, 52] and the proposal of separation algorithms (for instance, an overview on these works can be found in [50]).

4.4.2.2 Polynomial Mixing Models

As proposed by [51], the nonlinearity in the post-nonlinear mixing model can be represented using approximating functions allowing a wide class of nonlinearities to be described. More precisely, the nonlinear mapping g_p in (4.42) is defined by a second-order polynomial nonlinearity

$$\begin{aligned} g_p &: \mathbb{R} \to \mathbb{R} \\ &\quad x \to x + b_p\, x^2, \end{aligned} \tag{4.43}$$

governed by a unique parameter $b_p \in [0,1]$ which adjusts the weight of the quadratic term. According to this formulation, proposed in [18], the mixing model leads to a particular instance of the Linear–Quadratic model

$$x_p(t) = \sum_{r=1}^{R} A_{pr} s_r(t) + B_p\left(\sum_{r=1}^{R} A_{pr} s_r(t)\right)^2. \tag{4.44}$$

4.4.2.3 Linear–Quadratic Mixing Models

The second-order polynomial approximation can be extended to the more general form of the Linear–Quadratic (LQ) model which is given as

$$x_p(t) = \sum_{r=1}^{R} A_{pr} s_r(t) + \sum_{r=1}^{R}\sum_{\ell=r}^{R} B_{pr\ell} s_r(t)\, s_\ell(t), \tag{4.45}$$

where $b_{pr\ell}$ corresponds to the nonlinear mixing coefficients which are assumed to be non-negative. Linear–quadratic models actually arise in the context of remote sensing by hyperspectral imaging [60], especially when there are multiple reflections caused by the presence of buildings or trees in the observed urban surface [63, 64]. Other example is the processing of scanned images involving a show-through effect [65], the analysis of gas sensor array data [16, 66], and in fluorescence spectroscopy of highly concentrated solutions involving a screen effect [67].

4.4.2.4 Bilinear Mixing Models

These models can be obtained from the LQ model by canceling the *auto-term* coefficients $\{B_{(p,r,t)}, (\forall r = 1, \dots, R)\}$. It is thus expressed by

$$x_p(t) = \sum_{r=1}^{R} A_{pr} s_r(t) + \sum_{r=1}^{R-1} \sum_{\ell=r+1}^{R} B_{pr\ell} s_r(t) s_\ell(t). \tag{4.46}$$

Actually, the bilinear model can be written as a linear mixing model by adding $R(R-1)/2$ *extended* sources $\tilde{s}_{(r,\ell)}$ such that

$$\tilde{s}_{(r,\ell)}(t) = s_r(t)\, s_\ell(t) \quad (\forall r = 1, \dots, R-1, \forall \ell = r+1, \dots, R).$$

According to [68], a simplification of this model concerns the nonlinearity coefficients $b_{pr\ell}$, which was expressed as the product of the mixing coefficients appearing in the linear part of the model $b_{pr\ell} = a_{pr}\, a_{p\ell}$. One can see here a connection with the second-order polynomial mixture model.

4.5 Some Practical Issues on Algorithm Implementation

In Sections 4.2 and 4.3, the general guidelines for implementing a Bayesian source separation method were already discussed. However, before providing examples of the application of Bayesian source separation to actual data, we shall focus again on its implementation, but now by paying special attention to some issues that often arise in practice when one is considering a Monte Carlo-based solution. These issues will be discussed with the aid of a simple example for which a pseudo-code will be provided (Section 4.5.1).

As already discussed throughout the chapter, the idea behind Bayesian inference via Monte Carlo integration is to simulate, that is, to draw samples from the marginal posterior distributions $f(\boldsymbol{S}|\boldsymbol{X})$ and $f(\boldsymbol{A}|\boldsymbol{X})$, which are related to the sources and to the mixing system parameters, respectively. Therefore, the implementation of a Bayesian source separation algorithm requires a numerical computing environment that is able to generate random numbers from a given distribution.

Given the practical difficulty in generating random number in digital computers, one often resorts to pseudo-random generators [69]. These generators provide a practical way of simulating a uniform distribution via a deterministic machine. Moreover, from the well-known result that associates a random variable X of cumulative distribution function (CDF), denoted by $F(\cdot)$ – and given by $df(x)/dx$, where $f(x)$ represents the distribution function of X – with a uniform random variable U, as follows

$$X = F^{-1}(U), \tag{4.47}$$

it becomes possible to generate samples of X from a uniform distribution generator.

Unfortunately, the use of (4.47) to generate samples of X, which is known as the *inverse CDF method*, is not always possible. Indeed, to cite one limitation, there are distributions for which it is not possible to obtain an analytic expression for $F^{-1}(\cdot)$. Therefore, much effort has been put in the problem of generating random numbers for a given distribution and among the more elaborated approach in the use of Markov chains [30], as already mentioned in Section 4.2.3.

By considering either simple methods, such as the inverse CDF method, or more complex solutions, many numerical computing programs have built-in functions to simulate the usual distributions. For example, in the software Matlab, the function `normrnd(mu,sigma)` generates a random number drawn from a Gaussian distribution of mean `mu` and standard deviation `sigma`. Therefore, from a practical point of view, it becomes quite convenient to choose prior distributions that engender marginal posterior distributions for which a random generator is already implemented.[3] Otherwise, one must build a customized function that is usually based on more complex procedures such as hybridization of different sampling techniques, e.g. Metropolis–Hasting together with Gibbs sampler.

4.5.1 A Simple Example

Let us now present a simple example to illustrate how a Bayesian source separation method can be implemented from built-in functions. For the sake of simplicity, we shall consider a linear mixing model and that all hyperparameters are fixed, and, thus, only the mixing matrix coefficients and the source

3 Fortunately, one can often find implementations for the distributions presented in Table 4.1.

signals must be estimated. Finally, the posterior mean estimator based on a Monte Carlo integration is considered (see Section 4.2.3.3 for details). Therefore, by using the Markov chain Monte Carlo sampling approach based on the Gibbs sampler, the following iterative scheme must be set

1. For k=1,2,...
   1.a Simulate the source matrix $\boldsymbol{S}^{(k+1)}$ from
   $$f\left(\boldsymbol{S}|\boldsymbol{X},\boldsymbol{A}^{(k)}\right) \propto f\left(\boldsymbol{X}|\boldsymbol{S},\boldsymbol{A}^{(k)}\right) \times f\left(\boldsymbol{S}\right), \tag{4.48}$$
   1.b Simulate the mixing matrix $\boldsymbol{A}^{(k+1)}$ from
   $$f\left(\boldsymbol{A}|\boldsymbol{X},\boldsymbol{S}^{(k+1)}\right) \propto f\left(\boldsymbol{X}|\boldsymbol{S}^{(k+1)},\boldsymbol{A}\right) \times f\left(\boldsymbol{A}\right). \tag{4.49}$$
2. Compute estimates according to the Monte Carlo approximation given in (4.19).

As in Section 4.2.1, we assume that the noise samples are independent and identically distributed with a Gaussian distribution with zero-mean and variance σ_p^2, for all $p = 1, \ldots, P$. The likelihood term is given by

$$f(\boldsymbol{X}|\boldsymbol{S},\boldsymbol{A}) = \prod_{p=1}^{P}\prod_{t=1}^{T} f(x_p(t)|\boldsymbol{S},\boldsymbol{A}) = \prod_{p=1}^{P}\prod_{t=1}^{T} \mathcal{N}\left(\sum_{r=1}^{R} A_{pr}s_r(t), \sigma_p^2\right). \tag{4.50}$$

Moreover, we assume that the noise variance σ_p^2 of the p-th mixture is known.

Let us consider that each mixing coefficient can be modeled by Gaussian prior of mean μ_A and variance σ_A^2. Therefore, after some manipulation considering (4.49) and (4.50), one can check that

$$f\left(A_{pr}|\boldsymbol{X},\boldsymbol{S}^{(k+1)},\mathbf{A}_{pr}\right) \propto \mathcal{N}(\mu_{Post_{Apr}}, \sigma^2_{Post_{Apr}}). \tag{4.51}$$

where $\mathbf{A}_{pr}$ represents all the coefficients of $\mathbf{A}$ except A_{pr} and

$$\mu_{Post_{Apr}} = \frac{\mu_{L_{Apr}}\sigma_A^2 + \mu_A\sigma^2_{L_{Apr}}}{\sigma_A^2 + \sigma^2_{L_{Apr}}}, \tag{4.52}$$

$$\sigma^2_{Post_{Apr}} = \sqrt{\frac{\sigma_A^2\sigma^2_{L_{Apr}}}{\sigma_A^2 + \sigma^2_{L_{Apr}}}}. \tag{4.53}$$

The terms $\mu_{L_{Apr}}$ and $\sigma^2_{L_{Apr}}$ are related to the likelihood function and are given by:

$$\mu_{L_{Apr}} = \frac{1}{\sigma^2_{L_{Apr}}}\sum_{t=1}^{T} s_p(t)\left(x_p(t) - \sum_{\ell=1,\ell\neq r}^{R} A_{p\ell}\, s_\ell(t)\right), \tag{4.54}$$

and

$$\sigma^2_{L_{Apr}} = \frac{\sigma_p^2}{\sum_{t=1}^{T} s_r(t)}. \tag{4.55}$$

A nice aspect of (4.51) is that this conditional distribution is also given by a Gaussian function. Therefore, if a function such as `normrnd(mu, sigma)` is available, then the simulation of (4.51) is straightforward. Therefore, it becomes quite interesting to consider priors that engender a posterior distribution that is of the same family of the chosen prior distribution. These priors are said to be conjugate with the likelihood function and are often referred to as conjugate priors.

Unfortunately, there are many cases in which the prior chosen for the sources does not conjugate with a Gaussian likelihood function. For instance, consider that the sources can be modeled as a log-normal distribution of parameters μ_r and σ_r, such that:

$$f(s_r(t)|\mu_r, \sigma_r) \propto \frac{1}{s_r(t)} \exp\left(-\frac{(\ln(s_r(t)) - \mu_r)^2}{2\sigma_r^2}\right). \tag{4.56}$$

After some manipulation with (4.48) and (4.56), the following conditional *a posteriori* distribution can be obtained

$$f(s_r(t)|\mu_r, \sigma_r, \mathbf{X}, \mathbf{A}) \propto \exp\left(\sum_{p=1}^{P} -\frac{1}{2\sigma_p^2}\left(x_p(t) - \sum_{L=1,L\neq p}^{R} A_{\ell r)} s_\ell(t)\right)^2 - \frac{\left(\ln(s_r(t)) - \mu_r\right)^2}{2\sigma_r^2}\right)\frac{1}{s_r(t)}. \tag{4.57}$$

It is worth noticing that (4.57) does not correspond to an usual distribution function, and thus it is difficult to find built-in functions that simulate this distribution. In such cases, the user must design and implement a customized function to generate random numbers from the obtained posterior distribution. In the case of (4.57), this task can be achieved by considering, for instance, the Metropolis–Hastings algorithm (see [30] for details). We shall refer to this function as `postsourcesrnd(X, A, theta)`, where `theta` are extra parameters.

4.5.1.1 The Resulting Gibbs Sampler

The implementation of Gibbs sampler given in Eqs. (4.48) and (4.49) can be done by the pseudo-code expressed in Algorithm 4.1. In the sequel we shall detail each step of the algorithm.

Algorithm 4.1 The Gibbs sampler for separation of log-normal sources.

Require: $\boldsymbol{X}$, μ_r, σ_r^2, μ_A, σ_A^2, σ_p^2, N_{MC}, N_b, σ_{inst}^2
1: Initialize $\boldsymbol{S}^{(0)}$ and $\boldsymbol{A}^{(0)}$
2: **for** $k \leftarrow 0$ to N_{MC} **do**
3: $\quad\boldsymbol{A}^{(k+1)} \leftarrow$ `postcoeffrnd` $(\boldsymbol{X}, \boldsymbol{S}^{(k)}, \mu_A, \sigma_A^2, \sigma_p^2)$
4: $\quad\boldsymbol{S}^{(k+1)} \leftarrow$ `postsourcesrnd` $(\boldsymbol{X}, \boldsymbol{A}^{(k+1)}, \mu_r, \sigma_r^2, \sigma_p^2, \sigma_{inst}^2)$
5: **end for**;
6: $\boldsymbol{A}_{est} = \frac{1}{(N_{\mathrm{MC}}-N_b)} \sum_{\tau=N_b+1}^{N_{\mathrm{MC}}} \boldsymbol{A}^{(\tau)}$
7: $\boldsymbol{S}_{est} = \frac{1}{(N_{\mathrm{MC}}-N_b)} \sum_{\tau=N_b+1}^{N_{\mathrm{MC}}} \boldsymbol{S}^{(\tau)}$
8: **return** $\boldsymbol{A}_{est}$, $\boldsymbol{S}_{est}$

In addition to the mixtures $\boldsymbol{X}$, the inputs of Algorithm 4.1 comprise the parameters of the sources distribution (μ_r and σ_r^2) and the parameters of the mixing coefficients distribution (μ_A, σ_A^2). For the sake of simplicity, we consider that these parameters are known. However, when these parameters are unknown, one must also consider a hierarchical model and a sampling strategy for them, as discussed in Section 4.2.4. We also assume that the noise variance σ_p^2 is equal for each mixture and is known in this example. Again, a sampling strategy must be considered when σ_p^2 is unknown. The other inputs in Algorithm 4.1 are N_{MC}, N_b, and σ_{inst}^2. N_{MC} refers to the number of iterations executed by the Gibbs sampler and is used here as a stopping criterion. The parameter N_b is the burn-in phase (which is required so that the Markov chains can converge to the stationary distributions). Finally, σ_{inst}^2 is the variance of the instrumental distribution that is used by the Metropolis–Hastings [31] algorithm which draws samples for the sources, as will be detailed later.

Step 1 of Algorithm 4.1 corresponds to the initialization of the Markov chains related to the sources and to the mixing coefficients. This initialization can be done in a random fashion or by considering an initial estimation obtained by means of other methods.

The steps from 2 to 5 denote the iterative procedure of the Gibbs sampler. In Step 3, the function `postcoeffrnd` $(\boldsymbol{X}, \boldsymbol{S}^{(k)}, \mu_A, \sigma_A^2, \sigma_p^2)$ refers to the sampling step for the mixing coefficients. As discussed in Section 4.5.1, since, in our example, the chosen prior for the mixing coefficients was a Gaussian distribution and the mixing model is linear, the conditional *a posteriori* distribution is also a Gaussian distribution. Therefore, by considering an implementation via Matlab, this distribution could be simulated by the function `normrnd` $\left(\mu_{Post_{Apr}}, \sigma^2_{Post_{Apr}}\right)$.

Step 4 is the most difficult one. Indeed, as discussed in Section 4.5.1, sampling from (4.57) is not straightforward and would require strategies such as the Metropolis–Hastings (MH) algorithm. In Appendix 4.A, we detail the application of the MH algorithm in this case. Finally, Steps 6 and 7 are related to the estimation given by the MMSE strategy, which is given by Eq. (4.19). At the end, Algorithm 4.1 provides estimates for the sources and mixing coefficients, which are given by $\boldsymbol{A}_{est}$ and $\boldsymbol{S}_{est}$, respectively.

4.6 Applications to Case Studies in Chemical Sensing

4.6.1 Monitoring of Calcium Carbonate Crystallization Using Raman Spectroscopy

This section is inspired from [70], where a more detailed discussion on the application is given and a comparison with alternative mixture analysis approaches performed. Calcium carbonate is a chemical material used commercially for a large variety of applications such as filler for plastics or paper. Depending on operating conditions, calcium carbonate crystallizes as Calcite, Aragonite, or Vaterite. Calcite is the most thermodynamically stable of the three, followed by Aragonite or Vaterite. Globally, the formation of calcium carbonate by mixing two solutions containing calcium and carbonate ions takes place in two well-distinguished steps, respectively. The first step is the precipitation one. This step is very fast and provides a mixture of calcium carbonate polymorphs. The second step (a slow process) represents the phase transformation from the unstable polymorphs[4] to the stable one (Calcite). The physical properties of the crystallized product depend largely on the polymorphic composition, so it is necessary to quantify these polymorphs when they are mixed in order to explore favorable conditions for Calcite formation. The main objective of this experiment is to show how the Bayesian source separation based on non-negativity and sum-to-one constraints can be applied to the monitoring of calcium carbonate crystallization using Raman spectroscopy data.

4.6.1.1 Mixture Preparation and Data Acquisition

Calcium chloride and sodium carbonate, separately dissolved in sodium chloride solutions of the same concentration (5mol/L), were rapidly mixed

4 The ability of a chemical substance to crystallize with several types of structures, depending on a physical parameter, such as temperature, is known as polymorphism. Each particular form is said to be a polymorph.

to precipitate calcium carbonate. The temperature and the aging time are the most important factors that can affect the polymorphic composition. Raman spectra were collected on a Jobin-Yvon T64000 spectrometer equipped with an optical microscope, a threefold monochromator, and a nitrogen-cooled charge-coupled device (CCD) camera. The excitation was induced by a laser beam of argon Spectra Physic Laser Stability 2017 at a wavelength of 514.5 nm. The Raman spectra were collected at five points, which were randomly distributed throughout the mixture. The average of all spectra was considered as the Raman spectrum of the corresponding mixture for the considered temperature value and aging time. Raman spectra were collected two minutes after the beginning of the experiment for various temperatures ranging between 20 and 70°C in order to determine the influence of temperature on the polymorph precipitation. Moreover for each temperature, Raman spectra were collected at different time intervals for monitoring phase transformation. Finally, a total of $M = 37$ Raman spectra (see Figure 4.7) of $L = 477$ wavelengths have been obtained.

4.6.1.2 Mixture Analysis by Bayesian Source Separation

The separation is performed using a Bayesian approach incorporating the non-negativity of the sources signals using a Gamma distribution prior. Moreover, the constraint of the non-negativity and the sum-to-one of the

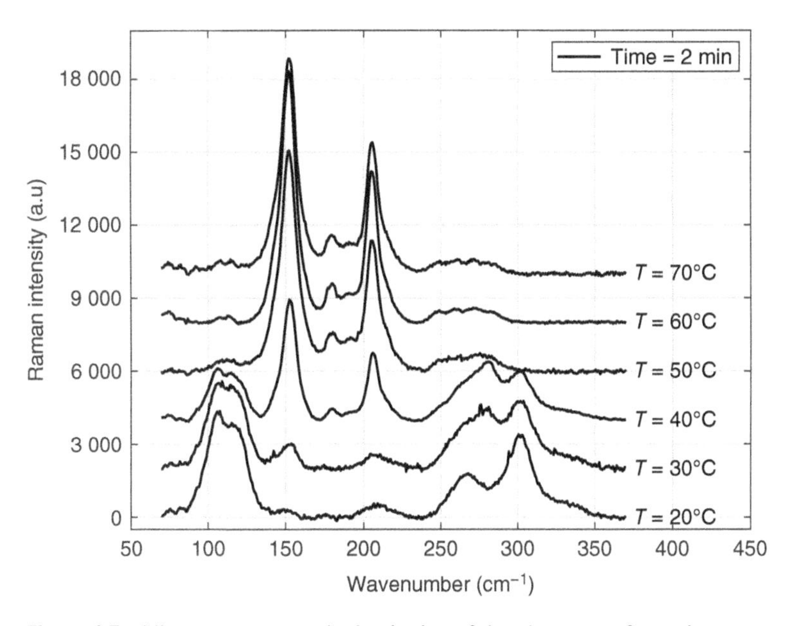

Figure 4.7 Mixture spectra at the beginning of the phase transformation.

mixing coefficients are encoded using a uniform prior. The separation algorithm, detailed in [12], consists in sampling the posterior distribution using an MCMC technique and approximating the MMSE estimators from the generated samples. The number of sources is fixed to three according to the physical knowledge on the phase transformation process and the iteration number of the MCMC algorithm is fixed to 1000, where the first 200 samples are discarded since they correspond to the burn-in period of the Gibbs sampler.

Figure 4.8 illustrates the estimated spectra. From a spectroscopic point of view and according to the positions of the vibrational peaks, the identification of the three components is very easy: the first source corresponds to Calcite, the second spectrum to Aragonite, and the third one to Vaterite. A measure of the dissimilarity between the estimated spectra and the measured pure spectra of the three components gives 4.56% for Calcite, 0.65% for Aragonite, and 4.76% for Vaterite. These results show that the proposed method can be applied successfully without imposing any additional prior information on the shape of the pure spectra.

The evolution of the polymorph proportions versus temperature is shown in Figure 4.9b. One can observe pure Vaterite at 20°C and a quite pure Aragonite is obtained at 60°C. However, between 20 and 60°C ternary

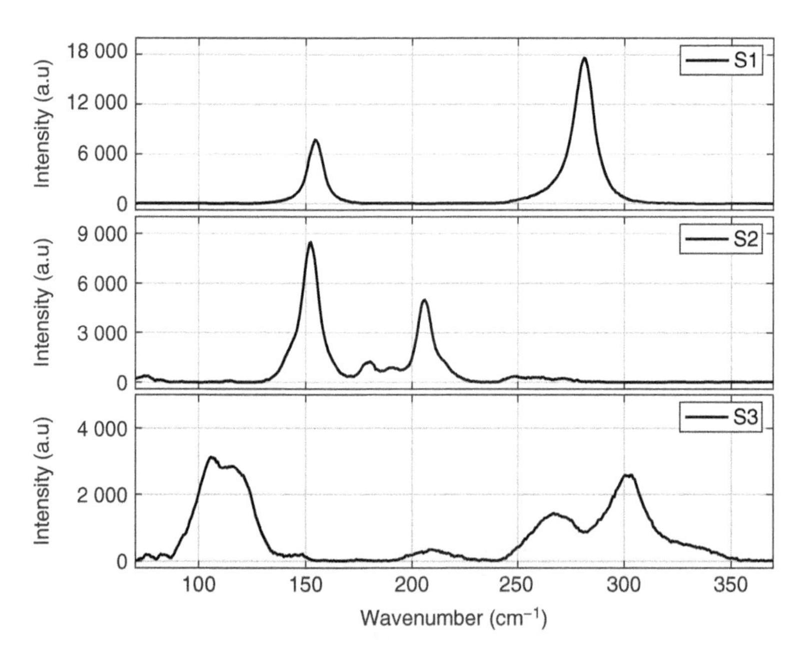

Figure 4.8 Estimated sources using a Bayesian source separation approach.

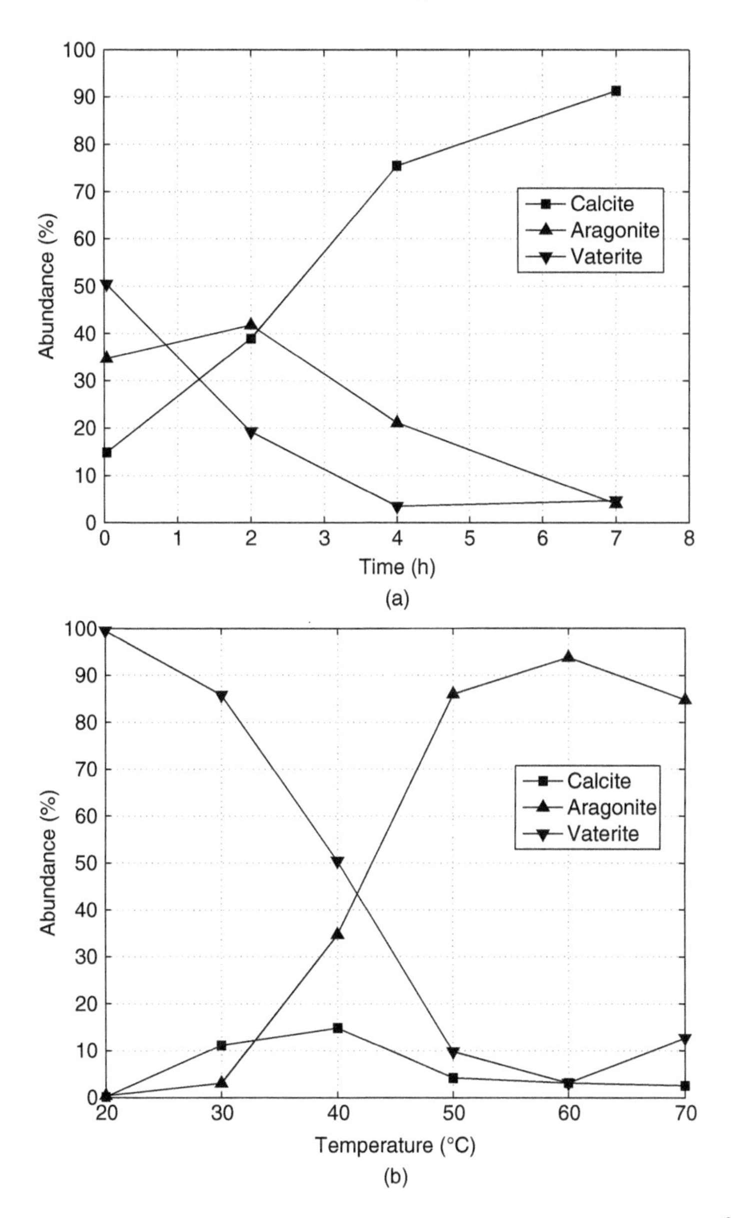

Figure 4.9 Evolution of the three component abundances: (a) for $T = 40°C$, (b) at the beginning of the phase transformation for different temperature values.

mixtures are observed. The abundance of Calcite is maximal at 40°C. Let us now consider the phase transformation evolution at this temperature value. The concentration profile (abundance) versus precipitation time at 40°C is reported in Figure 4.9a. At the beginning of the phase transformation (two minutes), the ternary mixture is composed of 50% Vaterite, 35% Aragonite, and 15% Calcite. After two hours, the Vaterite is transformed to Aragonite and Calcite. After seven hours, Vaterite and Aragonite are almost totally transformed to Calcite.

4.6.2 Dealing with Interference Issues in Ion-Selective Electrode Arrays

Measuring the activities of the ions potassium and ammonium are important in several domains of applications, such as food industry and clinical analysis [53]. However, these ions are among the most representative examples of interference that arise in ISE and, therefore, it is difficult to find potassium electrodes that do not respond to ammonium ions and vice versa.

In order to gain more insight into the interference between the ions potassium and ammonium, let us consider the data available in the ion-selective electrode array (ISEA) dataset [71] This dataset is publicly available and contains measurements acquired in several experiments in which data were acquired by ISE arrays. In two of these experiments, an ISE array was adopted to analyze the activities of potassium and ammonium (the ionic activities can be seen as a measure of effective concentration of a ion). In the context of blind source separation, the activities correspond to the sources of interest.

Since it was a controlled experiment, the actual activities were known and varied as shown in Figure 4.10, which depicts the concatenation of two experiments. The difference between the two experiments was the initial concentration of the potassium ion. In the first experiment, which corresponds to the first 80 samples of Figure 4.10), there was an injection of ammonium in an initial solution with a high concentration of potassium. In the second one, the same injection of ammonium was considered, but now in a solution with a low concentration of potassium.

The outputs of an array of two ISEs (one tailored to potassium and the other one to ammonium) are shown in Figure 4.10a. One can note that the response K^+-ISE is highly influenced by the activity of the ion NH_4^+, and vice versa. Moreover, as predicted by the Nicolsky–Eisenman equation (4.39), the nonlinear character of the mixing process is patent.

Performing source separation from the mixtures such as those exhibited in Figure 4.11 is not a straightforward task. First of all, the number of samples

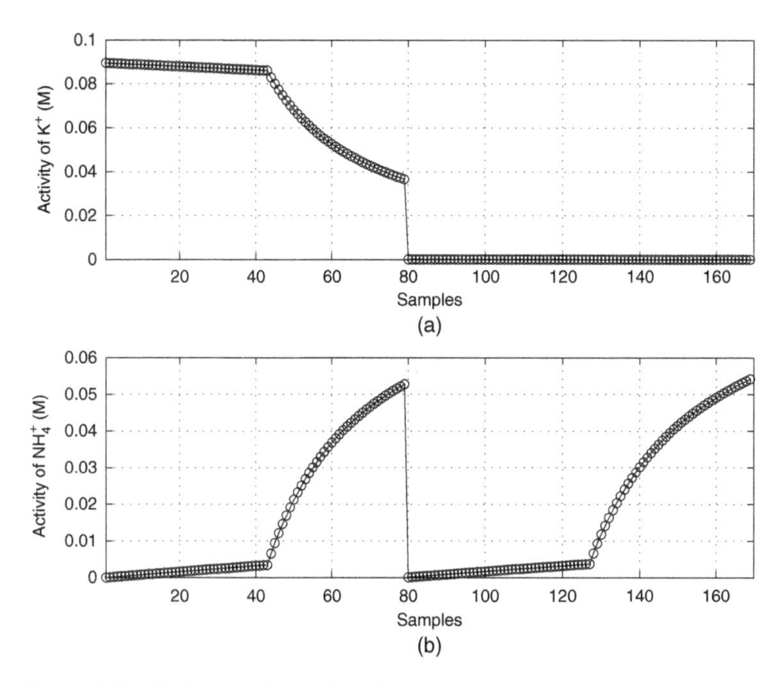

Figure 4.10 Evolution of activities of the potassium (a) and ammonium (b) sources.

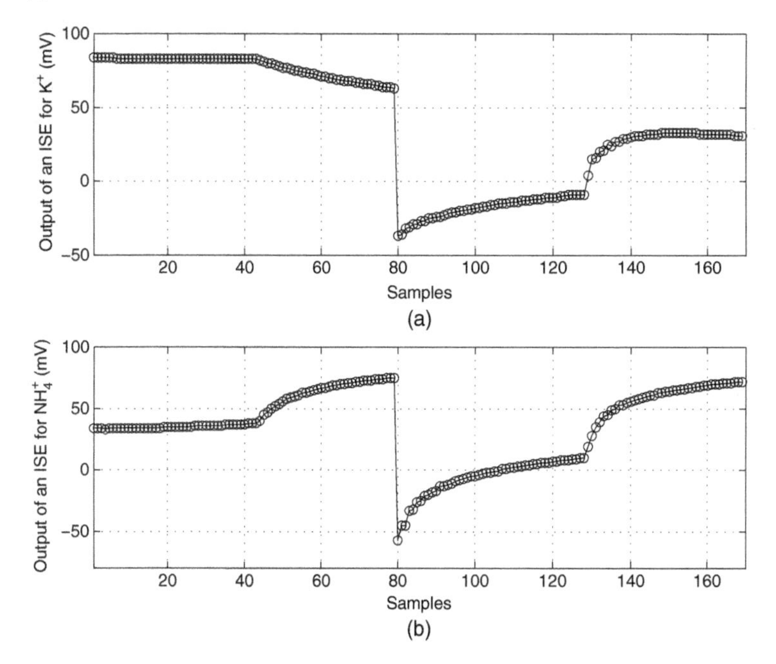

Figure 4.11 Acquired signals through an ISE array: responses of potassium electrode (a) and of ammonium electrode (b).

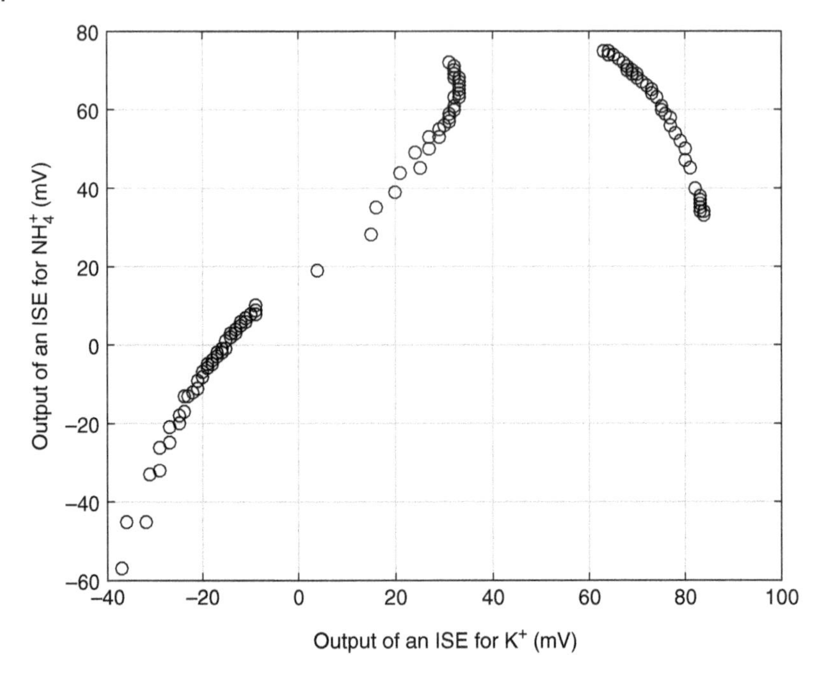

Figure 4.12 Scatter plot of the mixtures (data provided by the ISE array).

is reduced in this case (only 170), so the use of methods that rely on statistical estimators becomes limited. Another challenging aspect here is related to the sources. As can be noticed in Figure 4.11, the sources are strongly correlated and, therefore, in view of the discussion provided in Chapter 1, the application of BSS methods based on ICA is hopeless in this situation. In addition to these challenges, these mixtures have a strong nonlinear character, as can be observed in the scatter plot provided in Figure 4.12.

In view of such limitations, Bayesian methods arise as an interesting alternative, as discussed in [15]. In this paper, the authors proposed a Bayesian formulation that relies on model (4.39) and on several prior information on the problem. Due to the nonlinear model considered in the likelihood function, the resulting inference problem becomes difficult to be calculated and, therefore, the authors of [15] considered a strategy based on MCMC simulation. In the following, we shall discuss some elements of [15] in more detail in order to provide the reader an additional example on the use of Bayesian methods in a nonlinear scenario.

With respect to the priors, as is usual in physical–chemical sensing, a first natural information to be taken into account is the non-negativity of the sources – indeed, activities are always non-negative. As discussed in

Section 4.3.3.1, there are many alternatives to model non-negative data. In [15], the authors considered a log-normal distribution. A first interesting aspect of a log-normal distribution is related to the estimation of its parameters, which admits conjugate priors and, thus, simplifies inference via MCMC. Moreover, assuming a log-normal distribution means that the activities can be described by a normal distribution in the logarithmic scale, which is in accordance with ISE data analysis, as, often, one is interested in this scale.

Other information on the model (4.39) can also be incorporated to the solution in a Bayesian approach. For instance, the parameter d_p comes from the Nernst equation [53] and, therefore, can be expressed as a function of several constants, as follows

$$d_p = \frac{RT}{z_p F}, \tag{4.58}$$

where R and F denote the universal gas and Faraday constants, respectively; T is the temperature (in Kelvin); and z_p stands for the valence of the target ion. At room temperature, and considering the logarithm to the base 10 in model (4.39), the parameter d_p is given by $59/z_p$ mV. However, the observed value of d_p in practice may be very different to the theoretical value provided by (4.58), due to problems such as electrode quality or aging. In view of this observation, a Gaussian prior centered at $59/z_p$ mV can be considered for d_p [15]. Alternatively, a uniform prior can also be adopted since it is rare to observe in practice ISEs in which the slope d_p is either very small or high.

The other parameters of (4.39) can also be modeled based on some prior information. For instance, the selectivity coefficients a_{pr} are always non-negative and, very often, lie within the interval [0,1]. Therefore, a uniform distribution can be considered in this case. The offset terms e_p are also restricted to a given interval, thus also admitting a prior based on uniform distributions [15].

By assuming additive white Gaussian (AWG) noise, the likelihood function related to (4.39) is given by [15]:

$$f(\mathbf{X}|\mathbf{S}, \mathbf{A}, d_p, c_p, \sigma_p^2) = \prod_{t=1}^{T}\prod_{r=1}^{R} \mathcal{N}\left(c_p + d_p \log\left(\sum_{p=1}^{P} A_{pr} s_r(t)^{z_p/z_r}\right), \sigma_i^2\right). \tag{4.59}$$

The pronounced nonlinear nature of the likelihood function in this case makes inference difficult. In [15], this problem was tackled by a MCMC simulation scheme based on the Gibbs sampler and on the Metropolis–Hasting algorithm [30].

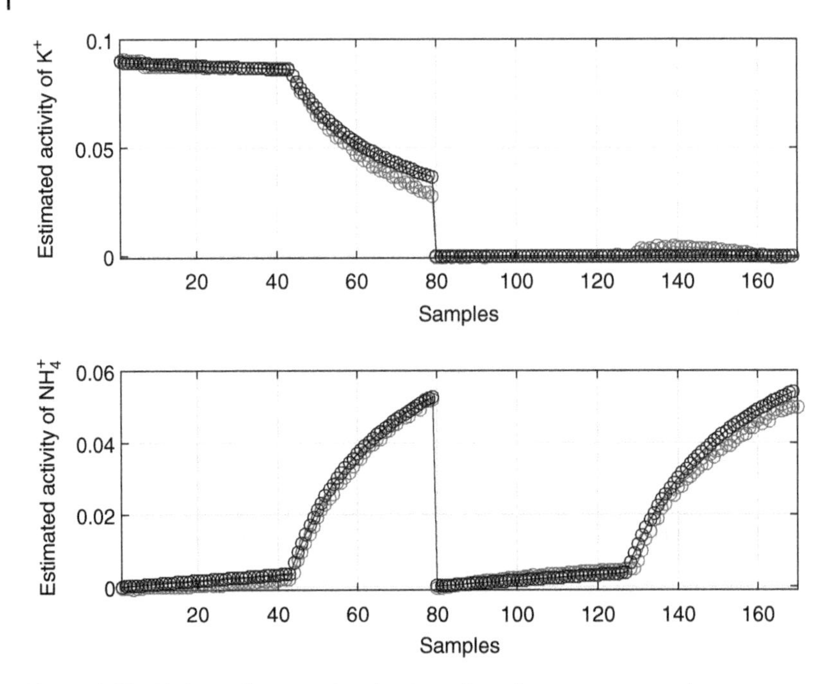

Figure 4.13 Estimated sources (gray) using a Bayesian source separation approach. The actual sources are depicted in black.

The application of the Bayesian approach developed in [15] on the mixtures shown in Figure 4.11 led to the results shown in Figure 4.13, which depicts the estimated sources. As can be seen in this figure, the Bayesian source separation was able to provide good estimates for sources.

It is worth mentioning that, after applying this BSS algorithm, it is necessary to perform a post-processing calibration process. Indeed, as discussed in Chapter 1, blind methods can retrieve the sources up to some ambiguities, which include a scale ambiguity. In other words, the application of BSS may lead, at best, to the estimation of the sources waveform. Of course, in many situations in physical–chemical analysis, the user is interested not only in the sources waveform but also in the exact values of the sources, which explains the need for a calibration step after performing source separation.

Since BSS methods also require calibration for processing data from ISE arrays, the question arises as to what are the real benefits of a BSS-based approach when compared to simple multivariate regression. This question was addressed in [15]. More precisely, the authors conducted numerical experiments showing that the conjunction BSS/post-calibration requires fewer calibration points to achieve similar performance compared to multivariate regression – practically, two calibration points per source are sufficient. For instance, in the results shown in Figure 4.13, only four calibration points were considered. In a practical scenario, such a feature is highly desirable since the process of obtaining calibration may be time consuming.

4.7 Conclusion

In this chapter, we provided an introduction to source separation methods which are based on the Bayesian approach. Special attention was given to the choice of prior distributions and the inference step through Monte Carlo sampling methods. Our discussion encompassed linear mixing models, as well as particular types of nonlinear mixing models which arise in chemical sensing problems. Finally, we also provided an example which showed how to implement a Bayesian method and a set of results obtained by considering actual data.

Of course, Bayesian methods also have limitations. A first one is the computational burden involved in the inference step (specially in the case relying on Monte Carlo sampling methods). A second challenge is that the choice of priors is not always evident given that in some cases there is no concrete information about the sources and the mixing coefficients.

Overall, the Bayesian approach is a quite useful methodology in source separation. A first interesting feature of Bayesian methods is the possibility of modeling priors by means of a probabilistic framework. Moreover, since they are based on a generative inference approach, Bayesian methods do not rely on strong hypotheses such as statistical independence between the sources. Such a feature is particularly helpful in chemical sensing since chemical sources do not always exhibit statistical independence. Finally, with respect to alternative separation methods, the Bayesian approach usually performs better in situations in which the number of samples is reduced, which is also usual in chemical sensing applications.

Appendix 4.A

Implementation of Function `postsourcesrnd` via Metropolis–Hasting Algorithm

The Metropolis–Hasting (MH) algorithm relies on the derivation of a Markov chain whose stationary distribution is the probability distribution one is interested to sample from, here denoted by $f(x)$. In the context of the example provided in Section 4.5.1, $f(x)$ is given by (4.57). Basically, the MH algorithm requires the definition of an instrumental distribution, denoted by $g(x;y)$. The instrumental distribution should be chosen among distributions which are easy to simulate. In our example, we consider a truncated Gaussian distribution (within the interval $[0,+\infty[$) whose mean is given by the current sample and the variance given by 0.05.

The idea behind the MH algorithm is to use the current sample drawn from $f(x)$, which is denoted here by $x^{(k)}$, in the generation of a candidate sample, here denoted by x^*, by means of the instrumental distribution $g(x;x^{(k)})$. Then, it becomes necessary to calculate the *acceptance probability*, which is given by

$$a = \min\left(1, \frac{f(x^*)g(x^{(k)};x^*)}{f(x^{(k)})g(x^*;x^{(k)})}\right).$$

If $u \leq a$, where u is drawn from a uniform distribution over [0,1], then the candidate x^* is accepted as the new sample $x^{(k+1)}$. Otherwise, the candidate x^* is not accepted, and the new sample is simply given by the current sample, i.e. $x^{(k+1)} = x^{(k)}$. In both cases, the complete procedure is performed again until the desired number of samples is obtained.

In order to illustrate how the MH algorithm can be coded in our example, we provide below the Matlab code related to the function `postsources-rnd`. The sampling in this example is done in a block basis in which, for a given source, all time samples are considered at the same time.

```
1  function Se = sampling_S_lognorm(X,Se,Ae,PSe,varEe)
2  % Input
3  %   X  -> Mixtures; Ae -> Current mixing coefficients
4  %   varEe -> Noise variance; PSe -> Source hyperparamenters
5
6  % Determination of number of sources, mixtures, and samples
7  [Ns,Nd] = size(Se);
8  Nc = size(X,1);
9
10 % Generating (for Ns sources) samples by means of the ...
       Metropolis-Hasting
11 % algorithm
12 % All the elements of each source are sampled in a block basis
13 sig2 = .005; % variance of the instrumental distribution
14 for jj = 1:Ns
15
16     x = Se(jj,:); % actual samples
17     xp = (trandn(x,sqrt(sig2)))'; % Generate candidate ...
           samples according a truncated Gaussian distribution
18
```

```
19      gxp = eval_trunc_gauss(xp,x,sqrt(sig2)); % evaluating ...
            the instrumental distribution for the candidate ...
            samples
20      gx = eval_trunc_gauss(x,xp,sqrt(sig2)); % evaluating ...
            the instrumental distribution for the actual samples
21      Rpxp_Rpx = eval_posterior_ratio(xp,x,jj,PSe,X,Ae, ...
            varEe,Se); % evaluating the posteriori ratio for ...
            the candidate and actual samples
22
23      % MH: acceptance or rejection
24      AP = min((Rpxp_Rpx.*gx)./(gxp +eps),1);  % Acceptance ...
            probability
25      Ind_vec_Acceptance = (rand(1,Nd)<=AP);    % Acceptance ...
            indexes
26      Se(jj,:) = Ind_vec_Acceptance.*xp + ...
            ~Ind_vec_Acceptance.*x; % Defining the new samples
27
28  end
29
30  function Rpxp_Rpx = ...
        eval_posterior_ratio(xp,x,jj,PS,X,Ae,varEe,Se)
31  % Evaluating posterior ratio for the MH algorithm
32
33  M = PS(jj,1); % Mean parameters of the lognormal distribution
34  V = PS(jj,2); % Variance parameters of the lognormal ...
        distribution
35
36  [Nc Nd] = size(X); % Determining number of mixtures and ...
        samples
37
38  Sauxx = Se;
39  Sauxxp = Se; Sauxxp(jj,:)= xp;
40
41  Rpxp_Rpx =  (x./xp).*exp( - sum( diag((1./(2*varEe)))*(( ...
        X-(Ae*Sauxxp) ).^2) ) ...
42      - (((log(xp) - M ).^2)/(2*(V))) ...
43      -( - sum( diag((1./(2*varEe)))*(( X- (Ae*Sauxx) ).^2) ) ...
44      - (((log(x) - M ).^2)/(2*(V)))   ) );
```

Listing 4.A.1 Matlab code for the implementation of the function `postsourcesrnd`.

References

1 Robert, C. (2001) *The Bayesian Choice*, Springer-Verlag, 2nd edn.

2 Roberts, S. (1998) Independent component analysis: source assessment and separation, a Bayesian approach. *IEE Proceedings on Vision, Image and Signal Processing*, 145 (3), 149–154.

3 Knuth, K. (1999) A Bayesian approach to source separation, in *Proceedings of International Workshop on Independent Component Analysis and Signal Separation (ICA'99)* (eds J.F. Cardoso, C. Jutten, and P. Loubaton), CRC Press Aussois, France, pp. 283–288.

4 Mohammad-Djafari, A. (1999) A Bayesian approach to source separation. *American Institute of Physics (AIP) proceedings*, 567, 221–244.

5 Rowe, D. (2003) *Multivariate Bayesian Statistics: Models for Source Separation and Signal Unmixing*, CRC Press, Boca Raton, FL.

6 Snoussi, H. and Idier, J. (2006) Bayesian blind separation of generalized hyperbolic processes in noisy and underdeterminate mixtures. *IEEE Transactions on Signal Processing*, 54 (9), 3257–3269.

7 Févotte, C. and Godsill, S. (2006) A Bayesian approach for blind separation of sparse sources. *IEEE Transactions on Audio, Speech and Language Processing*, 14 (6), 2174–2188.

8 Mohammad-Djafari, A. and Ichir, M. (2003) Wavelet domain blind image separation, in *SPIE, Mathematical Modeling, Wavelets X*.

9 Ochs, M.F., Stoyanova, R.S., Arias-Mendoza, F., and Brown, T.R. (1999) A new method for spectral decomposition using a bilinear Bayesian approach. *Journal of Magnetic Resonance*, 137, 161–176.

10 Miskin, J. and MacKay, D. (2001) Ensemble learning for blind source separation, in *Independent Component Analysis: Principles and Practice* (eds S. Roberts and R. Everson), Cambridge University Press, pp. 209–233.

11 Moussaoui, S., Brie, D., Mohammad-Djafari, A., and Carteret, C. (2006) Separation of non-negative mixture of non-negative sources using a Bayesian approach and MCMC sampling. *IEEE Transactions on Signal Processing*, 54 (11), 4133–4145.

12 Dobigeon, N., Moussaoui, S., Tourneret, J.Y., and Carteret, C. (2009) Bayesian separation of spectral sources under non-negativity and full additivity constraints. *Signal Processing*, 89 (12), 2657–2669.

13 Moussaoui, S., Hauksdottir, H., Schmidt, F., Jutten, C., Chanussot, J., Brie, D., Doute, S., and Benediksson, J. (2008) On the decomposition of Mars hyperspectral data by ICA and Bayesian positive source separation. *Neurocomputing*, 10, 2194–2208.

14 Dudok De Wit, T., Moussaoui, S., Guennou, C., Auchère, F., Cessateur, G., Kretzschmar, M., Vieira, L., and Goryaev, F.F. (2013) Coronal temperature maps from solar EUV images: a blind source separation approach. *Solar Physics*, 283 (1), 31–47.

15 Duarte, L., Jutten, C., and Moussaoui, S. (2009) A Bayesian nonlinear source separation method for smart ion-selective electrode arrays. *IEEE Sensors Journal*, 9 (12), 1763–1771.

16 Duarte, L., Jutten, C., and Moussaoui, S. (2011) Bayesian source separation of linear and linear-quadratic mixtures using truncated priors. *Journal of Signal Processing Systems*, 65, 311–323.

17 Halimi, A., Altmann, Y., Dobigeon, N., and Tourneret, J.Y. (2011) Nonlinear unmixing of hyperspectral images using a generalized bilinear model. *IEEE Transactions on Geoscience and Remote Sensing*, 49 (11), 4153–4162.

18 Altmann, Y., Halimi, A., Dobigeon, N., and Tourneret, J.Y. (2012) Supervised nonlinear spectral unmixing using a post-nonlinear mixing model for hyperspectral imagery. *IEEE Transactions on Image Processing*, 21 (6), 3017–3025.

19 Comon, P. (1994) Independent component analysis, a new concept? *Signal Processing*, 36 (3), 287–314.

20 Cardoso, J.F. (1998) Blind signal separation: statistical principles. *Proceedings of the IEEE*, 86 (10), 2009–2025.

21 Lee, A., Caron, F., Doucet, A., and Holmes, C. (2010) A hierarchical Bayesian framework for constructing sparsity-inducing priors. *arXiv preprint*.

22 Schmidt, M. (2009) Linearly constrained Bayesian matrix factorization for blind source separation, in *Proceedings of Advances in Neural Information Processing Systems (NIPS)*, vol. 22 (eds Y. Bengio, D. Schuurmans, J. Lafferty, C.K.I. Williams, and A. Culotta), Interface Foundation of North America, Inc., pp. 1624–1632.

23 Dobigeon, N., Moussaoui, S., Coulon, M., Tourneret, J.Y., and Hero, A.O. (2009) Joint Bayesian endmember extraction and linear unmixing for hyperspectral imagery. *IEEE Transaction on Signal Processing*, 57 (11), 4355–4368.

24 Tichý, O. and Šmídl, V. (2015) Bayesian blind source separation with unknown prior covariance, in *Proceedings of the Latent Variable Analysis and Signal Separation*, Springer, pp. 352–359.

25 Arngren, M., Schmidt, M.N., and Larsen, J. (2011) Unmixing of hyperspectral images using Bayesian nonnegative matrix factorization with volume prior. *Journal of Signal Processing Systems*, 65 (3), 479–496.

26 Demoment, G. (1989) Image reconstruction and restoration: overview of common estimation structures and problems. *IEEE Transactions on Acoustics Speech and Signal Processing*, 37 (12), 2024–2036.

27 Moussaoui, S., Brie, D., Caspary, O., and Mohammad-Djafari, A. (2004) A Bayesian method for positive source separation, in *Proceedings of IEEE International Conference on Acoustics, Speech, and Signal Processing*, vol. 5, Montreal, Canada, pp. V–485.

28 Celeux, G. and Diebolt, J. (1985) The SEM algorithm: a probabilistic teacher algorithm derived from the EM algorithm for mixture problem. *Computational Statistics Quarterly*, 2, 73–82.

29 Delyon, B., Lavielle, M., and Moulines, E. (1999) Convergence of a stochastic approximation version of the EM algorithm. *The Annals of Statistics*, 27, 94–128.

30 Gilks, W., Richardson, S., and Spiegehalter, D. (1995) *Markov Chain Monte Carlo in Practice*, Chapman & Hall, 1st edn.

31 Robert, C. (1999) *Monte Carlo Statistical Methods*, Springer-Verlag.

32 Doucet, A. and Wang, X. (2005) Monte Carlo methods for signal processing. *IEEE Signal Processing Magazine*, 22, 152–170.

33 Devroy, L. (1986) *Non-Uniform Random Variate Generation*, Springer-Verlag.

34 Robert, C.P. (2007) *The Bayesian Choice*, Springer.

35 Malinowski, E. (2000) *Factor Analysis in Chemistry*, John Willey & Sons, 3rd edn.

36 Lawton, W. and Sylvestre, E. (1971) Self-modeling curve resolution. *Technometrics*, 13, 617–633.

37 De Juan, A. and Tauler, R. (2006) Multivariate curve resolution (MCR) from 2000: progress in concepts and applications. *Critical Reviews in Analytical Chemistry*, 36, 163–176.

38 Paatero, P. and Tapper, U. (1994) Positive matrix factorization: a non-negative factor model with optimal utilization of error estimates of data values. *Environmetrics*, 5, 111–126.

39 Lee, D. and Seung, H. (1999) Learning the parts of objects by non–negative matrix factorization. *Nature*, 401, 788–791.

40 Févotte, C. and Idier, J. (2011) Algorithms for nonnegative matrix factorization with the beta-divergence. *Neural Computation*, 23 (9), 2421–2456.

41 Johnson, N.L., Kotz, S., and Balakrishnan, N. (1994) *Continuous Univariate Distributions*, John Wiley & Sons.

42 Eches, O., Dobigeon, N., and Tourneret, J.Y. (2011) Enhancing hyperspectral image unmixing with spatial correlations. *IEEE Transactions on Geoscience and Remote Sensing*, 49 (11), 4239–4247.

43 Betancourt, M.J. (2012) Cruising the simplex: Hamiltonian Monte Carlo and the Dirichlet distribution, in *Proceedings of the Maximum Entropy Bayesian Methods in Science and Engineering (MaxEnt), AIP Conference Proceedings*, vol. 1443, pp. 157–164.

44 Altmann, Y., Dobigeon, N., and Tourneret, J.Y. (2014) Unsupervised post-nonlinear unmixing of hyperspectral images using a Hamiltonian Monte Carlo algorithm. *IEEE Transactions on Image Processing*, 23 (6), 2663–2675.

45 Du, X., Zare, A., Gader, P., and Dranishnikov, D. (2014) Spatial and spectral unmixing using the beta compositional model. *IEEE Journal of Selected Topics in Applied Earth Observations and Remote Sensing*, 7 (6), 1994–2003.

46 Robert, C.P. and Soubiran, C. (1993) Estimation of mixture model through Bayesian sampling and prior feedback. *Test*, 2, 125–146.

47 McLachlan, G. and Peel, D. (2000) *Finite Mixture Models*, John Wiley & Sons.

48 Moussaoui, S., Carteret, C., Brie, D., and Mohammad-Djafari, A. (2006) Bayesian analysis of spectral mixture data using Markov chain Monte Carlo methods. *Chemometrics and Intelligent Laboratory Systems*, 81 (2), 137–148.

49 Jutten, C. and Karhunen, J. (2004) Advances in blind source separation (BSS) and independent component analysis (ICA) for nonlinear mixtures. *International Journal of Neural Systems*, 14, 267–292.

50 Deville, Y. and Duarte, L.T. (2015) An overview of blind source separation methods for linear-quadratic and post-nonlinear mixtures, in *Proceedings of the 12th International Conference on Latent Variable Analysis and Signal Separation (LVA/ICA 2015)*, Liberec, Czech Republic, pp. 155–167.

51 Taleb, A. and Jutten, C. (1999) Source separation in post-nonlinear mixtures. *IEEE Transactions on Signal Processing*, 47 (10), 2807–2820.

52 Achard, S. and Jutten, C. (2005) Identifiability of post-nonlinear mixtures. *IEEE Signal Processing Letters*, 85, 965–974.

53 Alegret, S. and Merkoci, A. (eds) (2007) *Comprehensive Analytical Chemistry: Electrochemical Sensor Analysis*, Elsevier.

54 Bakker, E. (2010) Generalized selectivity description for polymeric ion-selective electrodes based on the phase boundary potential model. *Journal of Electroanalytical Chemistry*, 639, 1–7.

55 Duarte, L.T. and Jutten, C. (2008) A nonlinear source separation approach to the Nicolsky-Eisenman model, in *Proceedings of the 16th European Signal Processing Conference*, EUSIPCO 2008.

56 Bioucas-Dias, J.M., Plaza, A., Dobigeon, N., Parente, M., Qian, D., Gader, P., and Chanussot, J. (2012) Hyperspectral unmixing overview: geometrical, statistical, and sparse regression-based approaches. *IEEE Journal of Selected Topics in Applied Earth Observations and Remote Sensing*, 5 (2), 354–379.

57 Hapke, B. (1981) Bidirectional reflectance spectroscopy 1. Theory. *Journal Geophysical Research*, 86 (B4), 3039–3054.

58 Hoffmann, J., Lubbers, D.T., and Heise, H.M. (1998) Applicability of the Kubelka–Munk theory for the evaluation of reflectance spectra demonstrated for haemoglobin-free perfused heart tissue. *Physics in Medicine and Biology*, 43, 3571–3587.

59 Close, R., Gader, P., Zare, A., Wilson, J., and Dranishnikov, D. (2012) Endmember extraction using the physics-based multi-mixture pixel model, in *SPIE Imaging Spectrometry XVII*, vol. 8515, San Diego, California, USA, pp. 85 150L–14.

60 Heylen, R., Parente, M., and Gader, P. (2014) A review of nonlinear hyperspectral unmixing methods. *IEEE Journal of Selected Topics in Applied Earth Observations and Remote Sensing*, 7 (6), 1844–1868.

61 Dobigeon, N., Tourneret, J.Y., Richard, C., Bermudez, J.C.M., McLaughlin, S., and Hero, A.O. (2014) Nonlinear unmixing of hyperspectral images: models and algorithms. *IEEE Signal Processing Magazine*, 31 (1), 89–94.

62 Korenberg, M. and Hunter, I. (1986) The identification of nonlinear biological systems: LNL cascade models. *Biological Cybernetics*, 55 (2–3), 125–134.

63 Meganem, I., Déliot, P., Briottet, X., Deville, Y., and Hosseini, S. (2014) Linear–quadratic mixing model for reflectances in urban environments. *IEEE Transactions on Geoscience and Remote Sensing*, 52, 544–558.

64 Meganem, I., Deville, Y., Hosseini, S., Déliot, P., and Briottet, X. (2014) Linear-quadratic blind source separation using NMF to unmix urban hyperspectral images. *IEEE Transactions on Signal Processing*, 62 (7), 1822–1833.

65 Almeida, M. and Almeida, L.B. (2012) Nonlinear separation of show-through image mixtures using a physical model trained with ICA. *Signal Processing*, 92, 872–884.

66 Yamazoe, N. and Shimanoe, K. (2008) Theory of power laws for semiconductor gas sensors. *Sensors and Actuators B: Chemical*, 128, 566–573.

67 Luciani, X., Mounier, S., Redon, R., and Bois, A. (2009) A simple correction method of inner filter effects affecting FEEM and its application to the PARAFAC decomposition. *Chemometrics and Intelligent Laboratory Systems*, 96 (2), 227–238.

68 Fan, W., Hu, B., Miller, J., and Li, M. (2009) Comparative study between a new nonlinear model and common linear model for analysing laboratory simulated-forest hyperspectral data. *International Journal of Remote Sensing*, 30 (11), 2951–2962.

69 Gentle, J.E. (2003) *Random Number Generation and Monte Carlo Methods*, Springer.

70 Carteret, C., Dandeu, A., Moussaoui, S., Muhr, H., Humbert, B., and Plasari, E. (2009) Polymorphism studied by lattice phonon Raman spectroscopy and statistical mixture analysis method. *Crystal Growth and Design*, 9, 807–812.

71 Duarte, L., Jutten, C., Temple-Boyer, P., Benyahia, A., and Launay, J. (2010) A dataset for the design of smart ion-selective electrode arrays for quantitative analysis. *IEEE Sensors Journal*, 10 (12), 1891–1892.

5

Geometrical Methods – Illustration with Hyperspectral Unmixing

José M. Bioucas-Dias[1] *and Wing-Kin Ma*[2]

[1] *Instituto de Telecomunicacoes, Instituto Superior Técnico, Lisboa, Portugal*
[2] *Department of Electronic Engineering, The Chinese University of Hong Kong, Hong Kong SAR, China*

5.1 Introduction

Many properties of physical and chemical data associated to source separation problems can be interpreted by means of geometrical tools. Indeed, in addition to non-negativity (see Chapters 3 and 2), constraints such as sum-to-one (sometimes called closure, full additivity, or column-stochastic) and sparsity provide nice geometrical features that can be used for deriving simple and efficient methods and algorithms for solving the source separation problem. Such properties are typical in chemistry, with reaction-based systems, where the proportions of the different species are non-negative and sum-to-one, in physics with Liquid Chromatography–Mass Spectrometry or in remote sensing by hyperspectral imaging, where the abundances of the different materials (called endmembers) of the ground surface are non-negative and sum-to-one in each pixel of the image. In this chapter, we will show how these properties lead to methods and algorithms based on a geometrical approach, and we will illustrate some of these methods in the context of hyperspectral unmixing. In fact, instead of presenting theoretical results based on geometrical properties of the data and presenting general algorithms with few applications, we preferred the chapter to be driven by the problem of hyperspectral unmixing. However, the application of this approach is not limited to this domain and could also be applied to many other applications.

The chapter is organized as follows. Section 5.2 is a short presentation of hyperspectral sensing. Section 5.3 addresses observation models in hyperspectral imaging including a discussion about data pre-processing.

Source Separation in Physical-Chemical Sensing, First Edition.
Edited by Christian Jutten, Leonardo Tomazeli Duarte, and Saïd Moussaoui.

Section 5.4 formulates the hypserspectral unmixing (HU) problem and, based on a geometric interpretation of the linear mixing model (LMM), provides a taxonomy of the blind and semi-blind HU problems. Sections 5.4.3.1 and 5.6 cover methods inspired by the geometric interpretation of HU, respectively, pure pixels search and simplex volume minimization. The former class includes non-negative matrix factorization. Section 5.5 addresses dictionary-based sparse regression (SR) to carry out semi-blind HU. Various types of structured sparsity-inducing regularizers are discussed along with their strengths and weaknesses. Sections 5.7 illustrates the potential and effectiveness of some of the presented methods with applications to hyperspectral unmixing in remote sensing and in chemical characterization of tablets. Finally, Section 5.8 ends the chapter with some concluding remarks.

The contents to be presented in this chapter are based on, or built on, those of the authors' overview articles in [1–3].

5.2 Hyperspectral Sensing

Hyperspectral sensing is concerned with the acquisition of the light reflected by surfaces or objects using a larger number of very high resolution spectral bands [4, 5]. Hyperspectral sensing, namely the imaging modality, termed *hyperspectral imaging*, has been increasingly used in remote sensing and in applications at laboratory scale (e.g. food safety, pharmaceutical process monitoring and quality control, biomedical, industrial, and biometric, and forensic) using small, commercial, high spatial, and spectral resolution instruments (see [1, 3] and references therein).

Figure 5.1 gives a partial indication of the relevance of the hyperspectral imaging applications, by displaying the published items per year and citations per year in the area of hyperspectral imaging. These results were obtained by searching the SCI-Expanded database of the ISI Web-Of-Science with the words (hyperspectral) and (imaging) in the title.

5.2.1 Hyperspectral Imaging

In hyperspectral imaging, also termed imaging spectroscopy [6], the sensor acquires a spectral vector of responses on hundreds or thousands of elements from every pixel in a given scene. The result is the so-called hyperspectral image (HSI). An equivalent viewpoint is the acquisition of a stack of images representing the radiance in the respective band (wavelength interval). Due to this interpretation, HSIs are also termed *hyperspectral data cubes*. These

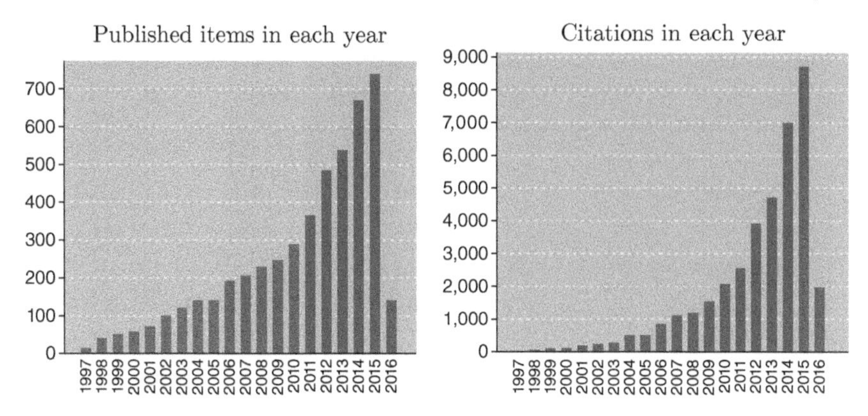

Figure 5.1 Paper counts per year in hyperspectral sensing by searching the SCI-Expanded database of the ISI Web-Of-Science with the words (hyperspectral) and (imaging) in the title. Search done on April 2016.

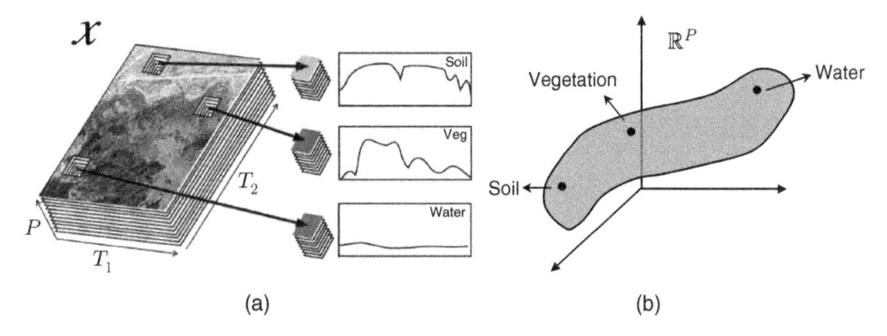

Figure 5.2 (a) Hyperspectral imaging concept. (b) Hyperspectral vectors represented in a low-dimensional manifold. Source: Bioucas-Dias et al. [3]/IEEE.

two viewpoints are illustrated in Figure 5.2a, where the HSI has P spectral bands and $T = T_1 \times T_2$ pixels. The curves plotted in the middle of the figures show the spectra of pixels containing soil, vegetation, and water.

The spectral reflectances of most materials are smooth curves and therefore the corresponding spectral vectors live in a low-dimensional manifold or a union of manifolds. This aspect is illustrated in Figure 5.2b, where the spectral vectors of soil, vegetation, and water are represented as $\mathbb{R}^P$ points on a surface. This characteristic has been actively exploited, namely, in dimensionality reduction, feature extraction, unmixing, classification, segmentation, and detection [1–4, 7]. Still related with the low dimensionality of the spectral information, the most recent trend is sparse and redundant modeling, which is reaching the areas of, e.g. restoration, unmixing, classification, segmentation, detection, and change detection (see, e.g. [8, 9] and references therein).

Table 5.1 Parameters of live hyperspectral sensors.

Parameter	HYDICE	AVIRIS	HYPERION	EnMAP	IASI
Altitude (Km)	1.6	20	705	653	817
Spatial reso (m)	0.75	20	30	30	V: 1–2 km H: 25 km
Spect reso (nm)	7–14	10	10	6.5–10	0.5 cm^{-1}
Coverage (μm)	0.4–2.5	0.4–2.5	0.4–2.5	0.4–2.5	3.62–15.5 (645-2760 cm^{-1})
Num bands	210	224	220	228	8461
Data cube size	200 × 320 × 210	512 × 614 × 224	660 × 256 × 220	1000 × 1000 × 228	765 × 120 × 8461

Table 5.1 displays spatial and spectral features of five well-known sensors used in Earth remote-sensing applications: two airborne (HYDICE and AVIRIS) and three spaceborne (HYPERION, EnMAP, and IASI). The spatial resolutions are higher for sensors carried by low-altitude platforms and vice versa. The spectral coverage of HYDICE, AVIRIS, HYPERION, and EnMAP corresponds to the visible, the near-infrared, and the shortwave infrared spectral bands, whereas the IASI covers the mid-infrared and the long-infrared bands. The number of bands is approximately 200 for HYDICE, AVIRIS, HYPERION, and EnMAP, with a spectral resolution of the order of 10 nm, and 8461 bands for IASI, with a resolution of 0.5 cm^{-1}. In any case, the spectral resolution is very high offering great potential to discriminate materials, in the case of the first four sensors, and to estimate physical parameters (temperature, moisture, and trace gases across the atmospheric column), in the case of the IASI sensor.

Since the number of incident photons per unit of area and per unit of time is limited, a high spectral resolution usually implies a low spatial resolution [10], and, as a consequence, the degradation of the discrimination power. For example, if a sensor is operating at a spatial resolution of 20 m over the Earth surface, it is very likely that most of pixels contain more than one material and thus the acquired spectral vectors are spectral mixtures.

5.2.2 Hyperspectral Unmixing

To recover the hyperspectral cameras' ability to discriminate materials based on their spectral responses, large research efforts have been devoted in the last two decades to develop effective Hyperspectral Unmixing (HU) algorithms [1–3, 11]. In addition to spectral mixing, other factors that add complexity to the analysis of hyperspectral data are its high dimensionality and correlation, large size, and degradation mechanisms associated to the measurement process (e.g. noise and atmosphere scattering and attenuation).

As explained above, this chapter is focused on HU, which is a class of mixture analysis problems conceived to revert the mixing process associated with the light scattering mechanisms and with the limited spatial resolution of the sensors. Blind HU, also known as unsupervised HU, is one of the most prominent research topics in signal processing for hyperspectral remote-sensing [1, 2, 10, 11]. It aims at identifying materials present in an observed scene, as well as their fractional abundances at each image pixel, by exploiting the high spectral resolution of the hyperspectral images. It is a blind source separation (BSS) problem from a signal processing (SP) viewpoint. Research on this topic started in the 1990s in geoscience and remote sensing enabled by technological advances in hyperspectral sensing at the time.

In recent years, blind HU has attracted much interest from other fields such as signal processing, machine learning, and optimization, and the subsequent cross-disciplinary research activities have made blind HU a vibrant topic. The resulting impact is not just on remote sensing – blind HU has provided a unique problem scenario that inspired researchers from different fields to devise novel blind SP methods. In fact, one may say that blind HU has established a new branch of BSS approaches not seen in classical BSS studies. In particular, the convex geometry (CG) concepts – discovered by early remote-sensing researchers through empirical observations [12–16] and refined by later research – are elegant and very different from statistical independence-based BSS approaches established in the SP field. Moreover, latest research on blind HU is rapidly adopting advanced techniques, such as those in sparse signal analysis and optimization. The present development of blind HU seems to be converging to a point where the lines between remote sensing-originated ideas and advanced SP and optimization concepts are no longer clear, and insights from both sides would be used to establish better methods.

In this chapter we will consider several key developments of HU related to the geometric approach, such as pure pixel search and convex geometry. We will also cover non-negative matrix factorization and dictionary-based SR. We will not cover techniques based on statistical inference, although readers should note that they also represent key developments in blind HU (see, e.g. [17–19]).

5.3 Hyperspectral Mixing Models

The signal recorded by a hyperspectral sensor at a given band and from a given pixel, letting alone the effects of the atmosphere, is a mixture of the "light" scattered by the constituent materials located in the respective pixel coverage. Figure 5.3 illustrates three types of mixtures owing to low spatial resolution of the sensor (a), multiple light scattering in a two-layer media (b), and presence of intimate mixtures (c). As a result, when mixing occurs, it is not anymore possible to determine which materials are present in the pixels directly from the respective measured spectral vectors. This is to say that the key feature of the hyperspectral sensors, which is its ability to discriminate materials based on their spectral responses, is compromised. This section addresses spectral mixing modeling and provides insights on the spectral unmixing inverse problems.

With the objective of recovering the ability to discriminate materials, an impressive amount of research work has been devoted to HU (see, e.g. [1, 2, 10], and the references therein). HU is, however, an ill-posed inverse problem. The difficulties begin with its formulation. Put in simple terms, given a measured spectral vector $\boldsymbol{x} \in \mathbb{R}^P$, HU aims at explaining $\boldsymbol{x}$ in terms of the spectral properties of the materials present in the respective pixel and of its distribution (spatial, temporal, etc.). A useful treatment of this problem cannot be given without a formal model, $\boldsymbol{x} = \mathcal{A}(\boldsymbol{s}, \boldsymbol{\theta})$, where $\mathcal{A}(\cdot)$ is the so-called forward (or direct) operator, linking the measurements $\boldsymbol{x}$ to the material signatures $\boldsymbol{s}$ and the scene parameters $\boldsymbol{\theta}$.

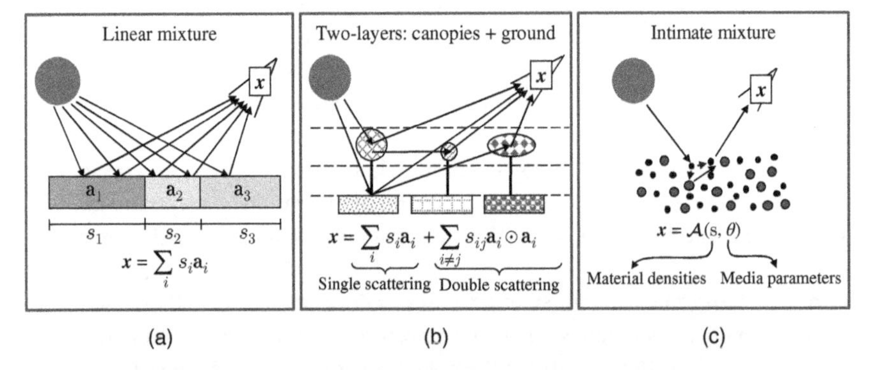

Figure 5.3 Schematic view of three types of spectral mixing. (a) linear mixing in a checkerboard-type surface. (b) Nonlinear (linear plus bilinear) mixing in a two-layer media. (c) Nonlinear mixing in an intimate (particulate) media.
Source: Bioucas-Dias et al. [1] and [3]/IEEE.

Radiative transfer theory (RTT) [20] is a mathematical model for the transfer of energy as photons interact with the materials in the scene, and thus to derive forward operators necessary to solve HU inverse problems. The core of the RTT is a differential equation describing radiance recorded by the sensor. It can be derived via the conservation of energy and the knowledge of the phase function, which represents the probability of light with a given propagation direction being scattered into a specified angle solid around a given scattering direction.

In general, the forward operator $\mathcal{A}$ is not invertible, unless we have partial knowledge of vector $\boldsymbol{\theta}$, which is often very hard to obtain. Three notable exceptions to these scenarios are the linear model (Figure 5.3a), the bilinear model (Figure 5.3b), and the Hapke model [1, 21] modeling the light scattering mechanism in intimate mixtures, as illustrated in Figure 5.3c. These are three approximations for the analytical solution to the RTT suitable to unsupervised applications, i.e. when no prior knowledge exists about the materials and their distributions.

The LMM is a good approximation when the mixing scale is macroscopic and the incident light interacts with just one material, as it is the case in checkerboard-type scenes [21] schematized in Figure 5.3a. The light from the materials is linearly mixed within the measuring instrument, owing to insufficient spatial resolution. Formally, for each pixel of the image, the measured spectral vector $\boldsymbol{x} := [x_1, \ldots, x_P]^\top$, holding the radiance at bands $p = 1, \ldots, P$, is expressed as

$$\boldsymbol{x} = \sum_{r=1}^{R} \boldsymbol{a}_r s_r, \tag{5.1}$$

where $\boldsymbol{a}_r$, for $r = 1, \ldots, R$, is the *spectral signature* of the r-th material, termed *endmember*, and s_r is the percentage that the rth material occupies inside the pixel, and termed *fractional abundance* or simply *abundance*. Inspired by the LMM, the HU problem is very often defined as the unsupervised, or blind, estimation of the endmembers and of the respective fractional abundances.

The LMM has been widely used in the past two decades to address HU problems. The reason is threefold: (i) despite its simplicity, LMM is an acceptable approximation for the light scattering in many real scenarios; (ii) under suitable conditions linked with $\mathcal{S} = \{\boldsymbol{s} \in \mathbb{R}^R \mid \boldsymbol{s} \geqslant 0, \boldsymbol{s}^\top \mathbf{1}_R = 1\}$, geometry of the dataset (i.e. the HSI), LMM yields well-posed inverse problems; (iii) under the LMM, HU is interpretable as a BSS problem or as a non-negative matrix factorization problem, which have been vastly researched in many SP areas.

In Section 5.4, we address in depth relevant aspects of HU under the LMM. Two classes of algorithms inspired by the LMM are the *pure pixel pursuit*, presented in Section 5.4.3.1, and the *minimum simplex volume estimation*, presented in Section 5.6. In the former class, the dataset contains at least one pure pixel of each material, that is, for all endmembers $r = 1, \dots, R$ there exists at least one pixel in the dataset such that the corresponding spectral vector $\boldsymbol{x} \in \mathbb{R}^T$ equals $\boldsymbol{a}_r$. In the latter class, not all materials have pure pixels, but there is a set of spectral vectors such that some components thereof are zero in a sense to be defined later.

In spite of the LMM attractiveness, researchers are beginning to expand more aggressively into the nonlinear mixing field to cope with the LMM limitations. A discussion of the nonlinear HU is, however, beyond the scope of this chapter. For a review on recent nonlinear HU methods, see, e.g. [11, 22]

5.4 Linear HU Problem Formulation

Under the LMM (5.1), a given measured hyperspectral vector $\boldsymbol{x}_t \in \mathbb{R}^P$, associated with the pixel $t \in \{1, \dots, T\}$, can be written as

$$\boldsymbol{x}_t = \sum_{r=1}^{R} \boldsymbol{a}_r s_{tr} + \boldsymbol{e}_t, \tag{5.2}$$

where $\boldsymbol{a}_r$, s_{tr}, and $\boldsymbol{e}_t$, for $r = 1, \dots, R$ and $t = 1, \dots, T$, are, respectively, the endmember signature, the fractional abundance (or just *abundance*) of the r-th endmember at pixel t, and the additive perturbation, due to, for example, model mismatches and additive noise, at pixel t. Henceforth, we will term $\boldsymbol{e}_t$ simply the noise. Because the abundances represent fractions, they must satisfy the so-called *abundance non-negativity constraint* (ANC) and *abundance sum constraint* (ASC):

$$\text{ANC:} \quad s_{tr} \geqslant 0, \quad \forall r = 1, \dots, R, \tag{5.3}$$

$$\text{ASC:} \quad \sum_{r=1}^{R} s_{tr} = 1. \tag{5.4}$$

The constraint set defined by the ANC and ASC is the unit $(R-1)$-simplex

$$\mathcal{S} = \{\boldsymbol{s} \in \mathbb{R}^R \mid \boldsymbol{s} \geqslant 0, \boldsymbol{s}^\top \mathbf{1}_R = 1\}, \tag{5.5}$$

where $\boldsymbol{a} \geqslant \mathbf{0}$ is to be understood in the component-wise sense and $\mathbf{1}_R$ denotes a column vector with R ones. By denoting the *abundance vector* of pixel t as $\boldsymbol{s}_t = [s_{t1}, \dots, s_{tR}]^\top$, the ANC and ASC may then be written compactly as $\boldsymbol{s}_t \in \mathcal{S}$.

Owing to spectral variability, the ASC is seldom observed in real applications. Nevertheless, because the spectral vectors are non-negative, it is always possible to build rescaled or projected versions thereof, belonging to an affine set [13], and thus satisfying the ASC. Another possibility, which also leads to a data set satisfying the ASC, is to project orthogonally the observed vectors into the $(R-1)$-dimensional affine set that best represents the observed data in the least squares (LS) sense. These two types of projections, used for example in the VCA algorithm [23], are studied in detail in [1], where they were termed, respectively, *projective projection* and *orthogonal projection*. The derivation of orthogonal projection was, meanwhile, carried in solid mathematical terms in [24]. We will get back to this topic in Section 5.4.1. Anyway, unless otherwise stated, we assume that the ASC holds true.

Let $\boldsymbol{A} = [\boldsymbol{a}_1, \dots, \boldsymbol{a}_R] \in \mathbb{R}^{P\times R}$ denote the *endmember matrix* (or *mixing matrix*) and suppose we are given a hyperspectral data set containing T spectral vectors of size P arranged in the matrix $\boldsymbol{X} = [\boldsymbol{x}_1, \dots, \boldsymbol{x}_T] \in \mathbb{R}^{P\times T}$. Define the *noise matrix* $\boldsymbol{E} = [\boldsymbol{e}_1, \dots, \boldsymbol{e}_T] \in \mathbb{R}^{P\times T}$, where $\boldsymbol{e}_t$ represents the *noise vector* of pixel t. With these definitions in place, the LMM, ANC, and ASC may be written as

$$\text{LMM:} \quad \boldsymbol{X} = \boldsymbol{A}\boldsymbol{S} + \boldsymbol{E}, \tag{5.6}$$

$$\text{ANC:} \quad \boldsymbol{S} \geqslant 0, \tag{5.7}$$

$$\text{ASC:} \quad \mathbf{1}_R^{\mathsf{T}}\boldsymbol{S} = \mathbf{1}_T^{\mathsf{T}}. \tag{5.8}$$

Figure 5.4 illustrates the LMM in the absence of noise, i.e. with $\boldsymbol{E} = \mathbf{0}$. The observed bands, shown in (a), are represented by the rows of $\boldsymbol{X}$; the endmember signatures, shown in (b), are the columns of $\boldsymbol{A}$; and the *abundance maps*, shown in (c), are represented by the rows of $\boldsymbol{S}$.

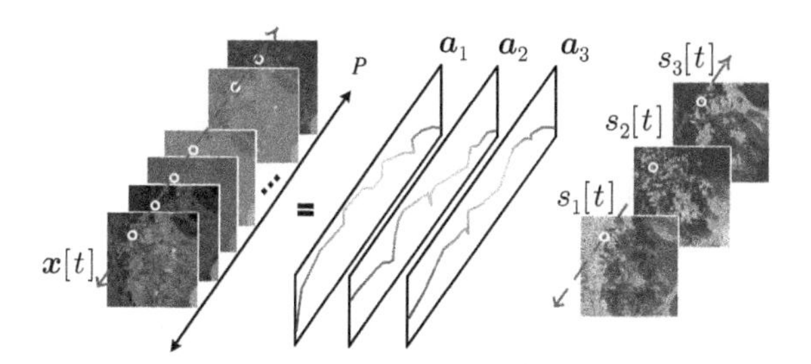

Figure 5.4 Schematic view of the LMM. The observed bands, shown in the left hand side, are represented by the rows of $\boldsymbol{X}$; the endmember signatures, shown in the middle, are the columns of $\boldsymbol{A}$; and the *abundance maps*, shown in the right hand side, are represented by the rows of $\boldsymbol{S}$. In this picture, we denote $\boldsymbol{x}[t] = \boldsymbol{x}_t$ and $\boldsymbol{s}[t] = \boldsymbol{s}_t$. Source: Wing-Kin Ma et al., [2]/IEEE.

Using the matrix representation (5.6), a possible formulation to the blind HU inverse problem, under the LMM, is

$$\begin{aligned} &\min_{\boldsymbol{A},\boldsymbol{S}} \quad \|\boldsymbol{X} - \boldsymbol{A}\boldsymbol{S}\|_F \\ &\text{s.t.} \quad \boldsymbol{S} \geqslant \boldsymbol{0}, \quad \boldsymbol{1}_R^T \boldsymbol{S} = \boldsymbol{1}_T, \end{aligned} \tag{5.9}$$

where $\|\boldsymbol{M}\|_F = \sqrt{\text{trace}\{\boldsymbol{M}\boldsymbol{M}^T\}}$ is the Frobenius norm of the matrix $\boldsymbol{M}$. The optimization problem (5.9) is interpretable both as a linear BSS problem or as a matrix factorization problem. In the former case, the independent component analysis (ICA) criterion [25, 26] comes to mind to separate sources (i.e. the fractional abundances). ICA has in fact been considered to solve spectral unmixing problems. ICA is, however, based on the assumption of mutually independent sources, which is not the case of hyperspectral data, since the sum of abundance fractions is constant, implying statistical dependence among them. This dependence compromises ICA applicability to hyperspectral data as shown in [27].

From the above discussion, we conclude that the ANC and the ASC put blind HU beyond the reach of the ICA-based methods. However, on the other hand, those constraints open the door to new approaches which exploit geometric properties of the dataset $\boldsymbol{X}$, linked with the way the columns of $\boldsymbol{X}$ are distributed in $\mathbb{R}^P$. These properties and the associated methods are studied in detail in the ensuing sections.

5.4.1 Preprocessing

Before unmixing, hyperspectral datasets usually undergo atmospheric calibration and dimension reduction (DR) [1, 28]. The atmospheric calibration step converts the measured radiance into reflectance, which is an intrinsic characteristic of the materials. However, the unmixing inverse problem can also be formulated in the radiance data, provided that the effects of atmosphere are pixel invariant. See [27] for details on the spectral radiance model.

5.4.1.1 Dimension Reduction (DR)

The DR step identifies the subspace where the spectral vectors live, often termed the *signal subspace*, and projects them onto this subspace. Given that the identified subspace is generally of much lower dimension than that of the spectral vectors, this projection yields considerable gains in algorithm performance and complexity, data storage, and noise reduction.

Let us assume that the columns of $\boldsymbol{A}$ are linearly independent and that $\boldsymbol{V} \in \mathbb{R}^{P \times R}$ is an orthonormal matrix (i.e. $\boldsymbol{V}^\mathsf{T}\boldsymbol{V} = \boldsymbol{I}$) whose columns are a basis for range$\{\boldsymbol{A}\}$, that is, the span of the columns of $\boldsymbol{A}$. We may then decompose $\boldsymbol{X}$ orthogonally as $\boldsymbol{X} = \boldsymbol{P}_A\boldsymbol{X} + \boldsymbol{P}_A^{\perp}\boldsymbol{X}$ where $\boldsymbol{P}_A = \boldsymbol{V}\boldsymbol{V}^{\mathsf{T}}$ and $\boldsymbol{P}_A^{\perp} = \boldsymbol{I} - \boldsymbol{P}_A$ denote, respectively, the projections matrices onto range$\{\boldsymbol{A}\}$ and onto the orthogonal complement of range$\{\boldsymbol{A}\}$. Since the component $\boldsymbol{P}_A^{\perp}\boldsymbol{X}$ does not depend on $\boldsymbol{A}$, it may be removed from $\boldsymbol{X}$ and the unmixing problem formulated with respect to $\boldsymbol{P}_A\boldsymbol{X}$, i.e. with respect to the projection of $\boldsymbol{X}$ onto range$\{\boldsymbol{A}\}$. Instead of working on the P-dimensional vectors resulting from that projection, we may work on the M-dimensional coordinates with respect to the basis $\boldsymbol{V}$ given by $\boldsymbol{X}_p = \boldsymbol{V}^{\mathsf{T}}\boldsymbol{P}_A\boldsymbol{X} = \boldsymbol{V}^T\boldsymbol{V}\boldsymbol{V}^{\mathsf{T}}\boldsymbol{X} = \boldsymbol{V}^{\mathsf{T}}\boldsymbol{X} \in \mathbb{R}^{R\times T}$, thus obtaining the LMM model

$$\boldsymbol{X}_p = \boldsymbol{A}_p\boldsymbol{S} + \boldsymbol{E}_p, \tag{5.10}$$

where $\boldsymbol{A}_p = \boldsymbol{V}^{\mathsf{T}}\boldsymbol{A} \in \mathbb{R}^{R\times R}$ and $\boldsymbol{E}_p = \boldsymbol{V}^{\mathsf{T}}\boldsymbol{E} \in \mathbb{R}^{R\times T}$. We conclude then that we may use the LMM (5.10) to infer $\boldsymbol{A}$ and $\mathbf{S}$, with two significant advantages over the LMM (5.6) when $R \ll P$, which is always the case in practical applications:

(a) The size of $\boldsymbol{X}_p \in \mathbb{R}^{R\times T}$ is much smaller than that of $\boldsymbol{X} \in \mathbb{R}^{P\times T}$.
(b) $\|\boldsymbol{E}_p\|_F \ll \|\boldsymbol{E}\|_F$ and then the signal-to-noise ratio (SNR) of model (5.10) is much higher than that of (5.6). When the components of $\boldsymbol{E}$ are independent and identically distributed (i.i.d.), we have $\|\boldsymbol{E}_p\|_F^2/\|\boldsymbol{E}\|_F^2 \simeq R/P \ll 1$.

In order to simplify the notation, we will drop the subscript R in (5.10) and will assume that $\boldsymbol{X} \in \mathbb{R}^{M\times T}$. This means that we will use the same notation for the unprojected LMM (5.6) and for the projected LMM (5.10). The distinction between the two cases will be made by the value of M, which takes the value $M = P$ if there is no DR and $M = R$ if there is DR. In any case, we assume that $M \leq T$.

5.4.2 Signal Subspace Identification

As seen above, the fundamental ingredient of dimension reduction is the knowledge of subspace spanned the mixing matrix $\boldsymbol{A} \in \mathbb{R}^{P\times R}$. Unsupervised subspace identification has been approached in many ways. A few representative algorithms are *maximum noise fraction* (MNF) [29], *noise adjusted principal components* (NAPC) [30], the optical real-time adaptive spectral identification system (ORASIS) [31], *virtual dimensionality*, [32], *hyperspectral signal identification by minimum error* (HySime) [33], and methods using random matrix theory [34–36].

Figure 5.5 illustrates the advantages of projecting the data on the signal subspace. The noise and the signal subspace were estimated with HySime

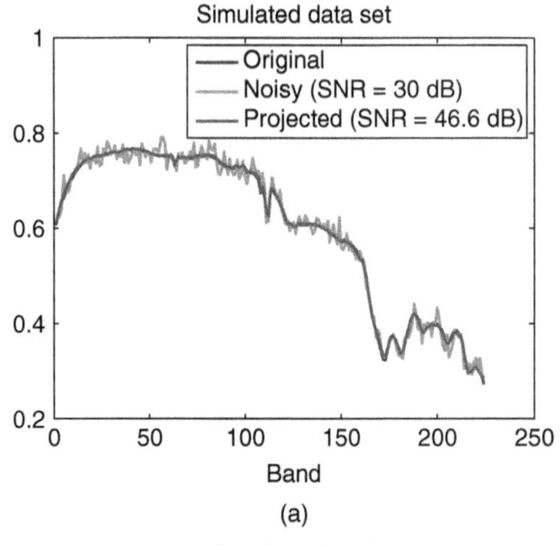

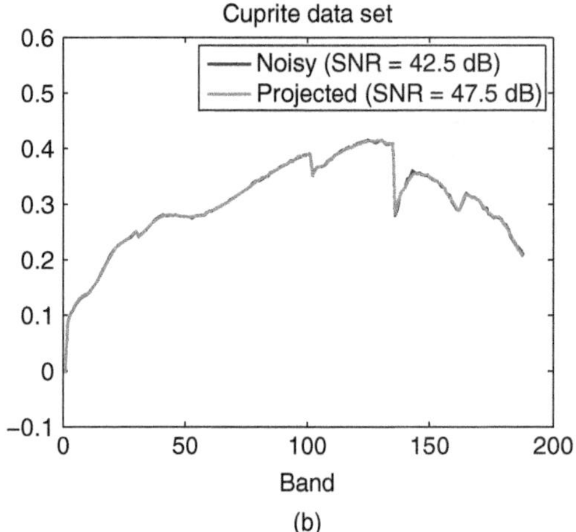

Figure 5.5 (a) Noisy and projected spectra from a simulated dataset. (b) Noisy and projected spectra from the real dataset cuprite. Source: Bioucas-Dias et al., [1]/IEEE.

[33]. The plot on the top shows a noisy spectral vector $\boldsymbol{x}$ and the corresponding projected spectra $\widehat{\boldsymbol{V}}\widehat{\boldsymbol{V}}^{\mathrm{T}}\boldsymbol{x}$, where the $\widehat{\boldsymbol{V}}$ was estimated by HySime. The simulated dataset has size $T = 5000$ and was generated according the LMM (5.6); the mixing matrix has $R = 5$ endmembers of size $P = 224$ randomly sampled from the United States Geological Survey (USGS) spectral library[1]; the fractional abundances are distributed uniformly on the four-unit simplex; the noise is i.i.d zero-mean Gaussian with SNR= $\|\boldsymbol{AS}\|_F^2/\|\boldsymbol{E}\|_F^2$ =30 dB. The subspace dimension was correctly identified. The SNR of the projected data set is 46.6dB, which is 16.6 dB $\simeq (P/R)$ dB above to that of the noisy dataset.

1 Available online from: http://speclab.cr.usgs.gov/spectral-lib.html

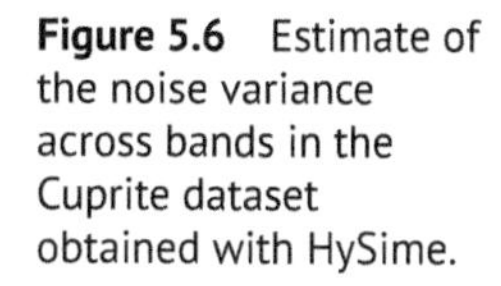

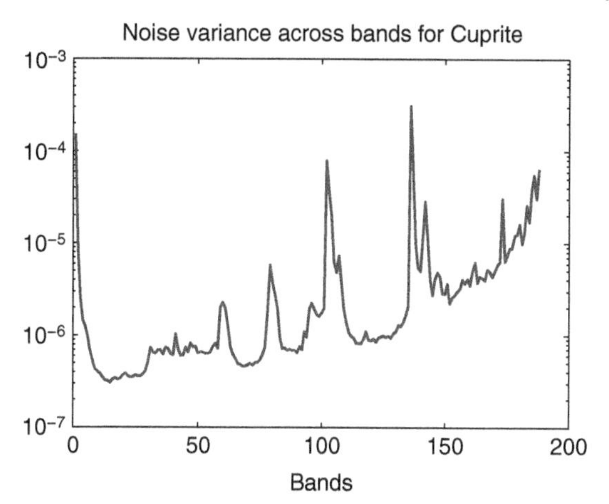

Figure 5.6 Estimate of the noise variance across bands in the Cuprite dataset obtained with HySime.

The plot on the bottom of Figure 5.5 is as in the top for a subset of well-known AVIRIS cuprite data cube[2] with size 250 lines by 191 columns by 188 bands (noisy bands due to water absorption were removed). The identified subspace has dimension 18. The SNR of the projected data set is 47.5dB, which is 5dB above to that of the noisy data set. Figure 5.6 shows the diagonal of the noise correlation matrix obtained with HySime. The noise power varies considerably across bands; the peaks of the noise variance are located at the extreme bands and at bands corresponding to the water absorption wavelengths. The colored nature of the noise explains the difference $(P/R)\,\mathrm{dB} - 5\,\mathrm{dB} \simeq 5\,\mathrm{dB}$.

A possible line of attack to further reduce the noise in the signal subspace is to exploit spectral and spatial contextual information. See [1] for details about this topic.

A final word of warning. Although the projection of the dataset onto the signal subspace often removes a large percentage of the noise, it does not improve the conditioning of the HU inverse problem, as this projection does not change the noise in the signal subspace.

5.4.2.1 Affine Set Estimation and Projection

Given the abundance vectors $\boldsymbol{s}_t \in \mathcal{S}$, for $t = 1, \ldots, T$, the clean spectral vectors $\boldsymbol{x}_t = \boldsymbol{A}\boldsymbol{s}_t \in \mathbb{R}^M$ are convex combinations of the columns of $\boldsymbol{A}$ and, therefore, live in the affine set $\mathrm{aff}\{\boldsymbol{A}\}$ of dimension $(R-1)$, assuming that $\mathrm{rank}\{\boldsymbol{A}\boldsymbol{S}\} = R$. Since any vector $\boldsymbol{x} \in \mathrm{aff}\{\boldsymbol{A}\}$ may be written as (see, e.g. [Ch. 2.1][37])

2 Available online from: `http://aviris.jpl.nasa.gov/data/free_data.html`

$$x = Cz + b, \tag{5.11}$$

where b is any vector in aff$\{A\}$, $C \in \mathbb{R}^{P\times(R-1)}$ is any matrix spanning the subspace aff$\{A\} - b$, and $z = C^{\dagger}(x - b)$, with $C^{\dagger} = (C^{\mathsf{T}}C)^{-1}C^{\mathsf{T}}$ denoting the pseudoinverse of C, we may use the representation (5.11) as an alternative to $x = As$.

A possible solution for (C, b) is

$$b = \frac{1}{T}\sum_{t=1}^{T} As_t, \quad C = [u_1, \ldots, u_{R-1}], \quad z = C^{\mathsf{T}}(x - b), \tag{5.12}$$

where u_i is the ith eigenvector, ordered by decreasing value of the respective eigenvalue, of the sample covariance matrix $ASS^{\mathsf{T}}A^{\mathsf{T}}/T - bb^{\mathsf{T}}$. In a real data set, the observed spectral vectors $x_t = As_t + e_t$ do not belong to an affine set due to the presence of additive noise and also because the abundances do not live in a unit simplex due to spectral variability. In this case, we may be interested in finding the affine set that best represents X, for example in LS sense, and project the observed data onto the estimated set. This affine fitting problem was addressed in [24] by solving the optimization

$$\min_{C\in\mathbb{R}^{P\times R-1},\ Z\in\mathbb{R}^{(R-1)\times T},\ b\in\mathbb{R}^{P}} \left\| X - CZ - b1_T^{\mathsf{T}} \right\|_F^2. \tag{5.13}$$

A solution of (5.13) similar to (5.12) is given by

$$b = \frac{1}{T}\sum_{n=1}^{T} x_t, \quad C = [u_1, \ldots, u_{R-1}], \quad Z = C^{\mathsf{T}}(X - b1_T^{\mathsf{T}}), \tag{5.14}$$

where u_i is the ith eigenvector, ordered by decreasing value of the respective eigenvalue, of the sample covariance matrix $XX^{\mathsf{T}}/T - bb^{T}$.

Figure 5.7 illustrates the affine set estimation and two types of projections, usually after DR, onto the estimated affine set, that further attenuates the noise and the spectral variability. The affine projection, illustrated by x_a and given by

$$X \leftarrow CC^{T}\left(X - b1_T^{\mathsf{T}}\right) + b1_T^{\mathsf{T}},$$

is the orthogonal projection of the columns of X onto the estimated affine set. The perspective projection, illustrated by y_p and given by

$$X \leftarrow \left[\frac{x_t}{x_t^{\mathsf{T}}u}, n = 1, \ldots, T\right], \quad u = \frac{(I - CC^{\mathsf{T}})b}{b^{\mathsf{T}}(I - CC^{\mathsf{T}})b},$$

is just a pixel-dependent scaling. We remark that vector u is orthogonal to range$\{C\}$ and its amplitude is such that $v^{\mathsf{T}}u = 1$ for any v in the identified

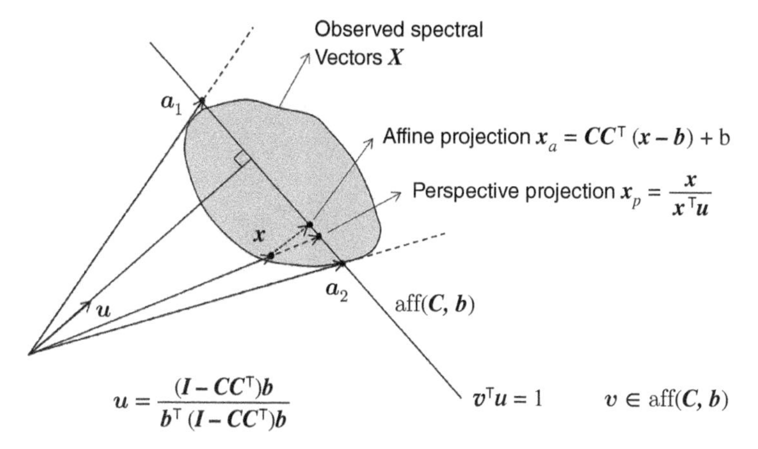

Figure 5.7 Illustration of the affine set estimation, affine projection, and perspective projection.

affine set. From Figure 5.7, we conclude that the affine projection may introduce a rotation in the projected vectors, whereas the perspective projection does not. This issue is studied in [1].

After DR and affine or perspective projection, we still assume that we have a LMM of the form (5.9) of which $\boldsymbol{A}$ and $\boldsymbol{S}$ are to be estimated. An alternative to the LMM (5.9) is to work with the matrix

$$\boldsymbol{Z} = \boldsymbol{C}^T\left(\boldsymbol{X} - \boldsymbol{b}\boldsymbol{1}_T^T\right) \tag{5.15}$$

By replacing $\boldsymbol{X} = \boldsymbol{A}\boldsymbol{S} + \boldsymbol{E}$ into (5.15), we may write

$$\begin{aligned} \text{LMM:} \quad & \boldsymbol{Z} = \boldsymbol{F}\boldsymbol{S} + \boldsymbol{E}_x, \\ \text{ANC:} \quad & \boldsymbol{S} \geqslant 0, \\ \text{ASC:} \quad & \boldsymbol{1}_R^T\boldsymbol{S} = \boldsymbol{1}_T^T, \end{aligned} \tag{5.16}$$

where

$$\boldsymbol{F} = \boldsymbol{C}^T(\boldsymbol{A} - \boldsymbol{b}\boldsymbol{1}_M^T) \in \mathbb{R}^{(M-1)\times M}, \tag{5.17}$$

$$\boldsymbol{E}_x = \boldsymbol{C}^T\boldsymbol{E}. \tag{5.18}$$

Notice that (5.16) is also a LMM. The objective now is to estimate $\boldsymbol{F}$ and, from it and the knowledge of $(\boldsymbol{C}, \boldsymbol{b})$, estimate $\boldsymbol{A}$.

A distinctive feature of model (5.16) is that the set conv$\{\boldsymbol{F}\}$ is a full-dimensional simplex, which may be useful to compute, for example, the volume of that set. Model (5.16) was adopted, for example, in [24, 38, 39]. Herein, and for uniform treatment, we mostly adopt the LMM (5.9), with references to (5.16), whenever necessary.

5.4.3 Classes of Linear HU Problems

In order to shed light on the blind HU problem, we now give an interpretation of problem (5.9) based on convex geometry (CG). The convex hull of set of vectors $\{\boldsymbol{a}_1, \dots, \boldsymbol{a}_R\} \subset \mathbb{R}^M$ is defined as

$$\mathrm{conv}\{\boldsymbol{a}_1, \dots, \boldsymbol{a}_R\} = \left\{ \boldsymbol{x} = \sum_{r=1}^{R} \boldsymbol{a}_r s_r \mid \boldsymbol{s} \in \mathbb{R}R, \mathbf{s}^{\mathsf{T}} \mathbf{1}_R = 1, \boldsymbol{s} \geqslant 0 \right\}. \tag{5.19}$$

The set $\mathrm{conv}\{\boldsymbol{a}_1, \dots, \boldsymbol{a}_R\}$, also denoted as $\mathrm{conv}\{\boldsymbol{A}\}$, with $\boldsymbol{A} = [\boldsymbol{a}_1, \dots, \boldsymbol{a}_R]$, is called a $(R-1)$-simplex in $\mathbb{R}^P$, or simply a simplex, if the set $\{\boldsymbol{a}_1, \dots, \boldsymbol{a}_R\}$ is affinely independent; that is, if the set $\{\boldsymbol{a}_2 - \boldsymbol{a}_1, \dots, \boldsymbol{a}_R - \boldsymbol{a}_1\}$ is linearly independent. We remark that the vertices of the simplex $\mathrm{conv}\{\boldsymbol{a}_1, \dots, \boldsymbol{a}_R\}$ are the vectors $\{\boldsymbol{a}_1, \dots, \boldsymbol{a}_R\}$.

Figure 5.8 illustrates a 2-simplex C for a hypothetical mixing matrix $\boldsymbol{A}$ containing three endmembers. The points in light gray denote non-pure spectral vectors, whereas the points in dark gray are pure spectral vectors, thus corresponding to the vertices of the simplex. Note that the inference of the mixing matrix $\boldsymbol{A}$ amounts to identifying the vertices of the simplex C. This geometrical point of view has been exploited by many unmixing algorithms, which can be mainly classified either as *pure pixel* or *non-pure pixel* based.

5.4.3.1 Datasets with Pure Pixels. Pure Pixel Pursuit

Very often, in blind HU, it is assumed that there is at least one pure spectral vector, or pixel, per endmember, the so-called *pure pixel assumption*. The geometric representation of this scenario is shown in Figure 5.9a; there is at least one spectral vector on each vertex of the simplex associated with the endmember matrix. In practice, there are scenarios where the pure pixel assumption holds. For example, imagine a scene that consists of water and soil. If there exist some local pixel regions that contain either water or soil only, then those regions contain pure pixels. Note that more than one pure pixel may exist in a dataset for a particular endmember.

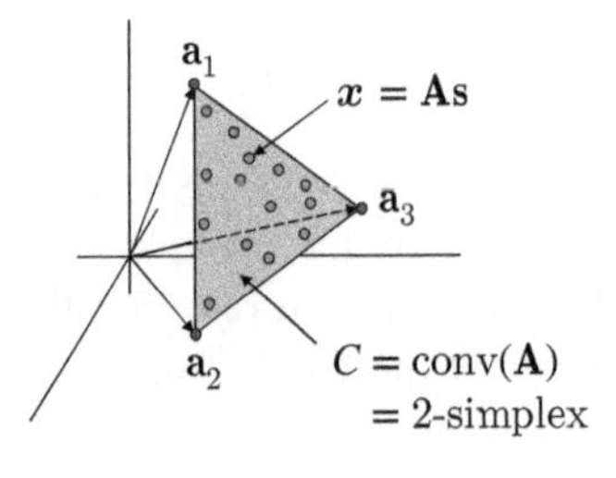

Figure 5.8 Illustration of the simplex set $C = \mathrm{conv}\{\boldsymbol{A}\}$, which is the convex hull of the columns of $\boldsymbol{A}$. Dark gray and light gray circles correspond, respectively, to pure and non-pure spectral samples (or pixels). Source: Bioucas-Dias et al., [1]/IEEE.

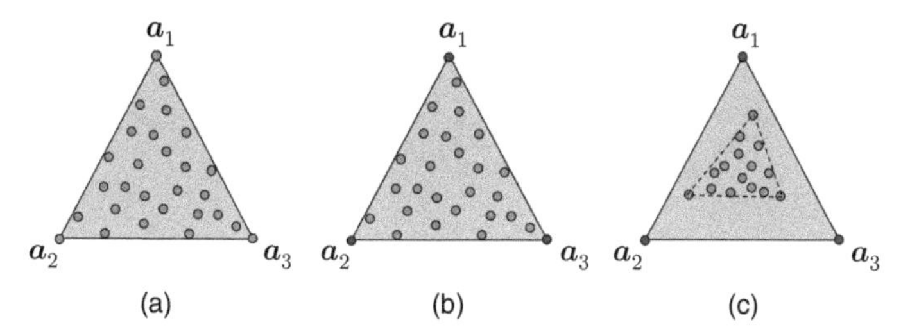

Figure 5.9 Three datasets with different distributions of spectral vectors. Light gray and dark gray circles correspond to observed samples and pure spectral vectors, respectively. (a) Existence of pure spectral vectors, (b) absence of pure spectral vectors but existence of pixels in the simplex facets, and (c) absence of pure spectral vectors and absence of pixels in the simplex facets, corresponding to a highly mixed scenario. Source: Bioucas-Dias et al., [1]/IEEE.

The pure pixel assumption has a clear conceptual meaning and is the key element in developing HU algorithms with light computational complexity. Most of the pure pixel-based algorithms search for pure pixels in the dataset by exploiting one of the following properties of the endmember signatures: (i) the extremes of the projection of the spectral vectors onto any subspace correspond to endmembers; (ii) the volume defined by any set of R spectral vectors is maximum when those are endmembers. Representative algorithms of class (i) are *pixel purity index* (PPI) [40], *vertex component analysis* (VCA) [23], simplex growing algorithm (SGA) [41], *successive volume maximization* (SVMAX) [38], and the *recursive algorithm for separable NMF* (RSSNMF) [42]; representative algorithms of class (ii) are N-FINDR [43], *iterative error analysis* (IEA), [44], *sequential maximum angle convex cone* (SMACC), and *alternating volume maximization*(AVMAX) [38].

5.4.3.2 Datasets Without Pure Pixels. Minimum Volume Simplex Estimation

Figure 5.9b,c, schematizes two datasets without pure pixels; the dataset in the middle does not contain pure pixels but contains at least $R - 1$ spectral vectors on each facet. The lack of pure pixels happens, for example, in a scene consisting of highly mixed materials, or if the spatial resolution of the hyperspectral camera is too low. In these datasets, the endmembers may by inferred by finding the facets of the simplex. A popular criterion to find the simplex facets is to fit a minimum volume (MV) simplex to the data; this rather simple and yet powerful idea, introduced by Craig in his seminal work [12, 13], underlies several geometrical-based unmixing algorithms. A similar idea was introduced by Perczel in the area of Chemometrics *et al.* [45, 46].

From an optimization point of view, the MV-based unmixing algorithms are typically formulated as

$$\min_{A,S} \quad \|X - AS\|_F^2 + \lambda \operatorname{vol}(A)$$
$$\text{subject to:} \quad S \geqslant 0, \quad 1_R^\top S = 1_T^\top, \tag{5.20}$$

where the regularizer $\operatorname{vol}(A) = \operatorname{volume}(\operatorname{conv}\{A\})$ promotes endmember matrices of "minimum volume" and $\lambda > 0$ is a regularization parameter setting the relative weight between the data term and the volume term.

We remark that the optimization (5.20) is a form of non-negative matrix factorization (NMF) [47]. The strong connection between MV HU and NMF HU has been actively exploited in the last decade by the signal processing and remote-sensing research communities, yielding a number of new NMF-based algorithms and theoretical results.

Most of the methods adopting the above formulation implement variations of the nonlinear block Gauss–Seidel iterative scheme minimizing (5.20) successively with respect to A and to S. This is the case of *iterative constrained endmembers* (ICE) algorithm [48] and of the *minimum volume transform non-negative matrix factorization* (MVC-NMF) [49], whose main differences are related with the way they define the regularizer $\operatorname{vol}(A)$. For variations of these ideas recently introduced, see [1]. The *sparsity-promoting ICE* (SPICE) [50] is an extension of the ICE algorithm that incorporates sparsity-promoting priors aiming at finding the number of endmembers.

Problem (5.20) is non-convex. Thus the solutions provided by greedy solvers are strongly dependent on the initialization. This handicap was mitigated in the minimum volume simplex analysis (MVSA) [51, 52], the *simplex identification via variable splitting and augmented Lagrangian* (SISAL) [53], the *minimum volume enclosing simplex* (MVES) [24], the *robust minimum volume enclosing simplex* (RMVES) [54], and the *robust minimum volume simplex analysis* (RMVSA) [52], by reformulating (5.20) with respect to A^{-1} instead of A.

5.4.3.3 Highly Mixed Datasets. Statistical Inference

The MV simplex shown in Figure 5.9c is strictly contained in the true one. This situation corresponds to a *highly mixed* dataset where there are no spectral vectors over, or near, some facets. For these classes of problems, the MV algorithms fail and we usually resort to the statistical framework, formulating HU as a statistical inference problem, usually adopting the Bayesian paradigm.

In the Bayesian approaches, formulated under the LMM, the posterior distribution of the parameters of interest is computed from the linear observation model (5.1) within a hierarchical Bayesian model, where conjugate prior distributions are chosen for some unknown parameters to account for physical constraints. The hyperparameters involved in the definition of the parameter priors are then assigned non-informative priors. Due to the complexity in obtaining closed-form expression for the posterior density, the parameters of interest, namely the mixing matrix and the fractional abundances, are, often, estimated from samples of the posterior density generated with Markov chain Monte Carlo (MCMC) techniques. See, e.g. [1, 55] and references therein.

A clear illustration of the potential of the Bayesian approach to cope with highly mixed data sets is provided by the DECA algorithm [18]; it models the abundance fractions as mixtures of Dirichlet densities. A cyclic minimization algorithm is developed where: (i) the number of Dirichlet modes is inferred based on the minimum description length (MDL) principle; (ii) a generalized expectation maximization (GEM) algorithm is derived to infer the model parameters.

Finally, we note that most of the matrix factorization methods referred to in Section 5.4.3.2 may also be formulated as Bayesian inference problems, with the advantage of attaching meaning to the model parameters and providing a principled framework to deal with them.

5.4.3.4 Hyperspectral Unmixing Through Sparse Regression (SR)

Hyperspectral unmixing via SR has recently been introduced with the objective of coping with highly mixed datasets, that is datasets lacking pure pixels or pixels on the facets of the simplex associated with the mixing matrix, such that MV methods may not be applied. In the SR formulation, it is assumed that the measured spectral vectors can be expressed as linear combinations of a small number of pure spectral signatures known in advance (e.g. spectra collected on the ground by a field spectro-radiometer) [8]. Unmixing then amounts to find the optimal subset of signatures in a (potentially very large) spectral library (*dictionary* in the SR jargon) that can best model each mixed pixel in the scene. In practice, this is a combinatorial problem, which calls for efficient linear SR techniques based on sparsity-inducing regularizers. Linear SR is an area of very active research with strong links to compressed sensing [56].

Works [8, 57] were seminal in sparse HU using spectral libraries. They introduced various pixel-based convex formulations for sparse HU and studied the viability of this new approach to HU. SR HU has attracted the attention of the researchers working in hyperspectral imaging. The current research efforts are mostly focused on diverse forms of structured sparsity [58, 59], and sensor array techniques to remove non-active signatures from the spectral libraries [60, 61].

5.4.3.5 Synopsis of the Linear HU Problems

Figure 5.10 shows a taxonomy of the linear HU problems, motivated by the spatial distribution of the dataset spectral samples. The sections where each class of problems are addressed are also indicated.

There are two main classes of problems, depending on the distribution of the dataset spectral vectors in the simplex corresponding to the endmember matrix $\boldsymbol{A}$, sometimes termed *the true simplex*: (i) identifiable mixtures, in which the datasets contain samples in all vertices of the simplex (i.e. pure pixels) or samples in the facets of the simplex from which the simplex may be inferred; (ii) non-identifiable mixtures, usually corresponding to highly mixed scenarios, in which the datasets do not contain enough samples neither in the vertices nor in the facets of the simplex allowing to infer the simplex.

We term HU problems involving identifiable mixtures as *blind*, since the unmixing may be carried out in an unsupervised way, whereas the non-identifiable HU problems are termed *semi-blind*, as the unmixing cannot be carried out without some sort of additional information.

As shown in Figure 5.10, the methods conceived for identifiable mixtures are developed in Section 5.4.3.1, named *Pure Pixel Pursuit*, and in Section 5.6, named *Minimum Volume Simplex Estimation*. The name "Pure Pixel Pursuit" is linked to the fact that the endmembers are searched for among the observed spectral samples, where the name "Minimum Volume Simplex Estimation" is linked to the fact that the endmembers are estimated by fitting a minimum volume simplex to the dataset. In the case of non-identifiable mixtures, we only address, in Section 5.5, sparse-based regression methods, which seek an optimal subset of signatures in a (potentially very large) spectral library that can best model each mixed pixel in the scene.

At this point, we would like to make a historical remark about the blind HU methods conceived for identifiable mixtures. We have previously

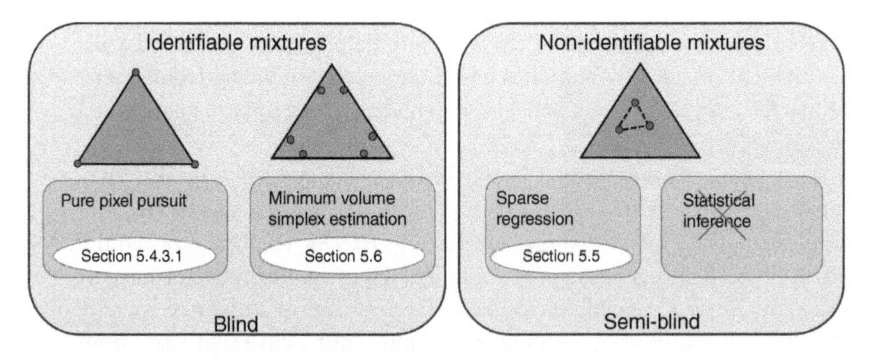

Figure 5.10 Taxonomy of the linear HU problems.

mentioned that blind HU may be easily handled under the pure pixel assumption. The pure pixel concept actually came from the study of CG of hyperspectral signals, where remote-sensing researchers examined the special geometric structure of hyperspectral signals and looked for automatic methods for endmember determination, that is, blind HU.

In fact, a vast majority of blind HU developments, if not all, are directly or intuitively related to concepts introduced in early CG studies, such as simplex volume minimization by Craig [13], simplex volume maximization by Winter [16], and the pure pixel search by Boardman *et al.* [15]. We give a historical review in the box "who discovered convex geometry (CG) for blind unmixing?"

Who Discovered Convex Geometry (CG) for Blind Unmixing?

In geoscience and remote sensing, the work by Craig in the early 1990s [12, 13] is widely recognized to be most seminal in introducing the notion of CG for hyperspectral signal analysis and unmixing. Craig's original work not only described simplex volume minimization, which turns out to become a key CG concept for blind HU, it also inspired other pioneers, such as Boardman who made notable early contributions to CG-based blind HU [14] and introduced pure pixel search [15]; and Winter who proposed the simplex volume maximization concept [16] which results in the popularized N-FINDR algorithm class. What is remarkable in these early studies is that they discovered such beautiful blind SP concepts through sharp empirical observations and strong intuitions, rather than through rigorous SP or mathematics.

Convex geometry is also an idea that has been discovered several times in different areas. The introduction of CG can be traced back to as early as 1964 by Imbrie and Van Andel [62]. The work by Imbrie and Van Andel belongs to another branch of geoscience studies wherein CG is used for analysis of compositional data in Earth science, such as mineral assemblages, grain-size distribution data, and geochemical and petrological data; see [63] for an overview. In fact, Imbrie's Q-mode analysis and the subsequent QMODEL by Klovan and Miesch [64] are conceptually identical to vertex or pure pixel search, although the methodology is different. Likewise, Full *et. al* already considered the same simplex volume minimization principle as Craig's in the 1980s [65]. CG has also been independently discovered in other fields such as chemometrics [46] and signal processing [66, 67].

(Continued)

(Continued)

In all the discoveries or rediscoveries mentioned above, the driving force that led researchers on different backgrounds to devise the same idea seems to be with the geometric elegance of CG and its powerful implications on solving blind unmixing problems.

5.5 Dictionary-Based Semiblind HU

This section describes semiblind HU by SR.

5.5.1 Sparse Regression

When performing blind HU, we generally assume no information on the spectral shapes of the true endmember signatures. The latter is not totally true. In geoscience and remote sensing, a tremendous amount of efforts has been spent on measuring and recording spectral samples of many different materials, which has resulted in spectral libraries for various research purposes. For example, the U.S. Geological Survey (USGS) library, which has taken over 20 years to assemble, contains more than 1300 spectral samples covering materials such as minerals, rocks, liquids, artificial materials, vegetations, and even microorganisms [68]. Such valuable knowledge base can be turned to blind HU purposes, or more precisely, semiblind HU.

A slight abuse of notations is required to explain the semiblind formulation. We redefine $\boldsymbol{A} = [\,\boldsymbol{a}_1, \dots, \boldsymbol{a}_K] \in \mathbb{R}^{M \times K}$ as a dictionary of K hyperspectral samples, where each $\boldsymbol{a}_i$ corresponds to one material (each $\boldsymbol{a}_i$ is also assumed to have been appropriately processed, e.g. atmospherically compensated). We assume that the dictionary $\boldsymbol{A}$ is known, obtained from an available spectral library, and that the true endmembers in each measured pixel $\boldsymbol{x}_t$ are covered by the dictionary. The measured pixels in the noiseless case (again, for tutorial purposes) can then be represented by

$$\boldsymbol{x}_t = \sum_{i \in S_n} \boldsymbol{a}_i s_{it}, \tag{5.21}$$

where $S_n \subseteq \{1, \dots, K\}$ is an index subset that indicates the materials present in the measured pixel $\boldsymbol{x}_t$, and $s_{it} > 0, i \in S_n$ are the corresponding abundances. In this representation, note that the sum-to-one constraint $\sum_{i \in S_n} s_{it} = 1$ may not hold; the measurement conditions of library samples and the actual scene are often different and this can introduce scaling inconsistencies between the library samples and true endmembers. By also letting $s_{it} = 0$ for all $i \notin S_n$ (5.21), can be written as

$$\boldsymbol{x}_t = \boldsymbol{A}\boldsymbol{s}_t, \tag{5.22}$$

where $\boldsymbol{s}_t = [\, s_{1t}, \ldots, s_{Kt} \,]^{\mathsf{T}} \in \mathbb{R}^K$ is now a *sparse* abundance vector. The problem now is to recover $\boldsymbol{s}_t$ from $\boldsymbol{x}_t$. This is not trivial because we often have $K > M$ and the corresponding system in (5.22) is underdetermined. However, we know beforehand that $\boldsymbol{s}_t$ have only a few non-zero components, since the number of materials present in one pixel is often very small, typically within 5 [1]. Hence, a natural formulation for the semiblind HU problem is to *find the sparsest* $\boldsymbol{s}_t$ *for the representation in (*5.22*).* This inference problem turns out to be identical to that investigated in compressive sensing (CS), where the objective is to recover a sparse representation of a signal on a given frame from compressive measurements [69]. This connection allows us to capitalize on the wealth of theoretical and algorithmic results available in the CS area.

The SR problem we describe above can be formulated as

$$\begin{aligned} &\min_{\boldsymbol{s}_t} \|\boldsymbol{s}_t\|_0 \\ &\text{s.t. } \boldsymbol{x}_t = \boldsymbol{A}\boldsymbol{s}_t, \end{aligned} \tag{5.23}$$

for each $n = 1, \ldots, T$, where $\|\boldsymbol{s}_t\|_0$ denotes the number of nonzero elements in $\boldsymbol{s}_t$. The above SR problem possesses provably good endmember identifiability. Specifically, problem (5.23) is known to have a unique solution if the true sparse abundance vector $\boldsymbol{s}_t$ satisfies

$$\|\boldsymbol{s}_t\|_0 < \frac{1}{2}\,\mathrm{spark}(\boldsymbol{A}), \tag{5.24}$$

where spark($\boldsymbol{A}$) is the smallest number of linearly dependent columns of $\boldsymbol{A}$ [70]. Since every $\boldsymbol{s}_t$ is highly sparse by nature (5.24), should hold in practice. The consequent implication is meaningful – the SR problem (5.23) can perfectly identify all the true endmembers in general.

While the SR approach sounds promising, there are challenges. Since problem (5.23) is NP-hard in general, it is natural to seek approximate solutions. Let us consider the popular ℓ_1 relaxation solution to problem (5.23):

$$\begin{aligned} &\min_{\boldsymbol{s}_t} \|\boldsymbol{s}_t\|_1 \\ &\text{s.t. } \boldsymbol{x}_t = \boldsymbol{A}\boldsymbol{s}_t, \end{aligned} \tag{5.25}$$

which is convex and has efficient solvers. The CS literature has a series of analysis results telling when problem (5.25) gives the same solution as problem (5.23), or simply sufficient conditions for exact recovery.

Those sufficient conditions usually depend on the conditioning of $\boldsymbol{A}$. For example, one sufficient exact recovery condition for problem (5.25) is $\|\boldsymbol{s}_t\|_0 < \frac{1}{2}(1 + \mu^{-1}(\boldsymbol{A}))$, where

$$\mu(\boldsymbol{A}) = \max_{1 \leq i,j \leq K,\, i \neq j} \frac{|\boldsymbol{a}_i^T \boldsymbol{a}_j|}{\|\boldsymbol{a}_i\|_2 \|\boldsymbol{a}_j\|_2}, \tag{5.26}$$

is called the mutual coherence of $\boldsymbol{A}$ [70]. Unfortunately, spectral libraries in practice are strongly correlated, yielding $\mu(\boldsymbol{A})$ almost being one [8]. A similar issue also occurs in other sufficient conditions, namely in the restricted isometry property (RIP) [69]. Thus, one may not obtain a desirable SR solution from a straight ℓ_1 relaxation application.

However, all is not lost. Recall that every $\boldsymbol{s}_t$ is, by nature, non-negative. Let us consider a *non-negative* ℓ_1 relaxation problem, which is problem (5.25) plus the non-negative constraint $\boldsymbol{s}_t \geqslant \boldsymbol{0}$. As it turns out, exploiting non-negativity helps a lot. There is a large amount of experimental evidence that indicates that non-negative ℓ_1 relaxation can yield useful unmixing results [1, 8, 71]. Also, non-negative ℓ_1 relaxation is theoretically proven to be able to give rather sparse solutions for certain classes of $\boldsymbol{A}$ [72, 73]. Although the above-noted theoretical result does not give a direct answer to exact recovery under highly correlated libraries, it gives good insight on the capability of non-negative ℓ_1 relaxation.

We can also combat the spectral library mutual coherence issue by using the multiple-measurement vector (MMV) formulation [74], which exploits the fact that in a given data set all the spectral vectors are generated by the same subset of library signatures, corresponding to the endmember signatures. Let $\boldsymbol{S} = [\ \boldsymbol{s}_1, \ldots, \boldsymbol{s}_T\ ] \in \mathbb{R}^{K \times T}$ and $\boldsymbol{X} = [\ \boldsymbol{x}_1, \ldots, \boldsymbol{x}_T\ ] \in \mathbb{R}^{M \times T}$, so that we can write $\boldsymbol{X} = \boldsymbol{A}\boldsymbol{S}$. Also, define $\|\boldsymbol{S}\|_{\text{row-0}}$ to be the number of nonzero rows in $\boldsymbol{S}$; i.e. $\|\boldsymbol{S}\|_{\text{row-0}} = |\text{rowsupp}(\boldsymbol{S})|$, $\text{rowsupp}(\boldsymbol{S}) = \{1 \leq i \leq K \mid \boldsymbol{s}^i \neq \boldsymbol{0}\}$, and $\boldsymbol{s}^i$ denotes the i-th row of $\boldsymbol{S}$. We consider a collaborative SR (CSR) problem [58]

$$\begin{aligned} &\min_{\boldsymbol{S}} \|\boldsymbol{S}\|_{\text{row-0}} \\ &\text{s.t. } \boldsymbol{X} = \boldsymbol{A}\boldsymbol{S}, \end{aligned} \tag{5.27}$$

where the rationale is to use the whole set of measured pixels, rather than one, to strengthen SR performance. It is interesting to note that $\|\boldsymbol{S}\|_{\text{row-0}}$ also represents the number of endmembers. Like the previous SR problem, we can apply a convex relaxation to CSR by replacing $\|\boldsymbol{S}\|_{\text{row-0}}$ in (5.27) by $\|\boldsymbol{S}\|_{2,1}$, where $\|\boldsymbol{S}\|_{p,q} = (\sum_{i=1}^{K} \|\boldsymbol{s}^i\|_p^q)^{1/q}$. In theory, there is no extra benefit in using the CSR or MMV formulation in the worst-case sense (think about a special and rather unrealistic case where $\boldsymbol{s}_1 = \cdots = \boldsymbol{s}_T$) [74]. However, an average analysis in [75] gives an implication that increasing the number of measurements (or pixels here) can significantly reduce the probability of recovery failure. In practice, this has been found to be so. Also, the non-negativity constraint $\boldsymbol{S} \geqslant \boldsymbol{0}$ can be incorporated in problem (5.27) to improve performance.

A practical SR or CSR solution should also cater for the presence of noise. For CSR, the following alternative convex relaxation formulation may be used to provide HU [58]

$$\min_{\boldsymbol{S} \geqslant \boldsymbol{0}} \ \|\boldsymbol{X} - \boldsymbol{A}\boldsymbol{S}\|_F^2 + \lambda \|\boldsymbol{S}\|_{2,1}, \tag{5.28}$$

for some constant $\lambda > 0$. The rationale is to seek an LS data fitting, rather than exact, with a sparse-promoting regularizer $\lambda\|\boldsymbol{S}\|_{2,1}$. It is important to note that while problem (5.28) is convex, it is a large-scale optimization problem.

A current state-of-the-art solver for problem (5.28) is the *collaborative sparse unmixing by variable splitting and augmented Lagrangian* (CLSUnSAL) provided in [58]. CLSUnSAL is an instance of the alternating direction method of multipliers (ADMM) methodology introduced in [76], termed CSALSA, conceived to solve optimization problems with and arbitrary number of convex terms.

The high spatial coherence of hyperspectral images has also been exploited to combat the spectral library mutual coherence issue. The SUnSAL-TV criterion and solver introduced in [77] exploits the contextual spatial information, by solving the optimization

$$\min_{\boldsymbol{S} \geqslant \boldsymbol{0}} \ \frac{1}{2}\|\boldsymbol{X} - \boldsymbol{A}\boldsymbol{S}\|_F^2 + \lambda \|\boldsymbol{S}\|_{1,1} + \lambda_{\mathrm{TV}} \ \mathrm{TV}(\boldsymbol{S}), \tag{5.29}$$

where $\mathrm{TV}(\boldsymbol{S})$ is the non-isotropic vector total variation of $\boldsymbol{S}$, which promotes piece-smooth abundance vectors $\boldsymbol{s}_t$, for $n = 1, \ldots, T$. As CLSUnSAL, SUnSAL-TV solver is implemented with CSALSA.

At this point readers may be wondering: How do we compare SR- and CG-based solutions? Simply speaking, CG relies on exploitation of simplex structures, while SR does not. To illustrate, consider the previous numerical example in Figure 5.11. In Figure 5.11c, we generated a heavily mixed (and noiseless) scenario where data do not possess simplex structures expected in CG. It is seen that even VolMin fails in this scenario. However, CSR, which was run under the USGS library with 498 spectral signatures, is seen to be able to identify the true endmembers perfectly. Note that the true endmember signatures were taken from the same library, which makes the setting slightly ideal. It would not be too surprising that if the library fails to cover all true endmember signatures (e.g. a new material), then SR solutions would fail. For further numerical results and real-data experiments, see [1, 8, 58]–[60].

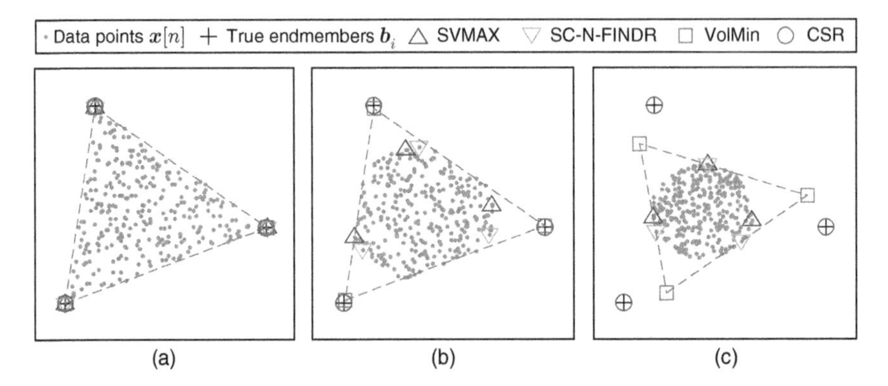

Figure 5.11 Numerical comparison of VolMax, VolMin, and sparse regression solutions. (a) Existence of pure spectral vectors, (b) absence of pure spectral vectors but sufficiently well spread pixels, and (c) absence of pure spectral vectors and a highly mixed scenario. Source: Wing-Kin Ma et al., [2]/IEEE.

5.5.2 Sensor Array Processing Meets Semiblind HU

MMV is a powerful concept which has been applied to estimation problems in statistical SP and sensor array processing [78]. Curiously, a classical concept originated from sensor array processing, namely, subspace methods, also finds its way to MMV research [79] – this provides yet another opportunity for semiblind HU [60].

The idea is simple for readers who are familiar with subspace methods or sensor array processing; or, see classical literature such as [80]. Consider the block model $\boldsymbol{X} = \boldsymbol{AS}$ (again, assuming no noise). Let $S = \text{rowsupp}(\boldsymbol{S})$ be the set of indices of active materials in the measured data $\boldsymbol{X}$, and $\boldsymbol{A}_S$ be a submatrix of $\boldsymbol{A}$ whose columns are $\{\boldsymbol{a}_i\}_{i\in S}$. Note that $\boldsymbol{A}_S$ is the true endmember matrix. Let us assume that $\{\boldsymbol{s}^i\}_{i\in S}$, the set of true abundance maps, is linearly independent; in practice this refers to situations where the abundance maps are sufficiently different. Then, one can easily deduce that $\text{range}\{\boldsymbol{X}\} = \text{range}\{\boldsymbol{A}_S\}$. The above expression implies that

$$\boldsymbol{P}_{\boldsymbol{X}}^{\perp}\boldsymbol{a}_k = \boldsymbol{0} \Longleftrightarrow k \in S, \tag{5.30}$$

for all $1 \leq k \leq K$, as far as $\{\boldsymbol{a}_k\} \cup \{\boldsymbol{a}_i\}_{i\in S\setminus\{k\}}$ is linearly independent for any $1 \leq k \leq K$. Since the latter holds for $|S| + 1 < \text{spark}(\boldsymbol{A})$, we have the following endmember identifiability condition for (5.30):

$$\|\boldsymbol{S}\|_{\text{row-0}} < \text{spark}(\boldsymbol{A}) - 1. \tag{5.31}$$

Remarkably, with the mild assumption of linear independence of $\{\boldsymbol{s}^i\}_{i\in S}$, we can achieve such provably good endmember identifiability by the simple subspace projection in (5.30).

In practice, the identification in (5.30) can be implemented by the classical multiple signal classification (MUSIC) method [80]; see [60] for implementation details.

Figure 5.11a shows a scenario where the pure pixel assumption holds. We see that both VolMax (via SVMAX or SC-N-FINDR) and VolMin perfectly identify the true endmembers. Figure 5.11b shows another scenario where pure pixels are missing. VolMax is seen to fail, while VolMin can still give accurate endmember estimates. Figure 5.11c shows a highly mixed scenario where both VolMax and VolMin fail, but the spare regression-based solution is accurate.

5.5.3 Further Discussion

There are a few more points to note.

(1) As a side advantage, the SR approach does not require knowledge of the number of endmembers N. Note that this does not apply to the subspace approach, which often requires knowledge of N to construct subspace projections.

(2) An interesting, but also elusive question is whether a given dictionary can truly cover the true endmembers. From an end user's viewpoint, it depends on the scene and whether one can preselect a reliable library for that scene specifically. Moreover, there are concurrent studies that consider learning the dictionary from the data, thereby circumventing these issues [71, 81, 82]. Dictionary learning is an active research topic. It is also related to NMF, to be described in Section 5.6.2. In addition, there has been interest in using the measured data $\boldsymbol{X}$ itself as the dictionary for MMV [83]. This self-dictionary MMV (SD-MMV) approach is related to pure pixel search. For example, both SPA and VCA can be derived from SD-MMV [84, 85].

5.6 Minimum Volume Simplex Estimation

As already discussed in Section 5.4.3.2, there are scenarios in which the pure pixel assumptions does not hold owing to, namely, highly mixed materials or low spatial resolution hyperspectral cameras. In this case, the use of pure pixel algorithms yields poor results.

Assuming that the dataset contains spectral vectors on each facet, as schematized in Figure 5.9b, we may try to find the endmembers by finding the the smallest simplex, i.e. the MV simplex, that contains the observed spectral vectors. The intuition underlying this criterion is that, if we have "enough" samples in the facets of the simplex not "far away" from the vertexes of the true simplex, then the MV simplex and the true one coincide.

In a real application with R materials, it is very likely that most pixels contain only a subset of the R materials. It is therefore likely that most pixels live in the boundary of the simplex, and hence it is conceivable that adjusting an MV simplex to the data yields good approximations of the true simplex. The fundamental question here is to know under what conditions the MV-based blind HU yields useful estimates. That is, what does it mean to have *enough samples in the facets not far away from the vertexes of the true simplex*? This identifiability question was recently addressed [86], in the noiseless case. Below we sketch the main result proved therein.

The MV criterion may be formulated as

$$\begin{aligned} &\min_{\boldsymbol{A},\boldsymbol{S}} \ \text{vol}(\boldsymbol{A}) \\ &\text{s.t.}\ \boldsymbol{X} = \boldsymbol{A}\boldsymbol{S}, \quad \boldsymbol{S} \in \mathcal{S}^T, \end{aligned} \tag{5.32}$$

or equivalently

$$\begin{aligned} &\min_{\boldsymbol{A}} \ \text{vol}(\boldsymbol{A}) \\ &\text{s.t.}\ \boldsymbol{x}_t \in \text{conv}\{\boldsymbol{a}_1, \dots, \boldsymbol{a}_R\}, \quad t = 1, \dots, T, \end{aligned} \tag{5.33}$$

which corresponds to the volume regularized matrix factorization optimization (5.20) when $\lambda \to \infty$ (equivalently, the noise power tends to zero). We call problems (5.32) and (5.33) *VolMin* for convenience. Figure 5.12a illustrates the concept of VolMin with pixels in the simplex facets defining the

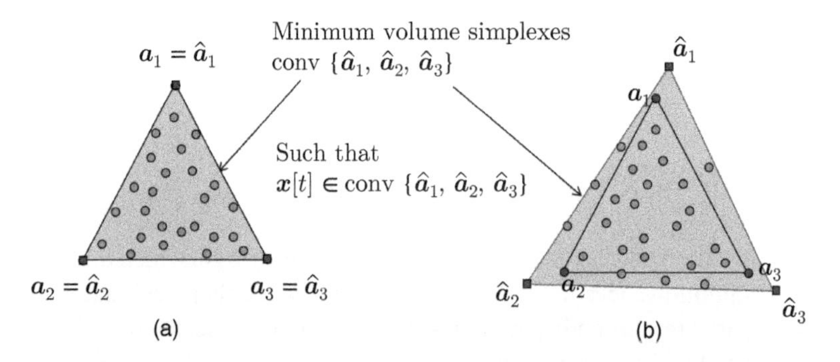

Figure 5.12 Simplex volume minimization. Light gray circles, dark gray circles, and dark gray squares represent, respectively, observed spectral vectors, endmembers, and estimated endmembers. (a) noiseless dataset without pure pixels; (b) noisy dataset without pure pixels.

true simplex, and Figure 5.12b illustrates a noisy scenario allowing only to infer an approximation of the true simplex.

Below, we present a result on MV identifiability recently proved in [86]. Let us denote $\mathbf{s}[t] = \mathbf{s}_t$. The result is as follows.

α-Pure Pixel Assumption

Assume that for every $i, j \in \{1, \dots, R\}$, $i \neq j$, there exists an index $n(i, j)$ such that

$$\mathbf{s}[n(i,j)] = \alpha_{ij}\mathbf{e}_i + (1 - \alpha_{ij})\mathbf{e}_j, \tag{5.34}$$

for some $\frac{1}{2} < \alpha_{ij} \leq 1$. Figure 5.13 schematizes a scenario with $R = 3$ and where the above assumption holds true.

VolMin Identifiability [86]

In the noiseless case and under the above assumption, the solution of the volMin optimization (5.32) is uniquely and exactly given by $[\mathbf{a}_1, \dots, \mathbf{a}_R]$ subject to ordering permutations if

1. $R = 3$ and $\alpha_{ij} > \dfrac{2}{3}$ for all i, j, or if
2. $R \geqslant 4$.

See [86] for a proof and for more details.

The above sufficient identifiability condition is restrictive in that it assumes that, for every two materials $i, j \in \{1, \dots, R\}$, there are pixels in the dataset satisfying (5.34). In practice, it has been systematically observed in a large array of numerical experiments that VolMin optimization provides useful results when for each of the $R - 1$ endmembers defining a facet there is a sample in that facet "close" thereof.

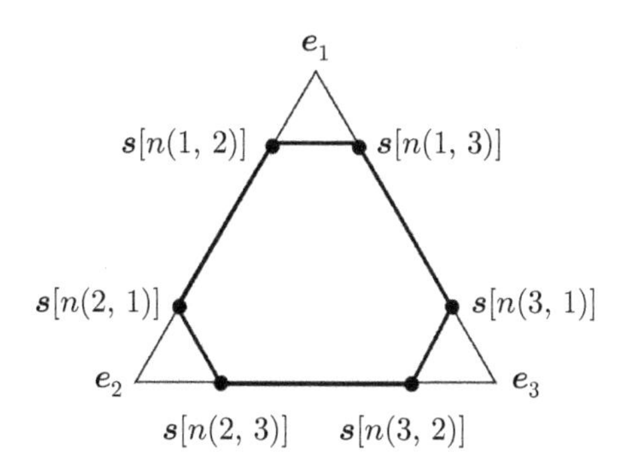

Figure 5.13 Illustration of abundances $\mathbf{s}[n(i,j)]$ satisfying the α-Pure Pixel Assumption (5.34), for $R = 3$ and with $2/3 < \alpha_{ij} \leq 1$. Source: Lin et al., [86]/IEEE.

5.6.1 VolMin Optimization

We now turn our attention to the simplex volume minimization (5.32), or simply VolMin, which was first pursued by Craig [13] and Boardman [14] in the blind HU context. VolMin uses a different approach to perform simplex fitting. Instead of using measured pixels as the endmember estimates, which is the case in VolMax, VolMin generates a set of "virtual" endmembers that do not belong to the measured pixels, and that can best cover the measured pixels. It performs simplex fitting by finding a simplex that encloses all the measured pixels, while yielding the minimum volume, as illustrated in Figures 5.12a and b. Consequently, VolMin can produce endmember estimates that are not measured pixels or their convex combinations, which is the case in VolMax.

Assuming that DR has been applied, such that $\boldsymbol{A} \in \mathbb{R}^{R\times R}$, and, denoting $\boldsymbol{Q} = \boldsymbol{A}^{-1}$, the optimization (5.32) may be reformulated as

$$\begin{aligned} &\max_{\boldsymbol{Q}} \ |\det(\boldsymbol{Q})| \\ &\text{s.t. } \boldsymbol{QX} \geqslant \boldsymbol{0}, \ \ \boldsymbol{1}^{\mathsf{T}}\boldsymbol{QX} = \boldsymbol{1}_T^{\mathsf{T}}, \end{aligned} \tag{5.35}$$

which is equivalent to

$$\begin{aligned} &\min_{\boldsymbol{Q}} \ -\log|\det(\boldsymbol{Q})| \\ &\text{s.t. } \boldsymbol{QX} \geqslant \boldsymbol{0}, \ \ \boldsymbol{1}^{\mathsf{T}}\boldsymbol{Q} = \boldsymbol{a}^{\mathsf{T}}, \end{aligned} \tag{5.36}$$

where $\boldsymbol{a}^{\mathsf{T}} = \boldsymbol{X}^{\mathsf{T}}(\boldsymbol{XX}^{\mathsf{T}})^{-1}$ (see [51, 53] for details). We remark that, although optimizations (5.35) and (5.36) are equivalent, the latter has much less equality constraints, which may have numerical advantages.

Another formulation of VolMin is as follows. Consider the LMM (5.16) after the DR in (5.15). The VolMin problem in (5.32) may be rewritten as:

$$\begin{aligned} &\min_{\boldsymbol{F}} \ \text{vol}(\boldsymbol{F}) \\ &\text{s.t. } \boldsymbol{z}_t \in \text{conv}\{\boldsymbol{f}_1, \ldots, \boldsymbol{f}_R\}, \quad t = 1, \ldots, T, \end{aligned} \tag{5.37}$$

We consider transforming the simplex to a polyhedron (see, e.g., [37, pp. 32–33]). To help understand the idea, an illustration is given in Figure 5.14. We see that a simplex can be equivalently represented by an intersection of halfspaces, i.e., a polyhedron. More precisely, the following equivalence holds for an affinely independent $\{\boldsymbol{b}_1, \ldots, \boldsymbol{b}_R\}$ [24]

$$\boldsymbol{v}_t \in \text{conv}\{\boldsymbol{b}_1, \ldots, \boldsymbol{b}_R\} \Longleftrightarrow \boldsymbol{Hv}_t - \boldsymbol{g} \geqslant \boldsymbol{0}, (\boldsymbol{Hv}_t - \boldsymbol{g})^{\mathsf{T}}\boldsymbol{1} \leq 1, \tag{5.38}$$

where the right hand side is a polyhedron, and

$$\boldsymbol{H} = [\, \boldsymbol{b}_1 - \boldsymbol{b}_R, \ldots, \boldsymbol{b}_{R-1} - \boldsymbol{b}_R \,]^{-1}, \quad \boldsymbol{g} = \boldsymbol{Hb}_R. \tag{5.39}$$

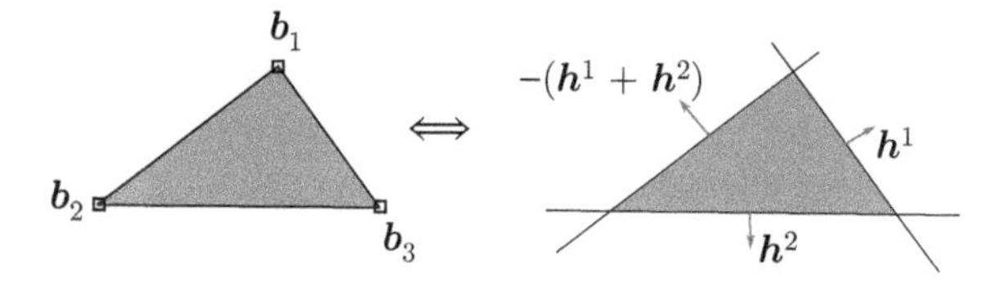

Figure 5.14 Transformation of a simplex to a polyhedron. Source: Wing-Kin Ma et al., [2]/IEEE.

By the change of variables in (5.39), and noting (5.38), we can recast problem (5.37) as

$$\max_{\boldsymbol{H},\mathbf{g}} \ |\det(\boldsymbol{H})|$$
$$\text{s.t. } \boldsymbol{H}\boldsymbol{z}_t - \mathbf{g} \geqslant \mathbf{0}, (\boldsymbol{H}\boldsymbol{z}_t - \mathbf{g})^{\mathsf{T}}\mathbf{1} \leq 1, \quad t = 1, \ldots, T. \tag{5.40}$$

The equivalent VolMin problem in (5.40) is arguably easier to handle than the original in (5.37). Specifically, the constraints in problem (5.40), which forms a data-enclosing polyhedron, are linear (and convex).

Work [24] introduces the VolMin criterion (5.40), jointly with the *minimum volume enclosing simplex* (MVES) algorithm to solve it. Work [51] introduces the VolMin criterion (5.36), jointly with the *minimum volume simplex analysis* (MVSA) to solve it. Work [51] introduces the VolMin criterion (5.36), jointly with the *minimum volume simplex analysis* (MVSA) to solve it. In both optimizations, the constraint sets are convex polytopes. However, the functions $-\log|\det(\boldsymbol{Q})|$ and $-|\det(\boldsymbol{Q})|$, for a generic matrix $\boldsymbol{Q}$, are nonconvex, yielding nonconvex optimizations. MVES and MVSA tackle this issue by successive convex approximation. MVES solves a sequence of linear subproblems by performing row-by-row optimization with respect to $\boldsymbol{Q}$, whereas MVSA solves a sequence of quadratically constrained subproblems. In both cases, the convex subproblems are solved effectively and efficiently using available solvers. In practical terms, it has been observed that VolMin is robust against lack of pure pixels, and the algorithms to solve (5.32) are faster and less prone to get stuck in poor local minima than those based on NMF. See [24, 51, 87] for the details and comparisons.

Readers are referred to [1, 18, 24, 38, 51, 53, 54] for more numerical comparisons and real-data experiments.

VolMin is generally recognized as a more powerful approach than VolMax, although VolMin requires numerical optimization and thus is more expensive to implement than VolMax. To support the former argument, we give a numerical result in Figure 5.16. The measured pixels are synthetically generated, following the same simulation setting as in [54]. The number of endmembers is $N = 8$. The factor ρ in the abscissa describes the pixel purity level. In short, $\rho = 1$ means that the pure pixel assumption holds; small ρ

means that all measured pixels are far away from the true endmembers. We see that in the noiseless case, VolMin via MVES shows very high endmember estimation accuracies even without pure pixels, while VolMax via SVMAX or SC-N-FINDR does not except for the pure-pixel case.

Contrarily to the lack of pure pixels, the presence of outliers and noise may affect significantly the solutions of VolMin. Figure 5.15a illustrates the effect of an outlier and Figure 5.15b illustrates the effect of noise. In the former case, a single outlier lying away from the true simplex yields a large error in the estimates of $\boldsymbol{a}_1$. In the latter case, the presence of noise in the samples lying close to the facets of the simplex degrades the estimate of the mixing matrix, mainly when the simplex is close to be degenerated (mixing matrix has a large condition number), a situation often seen in HU, where the angle between endmembers tends to be very small.

In the recent years, a number of robust MinVol algorithms have been developed with the objective of introducing robustness to outliers and noise. Three representative examples are the *simplex identification via split augmented Lagrangian* (SISAL) [53], the *robust minimum volume simplex analysis* (RMVES) [54], and the *robust minimum volume simplex analysis* (RMVSA) [52].

SISAL replaces the positivity constraints $\boldsymbol{QX} \geqslant \boldsymbol{0}$ in (5.36), forcing the abundances to be non-negative, by soft constraints enforced with the so-called hinge function. In this way, a few samples may have negative abundances and therefore lying outside the estimated simplex. The obtained problem is solved by a sequence of augmented Lagrangian method of multiplier (ADMM) optimizations. The resulting algorithm is very fast and able to solve problems far beyond the reach of the current state-of-the art algorithms [53].

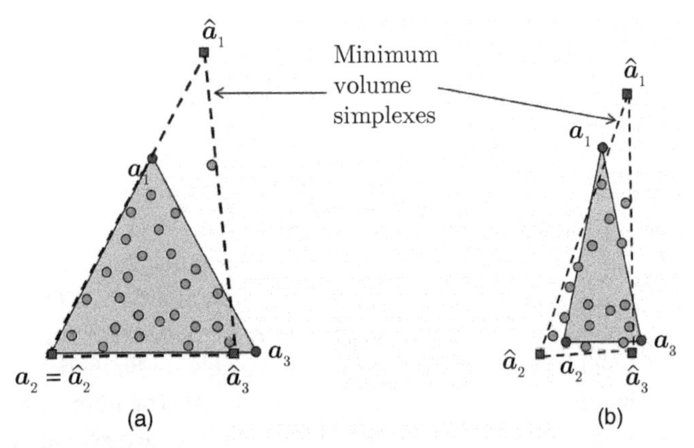

Figure 5.15 MV simplex estimation: (a) impact of an outlier; (b) impact of noise.

RMVES and RMVSA replace the constraints in (5.35) and (5.36) with chance constraints. These chance constraints in turn control the volume of the resulting simplex. Under the Gaussian noise assumption, the chance-constrained MVES problem can be formulated into a deterministic nonlinear program, which is handled by alternating optimization, in which each subproblem involved is handled by using sequential quadratic programming solvers. The difference RMVES and RMVSA concerns the way the quadratic programming is solved being faster in RMVSA. See [52, 54] for further details.

We complete this subsection with the following comments.

(1) As mentioned above, numerical evidence suggests that VolMin may be able to identify the true endmembers accurately in the absence of pure pixels. By analysis, it is known that in the noiseless case, the optimal solution of VolMin is uniquely the true endmembers' signatures *if* the *α-pure pixel assumption* holds (see page 229). This result extends the uniqueness under the *pure pixel assumption* proved in [24]. A proof for the no pure pixel case in the presence of noise is currently unavailable and is an open fundamental question.
(2) While VolMin is deterministic and geometric-based, it has a dual identity in stochastic maximum-likelihood (ML) estimation. Specifically, consider the noiseless case, and assume that every abundance vector $\mathbf{s}_t$ is i.i.d. uniformly distributed on the support of unit simplex $\mathcal{S}$. Then, it can be shown that the corresponding ML estimator is the same as the equivalent VolMin problem in (5.36) [18]. Note that the authors in [18] also consider a generalization where the abundance prior distribution is non-uniform.

5.6.2 Non-Negative Matrix Factorization

The objective of blind HU is to find two low-rank non-negative matrix factors, namely the spectral signatures matrix and the abundance matrix, which approximates well the observed dataset matrix. This is precisely the objective of algorithms of the non-negative matrix factorization (NMF) class. This section focuses on blind HU via NMF tailored to solve the optimization problems of the similar to (5.20), which incorporate not only ANC, ASC, and volume regularization, but also other regularization terms which may improve the conditioning of the blind HU in hand.

NMF was originally proposed as a linear DR tool for analyzing environmental data [88] and for data mining applications [47]. It is posed as a low-rank matrix approximation problem where, given a data matrix

$\boldsymbol{X} \in \mathbb{R}^{M\times T}$, the task is to find a pair of non-negative matrices $\boldsymbol{A} \in \mathbb{R}^{M\times R}$, $\boldsymbol{S} \in \mathbb{R}^{R\times T}$, with $R \leq \min\{M, T\}$, that solves

$$\min_{\boldsymbol{A}\geqslant\boldsymbol{0},\boldsymbol{S}\geqslant 0} \|\boldsymbol{X} - \boldsymbol{AS}\|_F^2. \tag{5.41}$$

In blind HU, the connection is that the NMF factors obtained, $\boldsymbol{A}$ and $\boldsymbol{S}$, can serve as estimates of the endmembers and abundances, respectively. However, there are two problems here. First, problem (5.41) is NP-hard in general [89]. For this reason, optimization schemes we see in the current NMF-based blind HU developments are rather pragmatic.We should however mention that lately, there are new theory-guided NMF developments in optimization [90, 91]. Second, NMF may not guarantee solution uniqueness. This is a serious issue to the blind HU application, since it means that an NMF solution may not necessarily be the true endmembers and abundances, even in the noiseless case.

In blind HU, NMF is modified to fit the problem better. Roughly speaking, we may unify many NMF-based blind HU developments under one formulation:

$$\min_{\boldsymbol{A}\geqslant\boldsymbol{0},\boldsymbol{S}\in S^T} \|\boldsymbol{X} - \boldsymbol{AS}\|_F^2 + \lambda g(\boldsymbol{A}) + \mu h(\boldsymbol{S}), \tag{5.42}$$

where g and h are regularizers, which vary from one work to another, and $\lambda, \mu > 0$ are regularization parameters. In particular, the addition of g and h is to make problem (5.42) more well-posed through exploitation of the problem natures. Also, for the same reason, we incorporate the unit simplex constraints on $\boldsymbol{S}$.

In (5.42), matrix $\boldsymbol{A}$ represents the spectral signatures of the endmembers, thus non-negative. This implies that the dataset matrix is $\boldsymbol{X} \in \mathbb{R}^{M\times T}$ has not undergone DR and thus $M = P$. We remark, however, that formulation (5.42) may also operate on the projected data after DR. In this case, we also have the LMM model $\boldsymbol{X} = \boldsymbol{AS} + \boldsymbol{E}$, with $\boldsymbol{X} \in \mathbb{R}^{R\times T}$ and $\boldsymbol{A} \in \mathbb{R}^{R\times R}$, which represent projection coefficients with values in $\mathbb{R}$. The original endmember matrix is recovered by computing $\boldsymbol{VA}$, where $\boldsymbol{V}$ is an orthogonal matrix containing a basis for the signal subspace. Therefore the constraint $\boldsymbol{A} \geqslant \boldsymbol{0}$ in (5.42) should be replaced with $\boldsymbol{VA} \geqslant \boldsymbol{0}$,

In the literature one can find a plethora of NMF-based blind HU algorithms – each work may use different g, h, modified constraints for simpler implementations (e.g. no constraints on $\boldsymbol{A}$), and a different optimization algorithm. Our intention here is not to give an extensive coverage of all these developments. Instead, we are interested in several representative NMF-based blind HU formulations, where we will see connections between NMF, CG and SR. A summary of those formulations is shown in Table 5.2.

Table 5.2 A summary of some NMF formulations.

Algorithm	$g(A)$, $h(S)$	Optimization schemes and remarks
MVC-NMF [49]	g: $\mathrm{vol}^2(\boldsymbol{C}^\dagger(\boldsymbol{A}-\boldsymbol{d}\boldsymbol{1}^T))$ h: 0	AO + one-step projected gradient
ICE [48]	g: $\sum_{r=1}^{R-1}\sum_{r'=r+1}^{R} \\|\boldsymbol{a}_r-\boldsymbol{a}'_r\\|_2^2$ h: 0	AO; unconstrained $\boldsymbol{A}$
DL [81]	g: 0 h: $\\|\boldsymbol{S}\\|_{1,1}$	AO + one-step projected gradient for $\boldsymbol{A}$; $\boldsymbol{S}\geqslant\boldsymbol{0}$
$L_{1/2}$-NMF [92]	g: 0 h: $\\|\boldsymbol{S}\\|_{1/2,1/2}^{1/2}$	AO + multiplicative update
APS [93]	g: 0 h: $\sum_{t=1}^{T}\sum_{t'\in\mathcal{N}_t} \\|\boldsymbol{s}_t-\boldsymbol{s}'_t\\|_1$ where $\mathcal{N}(n)$ is the neighborhood pixel index set of pixel t.	AO + one-step projected subgradient
SPICE [94]	g: $\sum_{r=1}^{R-1}\sum_{r'=r+1}^{R} \\|\boldsymbol{a}_r-\boldsymbol{a}'_r\\|_2^2$ h: $\sum_{r=1}^{R}\gamma_r\\|\boldsymbol{s}_r\\|_1$	AO; unconstrained $\boldsymbol{A}$; iteratively reweighted γ_i via $\gamma_r := 1/\\|[\boldsymbol{S}^{(k-1)}]_{r,1:L}\\|_1$, $1\leq r\leq R$
CoNMF [95]	g: $\sum_{r=1}^{R}\\|\boldsymbol{a}_r-\boldsymbol{\mu}_y\\|_2^2$ $h\sum_{r=1}^{R}\\|\boldsymbol{s}_r\\|_2^p$, $0<p\leq 1$	AO + one-step majorization minimization; unconstrained $\boldsymbol{A}$

Although we see many choices with the regularizers g and h, the philosophies behind the choices follow a few core principles. For the endmember regularizer g, the principle can be traced back to VolMin in CG. A classical example is minimum volume constrained NMF (MVC-NMF) [49]

$$\min_{\boldsymbol{A}\geqslant\boldsymbol{0},\boldsymbol{S}\in S^T} \|\boldsymbol{X}-\boldsymbol{A}\boldsymbol{S}\|_F^2+\lambda\,(\mathrm{vol}(\boldsymbol{B}))^2, \tag{5.43}$$

where $\boldsymbol{B}=\boldsymbol{C}^{\mathrm{T}}(\boldsymbol{A}-\boldsymbol{b}\boldsymbol{1}_M^{\mathrm{T}})$, with $(\boldsymbol{C},\boldsymbol{b})$ defining the affine set representing $\boldsymbol{X}$ (see Section 5.4.2.1 and expression (5.17)), and $\mathrm{vol}(\boldsymbol{B})$ standing for a volume of the generalized simplex $\mathrm{conv}\{\boldsymbol{B}\}$. MVC-NMF is essentially a variation of the VolMin formulation (see problem (5.32)) in the noisy case, with endmember non-negativity incorporated. As mentioned before, vol($\boldsymbol{B}$) is nonconvex. Iterated constrained endmember (ICE) [48] and sparsity-promoting ICE (SPICE) [94] avoid this issue by replacing $(\text{vol}(\boldsymbol{B}))^2$ with a convex surrogate, namely, $g(\boldsymbol{A}) = \sum_{r=1}^{R-1} \sum_{r'=r+1} R\|\boldsymbol{a}_r - \boldsymbol{a}'_r\|_2^2$, which is the sum of differences between vertices. A similar idea is also adopted in collaborative NMF (CoNMF) [95]; see Table 5.2.

As for the abundance regularizer h, the design principle usually follows that of sparsity. A good showcasing example, curiously, lies in dictionary learning (DL) [81]:

$$\min_{\boldsymbol{A}\geqslant\boldsymbol{0},\boldsymbol{S}\geqslant\boldsymbol{0}} \|\boldsymbol{X} - \boldsymbol{A}\boldsymbol{S}\|_F^2 + \mu\|\boldsymbol{S}\|_{1,1}; \tag{5.44}$$

note that $\|\boldsymbol{S}\|_{1,1} = \sum_{t=1}^{T} \sum_{r=1}^{R} |s_{tr}|$. The original idea of problem (5.44) is to learn the dictionary $\boldsymbol{A}$ by joint dictionary and sparse signal optimization, cf. Section 5.5 and particularly problem (5.28). However, problem (5.44) can also be seen as an NMF with sparse-promoting regularization.

Following the same spirit, $L_{1/2}$-NMF [92] uses a nonconvex, but stronger sparse-promoting regularizer based on the $\ell_{1/2}$ quasi-norm. Apart from sparsity, exploitation of spatial contextual information via TV regularization may also be used [93].

The aforementioned connection between DL and NMF provides an additional insight. In DL, the dictionary size is often set to be large and should be larger than the true number of endmembers; the number of endmembers is instead determined by the row sparsity of $\boldsymbol{S}$, i.e. $\|\boldsymbol{S}\|_{\text{row-0}}$. From an NMF-based blind HU perspective, this means that we can use row sparsity to provide joint endmember number, endmember and abundance estimation. More formally, consider a blind version of the MMV problem (5.28)

$$\min_{\boldsymbol{A}\geqslant\boldsymbol{0},\boldsymbol{S}\in\mathcal{S}^T} \|\boldsymbol{X} - \boldsymbol{A}\boldsymbol{S}\|_F^2 + \lambda g(\boldsymbol{A}) + \mu\|\boldsymbol{S}\|_{\text{row-0}}, \tag{5.45}$$

where the number of columns of $\boldsymbol{A}$, given by R, is now chosen to be a number greater than the true number of endmembers (say, by overestimating the latter), and we use $\|\boldsymbol{S}\|_{\text{row-0}}$ to represent the endmember number. SPICE is arguably the first algorithm that explores such opportunity [94]. In SPICE, the abundance regularizer can be expressed as $h(\boldsymbol{S}) = \sum_{i=1} R\gamma_i\|\boldsymbol{s}^i\|_1$ for some weights $\{\gamma_i\}$ that are iteratively updated; this regularizer is a convex surrogate of $\|\boldsymbol{S}\|_{\text{row-0}}$. CoNMF also aims at row sparsity, using a nonconvex surrogate $h(\boldsymbol{S}) = \sum_{i=1}^{K} \|\boldsymbol{s}^i\|_2^p$, $0 < p \leq 1$ [95].

We should also discuss optimization in NMF-based blind HU. Most NMF-based blind HU algorithms follow a two-block alternating optimization (AO) strategy, although their implementation details exhibit many differences. Two-block AO optimizes problem (5.42) with respect to either $\boldsymbol{A}$ and $\boldsymbol{S}$ alternatively. Specifically, it generates a sequence of iterates $\{(\boldsymbol{A}^{(k)}, \boldsymbol{S}^{(k)})\}_k$ via

$$\boldsymbol{A}^{(k)} = \arg\min_{\boldsymbol{A} \geqslant \boldsymbol{0}} \|\boldsymbol{X} - \boldsymbol{A}\boldsymbol{S}^{(k-1)}\|_F^2 + \lambda\, g(\boldsymbol{A}), \tag{5.46a}$$

$$\boldsymbol{S}^{(k)} = \arg\min_{\boldsymbol{S} \in S^T} \|\boldsymbol{X} - \boldsymbol{A}^{(k)}\boldsymbol{S}\|_F^2 + \mu\, h(\boldsymbol{S}). \tag{5.46b}$$

Note that if g and h are convex, then problems (5.46a)–(5.46b) are convex and hence can usually be solved efficiently. Moreover, every limit point of $\{(\boldsymbol{A}^{(k)}, \boldsymbol{S}^{(k)})\}_k$ is a stationary point of problem (5.42) under some fairly mild assumptions [96, 97]. A simple modification, termed *proximal alternating optimization* (PAO), consists in adding quadratic proximal terms to (5.46a) and (5.46b) as follows:

$$\boldsymbol{A}^{(k)} = \arg\min_{\boldsymbol{A} \geqslant \boldsymbol{0}} \|\boldsymbol{X} - \boldsymbol{A}\boldsymbol{S}^{(k-1)}\|_F^2 + \lambda\, g(\boldsymbol{A}) + \frac{\alpha_k}{2}\|\boldsymbol{A} - \boldsymbol{A}^{(k)}\|_F^2, \tag{5.47a}$$

$$\boldsymbol{S}^{(k)} = \arg\min_{\boldsymbol{S} \in S^T} \|\boldsymbol{X} - \boldsymbol{A}^{(k)}\boldsymbol{S}\|_F^2 + \mu\, h(\boldsymbol{S}) + \frac{\beta_k}{2}\|\boldsymbol{S} - \boldsymbol{S}^{(k)}\|_F^2, \tag{5.47b}$$

where $\alpha_k, \beta_k \in (r_-, r_+)$, with $0 < r_-, < r_+$, are sequences of positive numbers.

For practical reasons, most algorithms use cheap but inexact updates for (5.46a) and (5.46b), e.g. multiplicative update [92], one-step projected gradient or subgradient update [49, 81, 93], and one-step majorization minimization [95]. Convergence to a stationary point of these AO methods is however not guaranteed and is object of current considerable research efforts. A relevant inexact AO method worth referencing is the *proximal alternating linearized minimization* (PALM) [98] to the optimization (5.42). PALM is, in essence, similar to the PAO scheme (5.47a)–(5.47b) replacing the quadratic terms $\|\boldsymbol{X} - \boldsymbol{A}\boldsymbol{S}^{(k-1)}\|_F^2$ and $\|\boldsymbol{X} - \boldsymbol{A}^{(k)}\boldsymbol{S}\|_F^2$ with quadratic surrogates of the form $c_k\|\boldsymbol{G}_{k-1} - \boldsymbol{A}\|_F^2$ and $d_k\|\boldsymbol{H}_k - \boldsymbol{S}\|_F^2$. The resulting optimizations with respect to $\boldsymbol{A}$ and $\boldsymbol{S}$ are interpretable as proximity operators of $g(\boldsymbol{A}) + \iota_{\mathbb{R}_+^{M\times T}}(\boldsymbol{A})$ and $h(\boldsymbol{A}) + \iota_{S^T}(\boldsymbol{S})$, respectively, where ι_A stands for the indicator function in the set A. Under the conditions for g and h stated in the POA method and for suitable values of the sequences c_k and d_k, the sequence $\{(\boldsymbol{A}^{(k)}, \boldsymbol{S}^{(k)})\}_k$ generated by PALM converges to a stationary point of (5.42). See [98] for more details.

By numerical experience, many NMF-based blind HU algorithms work well under appropriate settings (e.g. using reasonable initializations which can be obtained for example with pure pixel algorithms such VCA, N-FINDR, or SVMAX).

To summarize, NMF is a versatile approach that has connections to both CG and SR. It leads to a fundamentally hard optimization problem, although practical solutions based on two-block AO usually offer good performance by experience. Also, before we finish, we should highlight that the more exciting developments of NMF-based blind HU lie in extensions to scenarios such as nonlinear HU [99], endmember variability [100], and multispectral and hyperspectral data fusion [101].

5.6.3 Illustrative Comparison of Geometrical Methods

In this subsection, we experimentally compare the results provided by the four algorithms (N-FINDR, VCA, MVC-NMF, and SISAL) in different situations. In Figure 5.16, in which pure pixels are present, unmixing results are perfect for all the methods. In Figures 5.17 and 5.18, when there is no longer pure pixel and with truncated (to 0.8) abundances, it is clear that methods based on pure pixel assumption like N-FINDR and VCA are not able to give

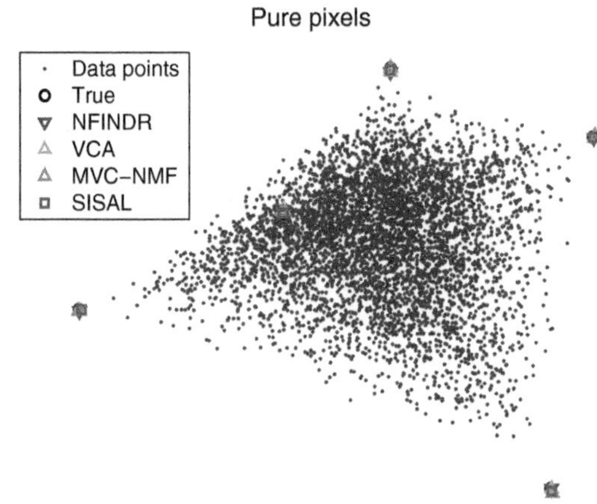

Figure 5.16 Unmixing results of N-FINDR, VCA, MVC-NMF, and SISAL on a simulated dataset with pure pixels ($R = 5$, SNR = 30 dB). Source: Bioucas-Dias et al., [1]/IEEE.

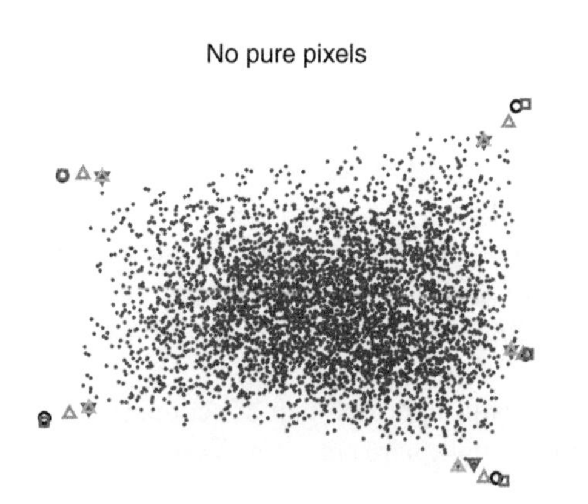

Figure 5.17 Unmixing results of N-FINDR, VCA, MVC-NMF, and SISAL on a simulated dataset without pure pixels ($R = 5$, SNR = 30 dB). Source: Bioucas-Dias et al., [1]/IEEE.

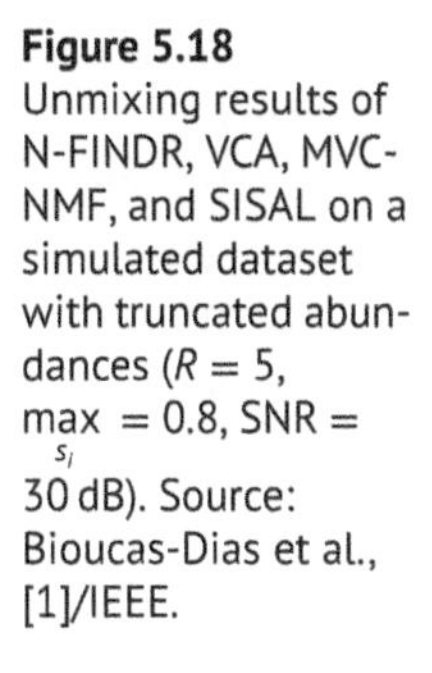

Figure 5.18 Unmixing results of N-FINDR, VCA, MVC-NMF, and SISAL on a simulated dataset with truncated abundances ($R = 5$, $\max_{s_i} = 0.8$, SNR = 30 dB). Source: Bioucas-Dias et al., [1]/IEEE.

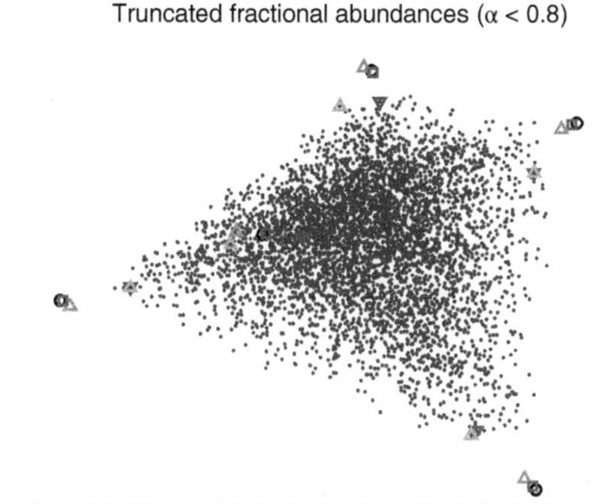

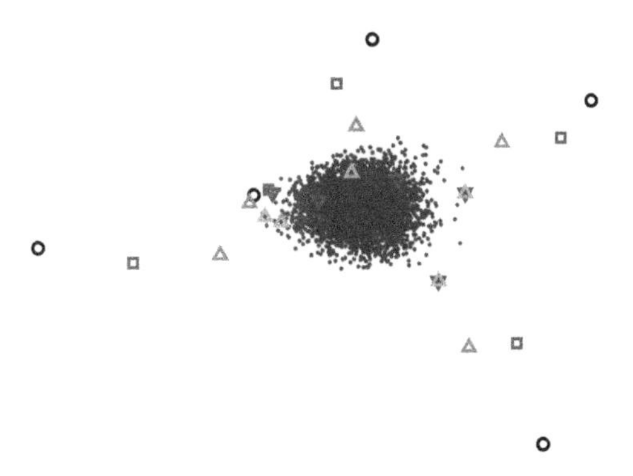

Figure 5.19 Unmixing results of N-FINDR, VCA, MVC-NMF, and SISAL on a highly mixed simulated dataset ($R = 5$, SNR = 30 dB). Source: Bioucas-Dias et al., [1]/IEEE.

an accurate factorization. Finally, in Figure 5.19, we show that all the algorithms fail, when the data are highly mixed and thus located in the middle of the simplex.

5.7 Applications

5.7.1 Unmixing Example Under the Pure Pixel Assumption

In this section, we illustrate part of the concepts presented before by unmixing the publicly available TERRAIN HSI[3] acquired by the HYDICE sensor

3 Data set available at `http://www.agc.army.mil/hypercube.`

[102] (see HYDICE parameters in Table 5.1). The low SNR bands due to water absorption were removed yielding a data set with 166 bands.

The TERRAIN HSI, shown in Figure 5.20a, was calibrated to reflectance, has size 500×307, and is mainly composed of soil, trees, grass, a lake, and shadows, disposed on a flat surface. The signal subspace was identified with the HySime [33] algorithm and the original data projected onto this subspace. The identified subspace dimension was 20. We have, however, discarded those orthogonal directions corresponding to SNR < 10 to avoid instability of the endmember identification (see [1] for more details). After this procedure, we ended up with a subspace of dimension 6.

The plots on Figure 5.20b show the identified endmember signatures with the VCA algorithm [23]. The corresponding pixels are referenced in the original image. They represent three types of soil, trees, grass, and a spectrum obtained in the lake, which we termed shade due to its low amplitude. The figure in (c) shows a scatterogram of the data set projected on the subspace defined by the first two subspace eigendirections determined by HySime. The endmembers identified by VCA and N-FINDR are also represented. The solution provided by the two algorithms is identical and, due to the height spatial resolution of the sensor, correspond to nearly pure pixels. Notice that there are endmembers in all in "extremes" of the scatterogram, which is coherent with the pure pixel hypothesis. The remaining parts of Figure 5.20 shows abundance fractions for soil 1, trees, and grass.

5.7.2 Unmixing via Sparse Regression

The work [58] introduces the collaborative sparsity approach in hypserspectral unmixing. Figure 5.21 shows a false color inference of $\mathbf{X} \in \mathbb{R}^{250\times 50}$ using a subset of the USGS library splib06 [4] with 250 signatures of size 220. The simulated ground-truth abundances contain 50 pixels and 4 endmembers randomly extracted from the library. The simulated measurements were contaminated with additive noise and SNR=30dB. The image on the (a) corresponds to the ℓ_1 relaxation computed with the SUnSAL algorithm [8], thus treating each pixel independently. The image on the (a) is the solution of the group sparse problem using the mixed $\ell_{2,1}$ norm and was computed with the CLSUnSAL algorithm introduced in [58]. The collaborative regularization yields a cleaner solution with many rows set to zero.

4 `http://speclab.cr.usgs.gov/spectral.lib06`

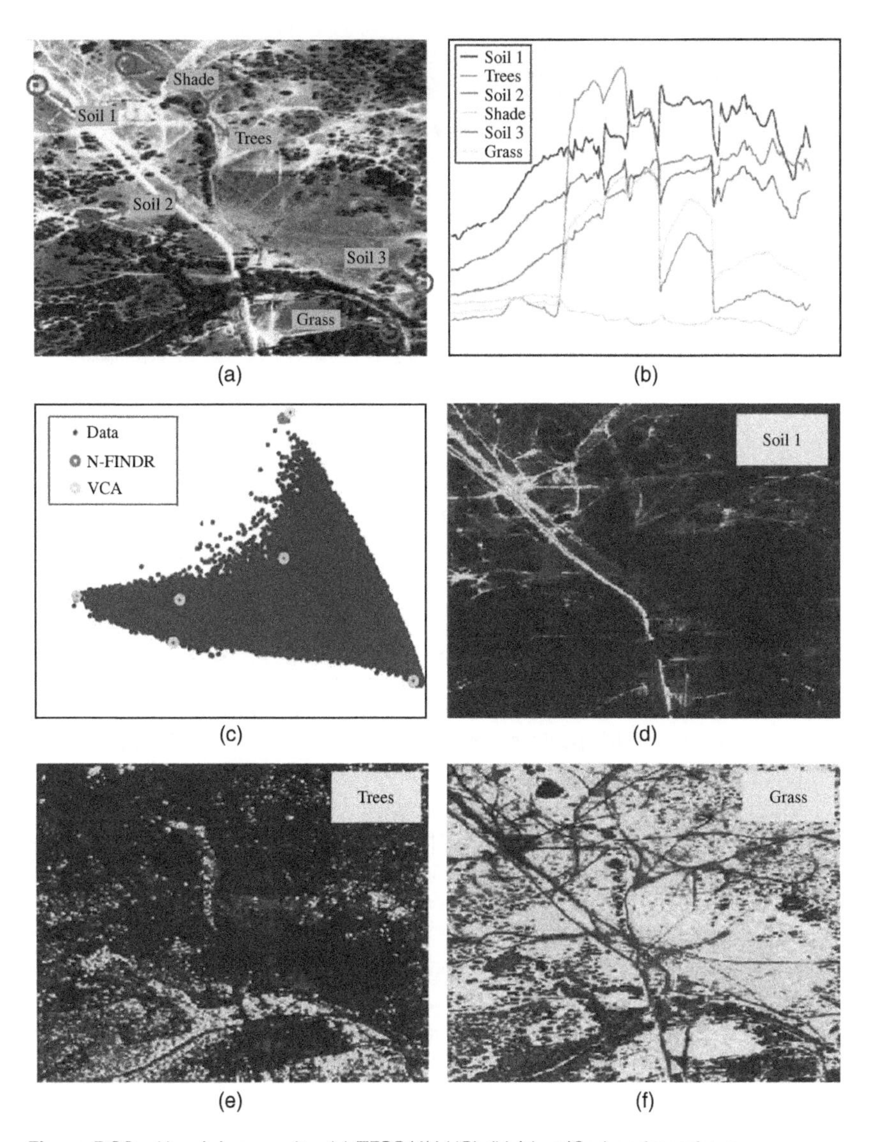

Figure 5.20 Unmixing results: (a) TERRAIN HSI, (b) identified endmembers, (c) data projection onto the subspace defined by the first two eigendirections, (d) soils abundance map, (e) tree abundance map, and (f) grass abundance map. Source: Bioucas-Dias et al., [3]/IEEE.

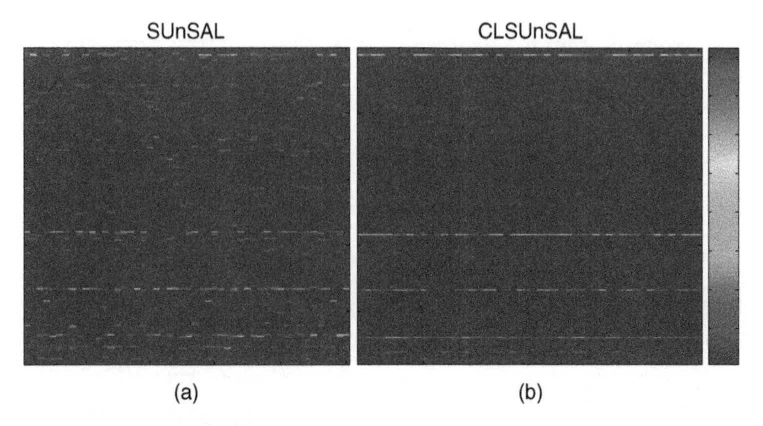

Figure 5.21 Sparse regression solutions **X** for 50 simulated spectral LMM generated with four endmembers and SNR=30dB and using a subset of the USGS library with 250 signatures. The horizontal axis represents pixels and the vertical axis represents endmembers. (a) Solution computed by the SUnSAL algorithm treating independently each pixel and (b) solution computed by CLSUnSAL algorithm, enforcing collaborative sparsity. Source: Bioucas-Dias et al. [3]/IEEE.

5.7.3 Near-Infrared Hyperspectral Unmixing of Pharmaceutical Tablets

Near-infrared (NIR) spectroscopy provides reflectance spectral images in the range 780 to 2500 nanometers. It can be applied in many domains. Recently, unmixing of such images has been used for characterizing pharmaceutical tablets and detecting counterfeit [87]. This emerging approach of NIR hyperspectral unmixing is interesting for a few reasons:

- it is fast and nondestructive,
- it can identify and quantify the composition of the counterfeit tablet, without prior knowledge on the materials,
- it can provide spatial information on the different materials, i.e. on the heterogeneity of the materials inside the tablets.

In [87], a few algorithms for hyperspectral unmixing has been evaluated on simulated and real data sets. Here, we only present results obtained on real data sets, extracted from a market survey of 55 counterfeit Heptodin^{TM} tablets [103]. For each tablet, the sensor provides a hyperspectral NIR image with 10^5 pixels and 126 spectral bands (spectral range from 1200 to 2450 nanometers, with 10 nanometers steps). The active pharmaceutical ingredient is lamivudine, but the counterfeit tablets contain microcrystalline cellulose (MCC), talc, and starch as main excipients.

Since the data don't contain pure pixel, algorithms based on pure pixel – like VCA – are not efficient. So, in the comparison, we use the classical NMF algorithm: multivariate curve resolution using alternative least square (MCR-ALS), and three algorithms based on minimum volume of the simplex: SISAL, MVSA, and MVES. Note that

- MCR-ALS does not always achieve a satisfactory solution, due to possible rotational ambiguity,
- although SISAL is computationally much faster and efficient, SISAL and MVSA are solving the same problem, and their results are the same and presented together in the figures.

After a first processing step providing a $(R-1)$-dimensional affine set, where $r = 4$ corresponds to the number of ingredients, we run the four algorithms on four counterfeit tablets. In Figure 5.22, we present the estimated

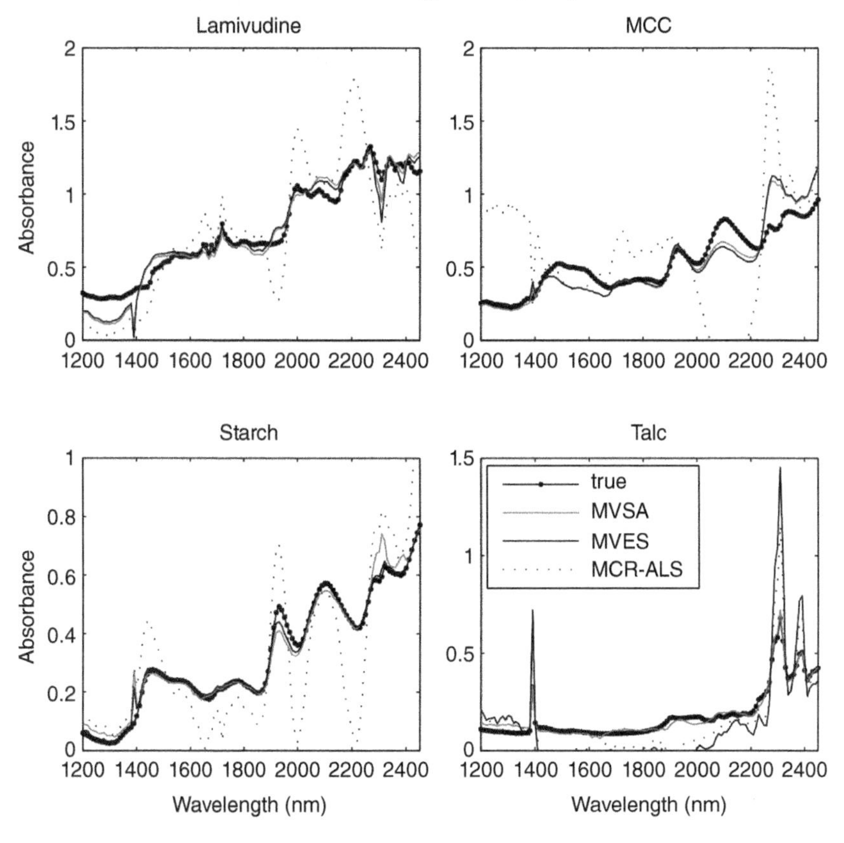

Figure 5.22 Estimated signatures of real data set (counterfeit tablets) by SISAL/MVSA, MVES, and MCR-ALS.

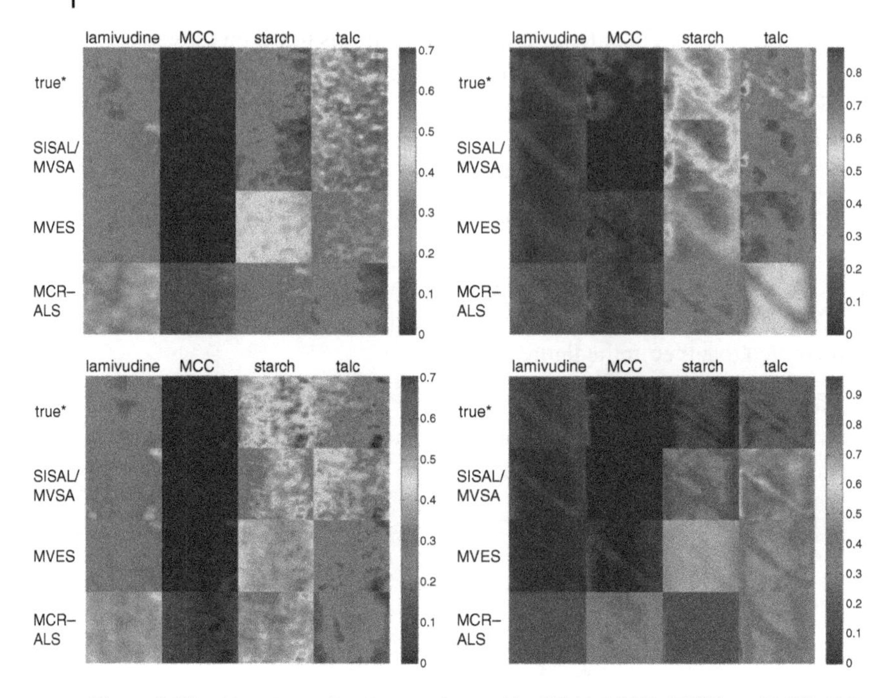

Figure 5.23 Abundance fractions estimated by SISAL/MVSA, MVES, and MCR-ALS, in each pixel of four counterfeit tablets (*the true images have been estimated using a library composed of the true NIR spectral signatures of the four components known to be present in the mixtures, i.e. lamivudine, MCC, starch, and talc). Source: Lopes et al. [87]/American Chemical Society.

spectra of the main ingredients. SISAL/MVSA obtains better results than MVES, while MCR-ALS achieves poor results. In Figure 5.23, we can compare the true spatial abundances of the four tablets for the four ingredients, and the abundances estimated by the different algorithms. We remark that the abundances estimated by SISAL/MVSA are close to the true ones and are better than those estimated by MVES. We also note that abundances estimated by MCR-ALS are quite whimsical. In these figures, the true values have been obtained based on a dictionary containing the true NIR spectral signature of the four ingredients.

5.8 Conclusions

In addition to non-negativity (see Chapter 3), sum-to-one (sometimes called closure, full additivity, or column-stochastic) and sparsity are usual properties encountered in the separation problem of physical or chemical sources. In this chapter, we showed that these natural properties lead to nice geometrical representations of the data set, using cones or simplex. These geometrical representations are inspiring various families of algorithms: a first one based on the source dominance (pure pixel) assumption, and a second one based on minimizing the simplex containing all the data. These ideas, and the suggested algorithms, have been intensively developed in hyperspectral unmixing for remote sensing and satellite imaging, but can be used in other applications like detection of counterfeit pharmaceutical tablets [87], in Liquid Chromatography–Mass Spectrum or with Positron Emission Tomography [104].

References

1 Bioucas-Dias, J., Plaza, A., Dobigeon, N., Parente, M., Du, Q., Gader, P., and Chanussot, J. (2012) Hyperspectral unmixing overview: geometrical, statistical, and sparse regression-based approaches. *IEEE Journal of Selected Topics and Applications in Earth Observation and Remote Sensing*, 5 (2), 354–379.

2 Ma, W.K., Bioucas-Dias, J., Chan, T.H., Gillis, N., Gader, P., Plaza, A., Ambikapathi, A., and Chi, C.Y. (2014) A signal processing perspective on hyperspectral unmixing: insights from remote sensing. *IEEE Signal Processing Magazine*, 31 (1), 67–81.

3 Bioucas-Dias, J., Plaza, A., Camps-Valls, G., Scheunders, P., Nasrabadi, N., and Chanussot, J. (2013) Hyperspectral remote sensing data analysis and future challenges. *IEEE Transactions on Geoscience and Remote Sensing*, 1 (2), 6–36.

4 Camps-Valls, G., Tuia, D., Gómez-Chova, L., Jiminéz, S., and Malo, J. (2011) *Remote Sensing Image Processing*, Morgan & Claypool Publishers.

5 Richards, J.A. and Jia, X. (2006) *Remote Sensing Digital Image Analysis: An Introduction*, Springer.

6 Goetz, A.F.H., Vane, G., Solomon, J.E., and Rock, B.N. (1985) Imaging spectrometry for Earth remote sensing. *Science*, 228, 1147–1153.

7 Heylen, R., Burazerovic, D., and Scheunders, P. (2011) Non-linear spectral unmixing by geodesic simplex volume maximization. *IEEE Journal of Selected Topics in Signal Processing*, 5 (3), 534–542.

8 Iordache, M.D., Bioucas-Dias, J., and Plaza, A. (2011) Sparse unmixing of hyperspectral data. *IEEE Transactions on Geoscience and Remote Sensing*, 49 (6), 2014–2039.

9 Eismann, M.T., Stocker, A.D., and Nasrabadi, N.M. (2009) Automated hyperspectral cueing for civilian search and rescue. *Proceedings of IEEE*, 97 (6), 1031–1055.

10 Keshava, N. and Mustard, J. (2002) Spectral unmixing. *IEEE Signal Processing Magazine*, 19 (1), 44–57.

11 Heylen, R., Parente, M., and Gader, P. (2014) A review of nonlinear hyperspectral unmixing methods. *IEEE Journal of Selected Topics and Applications in Earth Observation and Remote Sensing*, 7 (6), 1844–1868.

12 Craig, M.D. (1990) Unsupervised unmixing of remotely sensed images, in *Proceeding od Australasian Remote Sensing Conference*, Perth, Australia, pp. 324–330.

13 Craig, M.D. (1994) Minimum-volume transforms for remotely sensed data. *IEEE Transactions on Geoscience Remote Sensing*, 32 (3), 542–552.

14 Boardman, J.W. (1993) Automating spectral unmixing of AVIRIS data using convex geometry concepts, in *Proceeding of Summer 4th Annual JPL Airborne Geoscience Workshop*, vol. 1, pp. 11–14.

15 Boardman, J., Kruse, F., and Green, R. (1995) Mapping target signatures via partial unmixing of AVIRIS data, in *Proceeding of Summer 4th Annual JPL Airborne Geoscience Workshop*, vol. 1, Pasadena, CA, pp. 23–26.

16 Winter, M.E. (1999) N-FINDR: An algorithm for fast autonomous spectral end-member determination in hyperspectral data, in *Proceeding of SPIE Conference on Imaging Spectrometry*, Pasadena, CA, pp. 266–275.

17 Dobigeon, N., Moussaoui, S., Coulon, M., Tourneret, J.Y., and Hero, A.O. (2009) Joint Bayesian endmember extraction and linear unmixing for hyperspectral imagery. *IEEE Transactions on Signal Processing*, 57 (11), 4355–4368.

18 Nascimento, J. and Bioucas-Dias, J. (2012) Hyperspectral unmixing based on mixtures of Dirichlet components. *IEEE Transactions on Geoscience and Remote Sensing*, 50 (3), 863–878.

19 Themelis, K.E., Rontogiannis, A.A., and Koutroumbas, K.D. (2012) A novel hierarchical Bayesian approach for sparse semisupervised hyperspectral unmixing. *IEEE Transactions on Signal Processing*, 60 (2), 585–599.

20 Chandrasekhar, S. (1960) *Radiative Transfer*, Dover, New York.

21 Hapke, B. (1981) Bidirection reflectance spectroscopy. I. Theory. *Journal of Geophysical Research*, 86, 3039–3054.

22 Dobigeon, N., Tourneret, J.Y., Richard, C., Bermudez, J., McLaughlin, S., and Hero, A. (2014) Nonlinear unmixing of hyperspectral images: models and algorithms. *IEEE Journal of Selected Topics in Signal Processing*, 31 (1), 82–94.

23 Nascimento, J. and Bioucas-Dias, J. (2005) Vertex component analysis: a fast algorithm to unmix hyperspectral data. *IEEE Transactions on Geoscience and Remote Sensing*, 43 (4), 898–910.
24 Chan, T.H., Chi, C.Y., Huang, Y.M., and Ma, W.K. (2009) A convex analysis based minimum-volume enclosing simplex algorithm for hyperspectral unmixing. *IEEE Transactions on Signal Processing*, 57 (11), 4418–4432.
25 Hyvärinen, A., Karhunen, J., and Oja, E. (2004) *Independent Component Analysis*, vol. 46, John Wiley & Sons.
26 Comon, P. and Jutten, C. (2010) *Handbook of Blind Source Separation: Independent Component Analysis and Applications*, Academic Press.
27 Nascimento, J. and Bioucas-Dias, J. (2005) Does independent component analysis play a role in unmixing hyperspectral data? *IEEE Transactions on Geoscience and Remote Sensing*, 43 (1), 175–187.
28 Gao, B.C., Montes, M., Davis, C., and Goetz, A. (2009) Atmospheric correction algorithms for hyperspectral remote sensing data of land and ocean. *Remote Sensing of Environment*, 113, S17–S24.
29 Green, A., Berman, M., Switzer, P., and Craig, M.D. (1988) A transformation for ordering multispectral data in terms of image quality with implications for noise removal. *IEEE Transactions on Geoscience and Remote Sensing*, 26, 65–74.
30 Lee, J., Woodyatt, S., and Berman, M. (1990) Enhancement of high spectral resolution remote-sensing data by noise-adjusted principal components transform. *IEEE Transactions on Geoscience and Remote Sensing*, 28 (3), 295–304.
31 Bowles, J., Antoniades, J., Baumback, M., Grossmann, J., Haas, D., Palmadesso, P.J., and Stracka, J. (1997) Real-time analysis of hyperspectral data sets using NRL's ORASIS algorithm, in *Proceedings of SPIE Conference on Imaging Spectrometry III*, vol. 3118, pp. 38–45.
32 Chang, C.I. and Du, Q. (2004) Estimation of number of spectrally distinct signal sources in hyperspectral imagery. *IEEE Transactions on Geoscience and Remote Sensing*, 42 (3), 608–619.
33 Bioucas-Dias, J. and Nascimento, J. (2008) Hyperspectral subspace identification. *IEEE Transactions on Geoscience and Remote Sensing*, 46 (8), 2435–2445.
34 Kritchman, S. and Nadler, B. (2009) Non-parametric detection of the number of signals: hypothesis testing and random matrix theory. *IEEE Transactions on Signal Processing*, 57 (10), 3930–3941.
35 Cawse, K., Sears, M., Robin, A., Damelin, S., Wessels, K., Van den Bergh, F., and Mathieu, R. (2010) Using random matrix theory to determine the number of endmembers in a hyperspectral image, in

Proceedings of IEEE GRSS Workshop on Hyperspectral Image and Signal Processing: Evolution in Remote Sensing (WHISPERS), pp. 1–4.

36 Halimi, A., Honeine, P., Kharouf, M., Richard, C., and Tourneret, J.Y. (2016) Estimating the intrinsic dimension of hyperspectral images using a noise-whitened eigengap approach. *IEEE Transactions on Geoscience and Remote Sensing*, 54 (7), 3811–3821.

37 Boyd, S. and Vandenberghe, L. (2004) *Convex Optimization*, Cambridge University Press.

38 Chan, T.H., Ma, W.K., Ambikapathi, A., and Chi, C.Y. (2011) A simplex volume maximization framework for hyperspectral endmember extraction. *IEEE Transactions on Geoscience and Remote Sensing*, 49 (11), 4177–4193.

39 Chan, T.H., Ambikapathi, A., Ma, W.K., and Chi, C.Y. (2013) Robust affine set fitting and fast simplex volume max-min for hyperspectral endmember extraction. *IEEE Transactions on Geoscience and Remote Sensing*, 51 (7), 3982–3997.

40 Boardman, J. (1993) Automating spectral unmixing of AVIRIS data using convex geometry concepts, in *Proceedings of Annual JPL Airborne Geoscience Workshop*, vol. 1, pp. 11–14.

41 Chang, C.I., Wu, C.C., Liu, W., and Ouyang, Y.C. (2006) A new growing method for simplex-based endmember extraction algorithm. *IEEE Transactions on Geoscience and Remote Sensing*, 44 (10), 2804–2819.

42 Gillis, N. and Vavasis, S. (2013) Fast and robust recursive algorithms for separable nonnegative matrix factorization. *IEEE Transactions on Pattern Analysis and Machine Intelligence*, 36 (4), 698–714

43 Winter, M.E. (1999) N-FINDR: An algorithm for fast autonomous spectral endmember determination in hyperspectral data, in *Proceeding of SPIE Image Spectrometry V*, vol. 3753, pp. 266–277.

44 Neville, R.A., Staenz, K., Szeredi, T., Lefebvre, J., and Hauff, P. (1999) Automatic endmember extraction from hyperspectral data for mineral exploration, in *Proceedings of Canadian Symposium on Remote Sensing*, pp. 21–24.

45 Perczel, A., Hollósi, M., Tusnady, G., and Fasman, D. (1989) Convex constraint decomposition of circular dichroism curves of proteins. *Croatica Chimica Acta*, 62, 189–200.

46 Perczel, A., Hollósi, M., Tusnády, G., and Fasman, G.D. (1991) Convex constraint analysis: a natural deconvolution of circular dichroism curves of proteins. *Protein Engineering*, 4 (6), 669–679.

47 Lee, D. and Seung, H.S. (1999) Learning the parts of objects by non-negative matrix factorization. *Nature*, 401, 788–791.

48 Berman, M., Kiiveri, H., Lagerstrom, R., Ernst, A., Dunne, R., and Huntington, J.F. (2004) ICE: A statistical approach to identifying endmembers in hyperspectral images. *IEEE Transactions on Geoscience and Remote Sensing*, 42 (10), 2085–2095.

49 Miao, L. and Qi, H. (2007) Endmember extraction from highly mixed data using minimum volume constrained nonnegative matrix factorization. *IEEE Transactions on Geoscience and Remote Sensing*, 45 (3), 765–777.

50 Zare, A. and Gader, P. (2007) Sparsity promoting iterated constrained endmember detection for hyperspectral imagery. *IEEE Geoscience Remote Sensing Letters*, 4 (3), 446–450.

51 Li, J. and Bioucas-Dias, J. (2008) Minimum volume simplex analysis: a fast algorithm to unmix hyperspectral data, in *Proceedings of IEEE International Conference in Geoscience and Remote Sensing (IGARSS)*, vol. 3, pp. 250–253.

52 Li, J., Agathos, A., Zaharie, D., Bioucas-Dias, J., Plaza, A., and Li, X. (2015) Minimum volume simplex analysis: a fast algorithm for linear hyperspectral unmixing. *IEEE Transactions on Geoscience and Remote Sensing*, 53 (9), 5067–5082.

53 Bioucas-Dias, J. (2009) A variable splitting augmented Lagragian approach to linear spectral unmixing, in *Proceedings of IEEE GRSS Workshop on Hyperspectral Image and Signal Processing: Evolution in Remote Sensing (WHISPERS)*, pp. 1–4.

54 Ambikapathi, A., Chan, T.H., Ma, W.K., and Chi, C.Y. (2011) Chance-constrained robust minimum-volume enclosing simplex algorithm for hyperspectral unmixing. *IEEE Transactions on Geoscience and Remote Sensing*, 49 (11), 4194–4209.

55 Dobigeon, N., Moussaoui, S., Tourneret, J.Y., and Carteret, C. (2009) Bayesian separation of spectral sources under non-negativity and full additivity constraints. *Signal Processing*, 89 (12), 2657–2669.

56 Candès, E., Romberg, J., and Tao, T. (2006) Stable signal recovery from incomplete and inaccurate measurements. *Communications on Pure Applied Mathematics*, 59 (8), 1207–1223.

57 Bioucas-Dias, J. and Figueiredo, M. (2010) Alternating direction algorithms for constrained sparse regression: application to hyperspectral unmixing, in *Proceedings of IEEE GRSS Workshop on Hyperspectral Image and Signal Processing: Evolution in Remote Sensing (WHISPERS)*, Raykjavik, Iceland.

58 Iordache, M.D., Bioucas-Dias, J., and Plaza, A. (2013) Collaborative sparse regression for hyperspectral unmixing. *IEEE Transactions on Geoscience and Remote Sensing*, 52 (1), 341–354.

59 Iordache, M.D., Bioucas-Dias, J., and Plaza, A. (2012) Total variation spatial regularization for sparse hyperspectral unmixing. *IEEE Transactions on Geoscience and Remote Sensing*, 50 (11), 4484–4502.

60 Iordache, M.D., Bioucas-Dias, J., Plaza, A., and Somers, B. (2014) MUSIC-CSR: Hyperspectral unmixing via multiple signal classification and collaborative sparse regression. *IEEE Transactions on Geoscience and Remote Sensing*, 52 (7), 4364–4382.

61 Fu, X., Ma, W.K., Bioucas-Dias, J., and Chan, T.H. (2016) Semiblind hyperspectral unmixing in the presence of spectral library mismatches. *IEEE Transactions on Geoscience and Remote Sensing*, 54 (9), 5171–5184.

62 Imbrie, J. and Van Andel, T.H. (1964) Vector analysis of heavy-mineral data. *Geological Society of America Bulletin*, 75, 1131–1156.

63 Ehrlich, R. and Full, W.E. (1987) Sorting out geology—unmixing mixtures, in *Use and Abuse of Statistical Methods in the Earch Sciences* (ed. W. Size), Oxford University Press, pp. 33–46.

64 Klovan, J.E. and Miesch, A.T. (1976) Extended CABFAC and QMODEL computer programs for Q-mode factor analysis of compositional data. *Computers & Geosciences*, 1, 161–178.

65 Full, W.E., Ehrlich, R., and Klovan, J.E. (1981) EXTENDED QMODEL—Objective definition of external endmembers in the analysis of mixtures. *Mathematical Geology*, 13 (4), 331–344.

66 Naanaa, W. and Nuzillard, J.M. (2005) Blind source separation of positive and partially correlated data. *Signal Processing*, 85 (9), 1711–1722.

67 Chan, T.H., Ma, W.K., Chi, C.Y., and Wang, Y. (2008) A convex analysis framework for blind separation of non-negative sources. *IEEE Transactions on Signal Processing*, 56 (10), 5120–5134.

68 Clark, R., Swayze, G., Wise, R., Livo, E., Hoefen, T., Kokaly, R., and Sutley, S. (2007), USGS digital spectral library splib06a: U.S. Geological Survey, Digital Data Series 231, http://speclab.cr.usgs.gov/spectral.lib06.

69 Baraniuk, R.G. (2007) Compressive sensing [lecture notes]. *IEEE Signal Processing Magazine*, 24 (4), 118–121.

70 Donoho, D.L. and Elad, M. (2003) Optimally sparse representation in general (nonorthogonal) dictionaries via l1 minimization. *Proceedings of the National Academy of Sciences of the United States of America*, 100 (5), 2197–2202.

71 Greer, J.B. (2012) Sparse demixing of hyperspectral images. *IEEE Transactions on Image Processing*, 21 (1), 219–228.

72 Donoho, D.L. and Tanner, J. (2005) Sparse nonnegative solution of underdetermined linear equations by linear programming. *Proceedings* *of the National Academy of Sciences of the United States of America*, 102 (27), 9446–9451.

73 Itoh, Y., Duarte, M., and Parente, M. (2016) Perfect recovery conditions for non-negative sparse modeling. *IEEE Transactions on Signal Processing*, 65 (1), 69–80.

74 Chen, J. and Huo, X. (2006) Theoretical results on sparse representations of multiple-measurement vectors. *IEEE Transactions on Signal Processing*, 54 (12), 4634–4643.

75 Eldar, Y.C. and Rauhut, H. (2010) Average case analysis of multichannel sparse recovery using convex relaxation. *IEEE Transactions on Information Theory*, 56 (1), 505–519.

76 Afonso, M.V., Bioucas-Dias, J.M., and Figueiredo, M.A. (2011) An augmented Lagrangian approach to the constrained optimization formulation of imaging inverse problems. *IEEE Transactions on Image Processing*, 20 (3), 681–695.

77 Iordache, M., Bioucas-Dias, J., and Plaza, A. (2012) Total variation spatial regularization for sparse hyperspectral unmixing. *IEEE Transactions on Geoscience and Remote Sensing*, 50 (11), 4484–4502.

78 Malioutov, D., Cetin, M., and Willsky, A. (2005) A sparse signal reconstruction perspective for source localization with sensor arrays. *IEEE Transactions on Signal Processing*, 53 (8), 3010–3022.

79 Kim, J.M., Lee, O.K., and Ye, J.C. (2012) Compressive MUSIC: revisiting the link between compressive sensing and array signal processing. *IEEE Transactions on Information Theory*, 58 (1), 278–301.

80 Van Trees, H.L. (2002) *Optimum Array Processing, Part IV of Detection, Estimation, and Modulation Theory*, John Wiley & Sons.

81 Charles, A., Olshausen, B., and Rozell, C. (2011) Learning sparse codes for hyperspectral imagery. *IEEE Journal of Selected Topics in Signal Processing*, 5 (5), 963–978.

82 Castrodad, A., Xing, Z., Greer, J.B., Bosch, E., Carin, L., and Sapiro, G. (2011) Learning discriminative sparse representations for modeling, source separation, and mapping of hyperspectral imagery. *IEEE Transactions on Geoscience and Remote Sensing*, 49 (11), 4263–4281.

83 Esser, E., Moller, M., Osher, S., Sapiro, G., and Xin, J. (2012) A convex model for nonnegative matrix factorization and dimensionality reduction on physical space. *IEEE Transactions on Image Processing*, 21 (7), 3239–3252.

84 Fu, X., Ma, W.K., Chan, T.H., Bioucas-Dias, J.M., and Iordache, M.D. (2013) Greedy algorithms for pure pixels identification in hyperspectral unmixing: a multiple-measurement vector viewpoint. *Proceedings of EUSIPCO*.

85 Fu, X., Ma, W.K., Chan, T.H., and Bioucas-Dias, J. (2015) Self-dictionary sparse regression for hyperspectral unmixing: greedy pursuit and pure pixel search are related. *IEEE Journal of Selected Topics in Signal Processing*, 9 (6), 1128–1141.

86 Lin, C.H., Ma, W.K., Li, W.C., Chi, C.Y., and Ambikapathi, A. (2015) Identifiability of the simplex volume minimization criterion for blind hyperspectral unmixing: the no-pure-pixel case. *IEEE Transactions on Geoscience and Remote Sensing*, 53 (10), 5530–5546.

87 Lopes, M., Wolff, J., Bioucas-Dias, J., and Figueiredo, M. (2010) NIR hyperspectral unmixing based on a minimum volume criterion for fast and accurate chemical characterization of counterfeit tablets. *Analytical Chemistry*, 82 (4), 1462–1469.

88 Paatero, P. and Tapper, U. (1994) Positive matrix factorization: a non-negative factor model with optimal utilization of error estimates of data values. *Environmetrics*, 5 (2), 111–126.

89 Vavasis, S.A. (2009) On the complexity of nonnegative matrix factorization. *SIAM Journal on Optimization*, 20 (3), 1364–1377.

90 Gillis, N. and Plemmons, R. (2012) Sparse nonnegative matrix underapproximation and its application to hyperspectral image analysis. *Linear Algebra and Applications*, 438 (10), 3991–4007.

91 Arora, S., Ge, R., Kannan, R., and Moitra, A. (2012) Computing a nonnegative matrix factorization–provably, in *Proceedings of the 44th Symposium on Theory of Computing*, ACM, pp. 145–162.

92 Qian, Y., Jia, S., Zhou, J., and Robles-Kelly, A. (2011) Hyperspectral unmixing via $L_{1/2}$ sparsity-constrained nonnegative matrix factorization. *IEEE Transactions on Geoscience and Remote Sensing*, 49 (11), 4282–4297.

93 Zymnis, A., Kim, S.J., Skaf, J., Parente, M., and Boyd, S. (2007) Hyperspectral image unmixing via alternating projected subgradients, in *Proceedings of Asilomar Conference in Signals Systems and Computers.*

94 Zare, A. and Gader, P. (2007) Sparsity promoting iterated constrained endmember detection in hyperspectral imagery. *IEEE Geoscience and Remote Sensing Letters*, 4 (3), 446–450.

95 Li, J., Bioucas-Dias, J.M., and Plaza, A. (2012) Collaborative nonnegative matrix factorization for remotely sensed hyperspectral unmixing, in *Proceedings of IEEE IGARSS.*

96 Grippo, L. and Sciandrone, M. (2000) On the convergence of the block nonlinear Gauss-Seidel method under convex constraints. *Operational Research Letters*, 26 (3), 127–136.

97 Tseng, P. (2001) Convergence of a block coordinate descent method for nondifferentiable minimization. *Journal of Optimization Theory and Applications*, 109 (3), 475–494.

98 Bolte, J., Sabach, S., and Teboulle, M. (2014) Proximal alternating linearized minimization for nonconvex and nonsmooth problems. *Mathematical Programming*, 146 (1-2), 459–494.

99 Gader, P., Dranishnikov, D., Zare, A., and Chanussot, J. (2012) A sparsity promoting bilinear unmixing model, in *Proceedings of IEEE GRSS Workshop on Hyperspectral Image and Signal Processing: Evolution in Remote Sensing (WHISPERS)*.

100 Zare, A. and Gader, P. (2010) PCE: Piecewise convex endmember detection. *IEEE Transactions on Geoscience and Remote Sensing*, 48 (6), 2620–2632.

101 Yokoya, N., Yair, T., and Iwasaki, A. (2012) Coupled nonnegative matrix factorizaion unmixing for hyperspectral and multispectral data fusion. *IEEE Transactions on Geoscience and Remote Sensing*, 50 (2), 528–537.

102 Basedow, R., Carmer, D., and Anderson, M. (1995) Hydice system: implementation and performance, in *SPIE's 1995 Symposium on OE/Aerospace Sensing and Dual Use Photonics*, International Society for Optics and Photonics, pp. 258–267.

103 Lopes, M.B. and Wolff, J.C. (2009) Investigation into classification/-sourcing of suspect counterfeit heptodin^{TM} tablets by near infrared chemical imaging. *Analytica Chimica Acta*, 633 (1), 149–155.

104 Ouedraogo, W.S.B., Souloumiac, A., Jaidane, M., and Jutten, C. (2014) Non-negative blind source separation algorithm based on minimum aperture simplicial cone. *IEEE Transactions on Signal Processing*, 62 (2), 376–389.

6

Tensor Decompositions: Principles and Application to Food Sciences

Jérémy Cohen[1], Rasmus Bro[2], and Pierre Comon[3]

[1] *CNRS, CREATIS, Bâtiment Léonard de Vinci 21 Av. Jean Capelle O, F-69100 Villeurbanne, Lyon, France*
[2] *Department of Food Science, University of Copenhagen, Rolighedsvej 26, 1958 Frederiksberg, Denmark*
[3] *CNRS, GIPSA-lab, Université Grenoble Alpes, 11 rue des Mathematiques, BP.46, F-38402 St Martin d'Héres, France*

6.1 Introduction

Most graduate students fear the concept of tensor, as it reminds them of intricate astrophysics, material science, differential calculus or multilinear algebra formalism. Regarding the fields of signal processing and data science, while it is true that using tensor methods requires understanding at least linear algebra and convex optimization, which are both rich applied mathematics domains, the authors believe that most of this fear about tensors is unjustified in the context of engineering. Indeed, for data scientists, tensors are simply arrays that may have more than two indices, and most of the discussion in this chapter will actually be a generalization of well-understood techniques for matrices to such arrays.

Before entering the technical details of tensor algebra and tensor decomposition methods for analyzing data sets, we shall begin this chapter with a friendly introduction to tensors, so that the reader can hopefully get rid of any anxiety about tensors.

6.1.1 A Simplified Definition

Let us start with a definition of what a tensor is, within the scope of this chapter.

Definition 6.1 *A (real) tensor $\mathcal{T}$ is an element of $\mathbb{R}^{n_1 \times \cdots \times n_d}$ where n_i are integers greater than or equal to 2. Integer d is called the order of the tensor.*

Source Separation in Physical-Chemical Sensing, First Edition.
Edited by Christian Jutten, Leonardo Tomazeli Duarte, and Saïd Moussaoui.

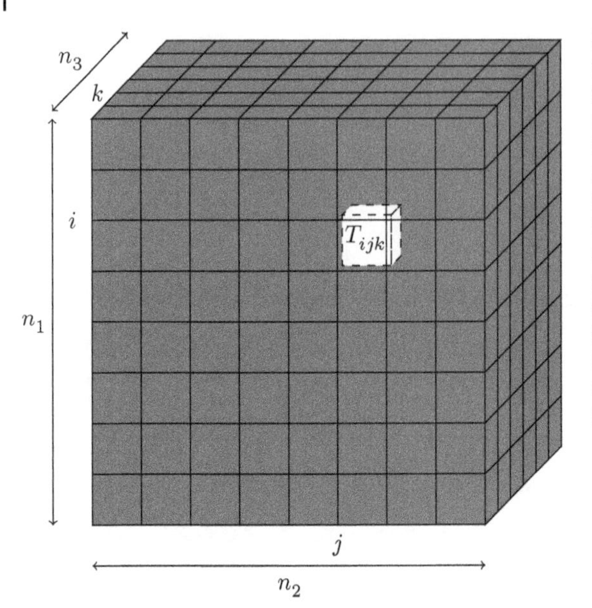

Figure 6.1 A third-order real tensor is nothing more than a three-way array of size $n_1 \times n_2 \times n_3$, that contains real numbers $T_{i,j,k}$ at each entry (i, j, k). To rephrase using a different vocabulary, this real tensor T has three modes of sizes n_1, n_2, and n_3, respectively.

In other words, we consider tensors as arrays containing real numbers, and a d-order tensor is a real array with exactly d dimensions, which we call modes. This means that real matrices are second-order tensors, i.e. matrices are tensors with only two modes, while vectors of $\mathbb{R}^n$ are first-order tensors. A third-order tensor is depicted in Figure 6.1. As a side note, there exist definitions of tensors which are much more general than this one, the most general one in the mathematical sense involving monoidal category theory [1], which is far outside the scope of this chapter. However, an interested reader may turn to [2–4] for other descriptions of tensors.

The study of tensors as a data structure is necessary in many applications where such d-dimensional arrays emerge as efficient representations for the studied phenomenon. For instance, it may be that sensors directly output tensors, e.g. a video or a 3D image. It is also often the case that data are collected along several experimental variables, which become modes in a measurement tensor. This is the case with fluorescence spectroscopy; see the end of this introduction for a description of how fluorescence spectroscopy measurements lead to data being contained naturally in a tensor. Finally, it is sometimes profitable to augment a dataset to obtain a tensor. An example of such augmentation is obtained by stacking shifted versions of gray-scale images. The obtained data set is a collection of matrices, i.e. a third-order tensor.

6.1.2 Separability: A Key Concept for Tensor Decomposition Model

We have sketched what a tensor is and how one may end up manipulating such an object. However, we have yet to define the type of information that we want to extract from tensors.

Recalling from Definition 6.1 that modes are the various ways/dimensions of a tensor, this chapter describes some tools that exploit mode-wise information to explain the content of a tensor. The main mathematical concept that formalizes this idea of collecting mode-wise information to describe the whole tensor is "separability." By definition, a separable function $f(x, y, z)$ verifies the following equality:

$$f(x, y, z) = f_1(x)f_2(y)f_3(z) \tag{6.1}$$

for some functions f_1, f_2, f_3. Of course, very few functions are of this form. A separable function f is therefore entirely described by a triplet of functions f_i, each depending only on a single variable. Therefore, they are desirable to describe multivariate patterns in an understandable, mode-wise manner. We shall see how to do that in Sections 6.2.2 and 6.2.8.

The link between separable functions and tensors is the following: a tensor can be seen as an array collecting values of a sampled multivariate function. Indeed, one can always define a function $f(x, y, z)$ such that

$$T_{ijk} = f(x_i, y_j, z_k). \tag{6.2}$$

If this function f is separable as defined in (6.1), then the corresponding tensor has received several names, and in particular *simple* tensor, or *decomposable* tensor in the early literature. In the remainder, we shall assume the terminology of *separable tensor*, which is more intuitive, and directly related to (6.1). This leads to the following definition:

Definition 6.2 *A separable tensor is a tensor from $\mathbb{R}^{n_1 \times n_2 \times n_3}$ whose general term $\mathcal{T}$ verifies the following equality:*

$$\mathcal{T}_{i,j,k} = a_i b_j c_k \tag{6.3}$$

where $\mathbf{a}, \mathbf{b}, \mathbf{c}$ are real vectors from $\mathbb{R}^{n_1}$, $\mathbb{R}^{n_2}$, and $\mathbb{R}^{n_3}$.

Separable tensors are thus a formal description of what was referred to in the beginning of this introduction as "patterns explaining the data across all modes." We shall see in Section 6.2.2 the definition of tensor rank; it will then be clear that separable tensors are nothing else but *rank-one* tensors. Because rank-one tensors make up for only one simple pattern, a more complex data tensor should be composed of several of them. This is the rationale under the Canonical Polyadic Decomposition (CPD), which is more formally described in Section 6.2.2. Taking this reasoning one step further, a tensor decomposition model always writes a tensor as a sum of separable terms. Therefore, tensor decomposition models always aim at expressing global information contained in a tensor using mode-wise descriptors.

6.1.3 The Fluorescence Excitation Emission Matrix (FEEM)

To conclude this introduction, we explore how fluorescence spectroscopy measurements can lead to a tensor which has a CPD structure, and how this structure can be exploited in practice. Further details are provided in Section 6.6 dedicated to applications of tensor decompositions in food sciences.

Fluorescence spectroscopy takes advantage of the fact that in food sciences, many interesting compounds have their own fluorescent response to stimulation. To be more precise, these compounds react only in a specific range of stimulation wavelengths at various intensities. Fluorophores respond to light excitation by emitting a light that has a spread spectrum, called the emission spectrum. The emitted intensity consequently depends on excitation and emission wavelengths.

Given one sample of possible several fluorescent chemicals, and stimulating the sample with a light of varying wavelengths, a matrix of data is obtained. The two modes of the matrix are the two experimental variables, i.e. the excitation wavelength λ_{ex} and the emission wavelength λ_{em}, and the elements of the matrix are the measured intensity values of the light that the sample outputs. An example is provided in Figure 6.2.

Such a matrix is called a FEEM (Fluorescence Emission Excitation Matrix). Note that FEEM does not restrict to cases where only one compound is present in the solution. But in the case of a single fluorophore, the fluorescence phenomenon is separable with respect to excitation and emission wavelengths, and therefore the general term of a FEEM that contains the spectra of only one compound can be expressed as follows: $\boldsymbol{M}_{\lambda_{ex},\lambda_{em}} = a_{\lambda_{ex}} b_{\lambda_{em}}$, where $\mathbf{a}$ is the excitation spectrum and $\mathbf{b}$ is the emission spectrum. In other words, $\boldsymbol{M}$ is a rank-one matrix, i.e. a rank-one tensor of order 2.

Now suppose that instead of a single sample, one has several samples of the same single chemical, but at various concentrations in the solvent. The fluorescence phenomenon being linear with respect to concentration (using a first-order approximation valid at low concentrations), repeating the above reasoning on each sample yields a separable third-order tensor $\mathcal{T}_{\lambda_{ex},\lambda_{em},k} = a_{\lambda_{ex}} b_{\lambda_{em}} c_k$, where k is the sample index and $\mathbf{c}$ contains the relative

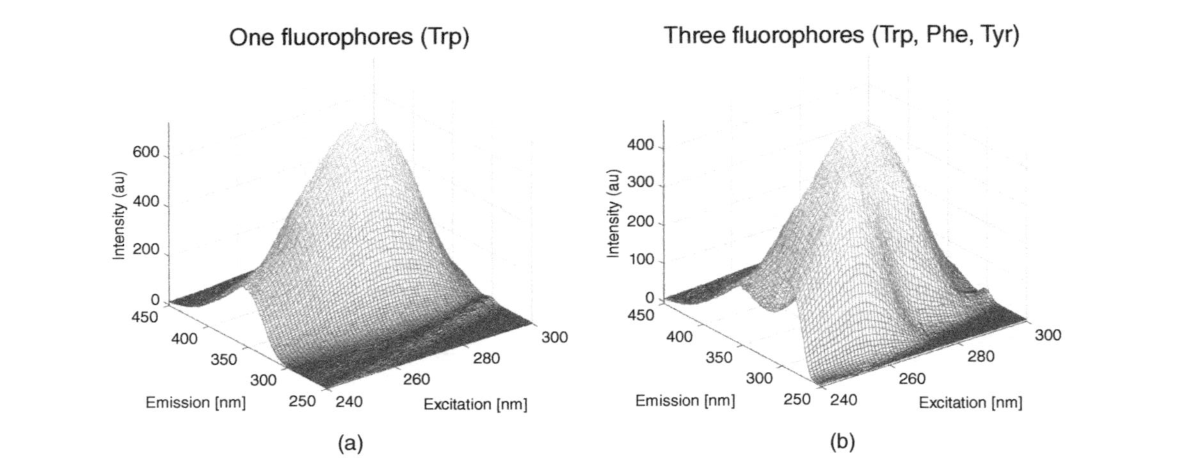

Figure 6.2 (a): A fluorescence Excitation-Emission Matrix (EEM) of a single fluorophore, the amino acid tryptophan. (b): A sample containing three different amino acids; each with a unique contribution (peak) to the EEM. The three amino acids are tryptophan, tyrosine, and phenylalanine. The (a) matrix is separable up to noise, while the (b) matrix is a sum of three separable matrices, also up to noise.

concentrations with respect to the first sample. In other words, ideally a fluorophore gives a simple, rank-one tensor.

At this stage, the reader can understand why studying tensors models is crucial in fluorescence spectroscopy. Indeed, given a tensor $\mathcal{T}$, if an informed user knows that this tensor is a collection of FEEM measured on solutions containing a single compound, then finding $\mathbf{a}$, $\mathbf{b}$, and $\mathbf{c}$ means estimating, from the data tensor, the spectra, and concentrations of the compound. Computing these parameters is typically achieved by solving an optimization problem, as discussed further in Section 6.5.

Usually, studied samples contain more than one chemical compound. The fluorescence phenomenon being additive at low concentrations, a tensor obtained by measuring a mixture is the sum of the separable tensors that would be obtained if each compound in the mixture was measured separately. Using a more technical vocabulary later defined in the chapter in Section 6.2, we shall see that a tensor of fluorescence spectroscopy measurements follows an approximate CPD model, with as many separable terms as there are compounds in the mixture. Again, identifying the parameters of this model can be done by solving an optimization problem, which leads to estimating the various spectra and relative concentrations in the mixture.

To summarize, we have seen that when studying tensors of fluorescence spectroscopy measurements, the data are naturally contained in a tensor that can be written as the sum of a small number of separable (i.e. rank-one) terms. It will be shown later that the CPD is a source separation technique which extracts on each mode a pre-defined number of sources from data stored in a tensor. Interestingly, these sources do not need to be independent in the statistical sense, as opposed to independent component analysis (ICA) for instance (see Section 1.3). In that sense, CPD and other tensor decomposition models are similar to Non-negative Matrix Factorization, presented in Chapter 3 of this book.

6.1.4 Structure of the Chapter

In the remainder of this chapter, we first define more formally a few tensor decompositions (including the CPD), and introduce mild conditions that ensure the underlying sources can be extracted from a tensor, see Section 6.2. In Sections 6.3 and 6.4, we then introduce the concepts of constrained decomposition and coupled decompositions, which are particular tensor-based models inspired from the CPD and actually compute tensor decompositions. Finally, in Section 6.6, we show in more depth how these methods can be employed to extract relevant information from various measurement techniques used in food sciences.

6.1.4.1 Note

For simplicity reasons, we will only work with third-order tensors. However, all the models introduced after this point can trivially be extended to higher orders. Third-order tensors are convenient for notations, and are also more common than tensors of order four and higher in practice.

6.1.4.2 Other Introductions

Survey papers have been proposed in the literature that focus on various aspects of tensors. If the reader wishes to have other accessible introductions to tensors before or after reading the remainder of this chapter, we can recommend these references: [4–7].

6.2 Tensor Decompositions

6.2.1 Tensor-Based Method, the Matrix Case

A convenient way to start addressing the tensor decomposition subject is to look first at the matrix case. Any matrix $\boldsymbol{M}$ of size $n_1 \times n_2$ can be written in the canonical basis as $\boldsymbol{M} = \sum_{ij} M_{ij} \boldsymbol{E}(i,j)$, where $\boldsymbol{E}(i,j)$ is the matrix having only one nonzero entry at position (i,j). This is a trivial decomposition, which only shows that the linear space of $n_1 \times n_2$ matrices is of dimension $n_1 n_2$; it is not so much useful otherwise. In other words, this trivial element-wise decomposition has $n_1\ n_2$ parameters (one for each element in the sum), so it is not parsimonious and is seldom used for extracting information.

Now, looking for more parsimonious representations and bearing in mind the separability property emphasized in the introduction, we can decompose a matrix $\boldsymbol{M}$ into a sum of rank-one terms as $\boldsymbol{M} = \sum_{r=1}^{R} \sigma_r \boldsymbol{D}_r$, where $\boldsymbol{D}_r = \boldsymbol{u}_r \boldsymbol{v}_r^\mathsf{T}$ are unit-norm rank-one matrices, i.e., vectors $\boldsymbol{u}_r$ and $\boldsymbol{v}_r$ are of unit-norm.

Stacking $\boldsymbol{u}_r$ and $\boldsymbol{v}_r$ vertically in matrices $\boldsymbol{U}$ and $\boldsymbol{V}$ of sizes $n_1 \times R$ and $n_2 \times R$, respectively, and setting $\boldsymbol{\Sigma}$ as a $R \times R$ diagonal matrix containing the values σ_r, this decomposition can be rewritten in compact form as $\boldsymbol{M} = \boldsymbol{U}\boldsymbol{\Sigma}\boldsymbol{V}^\mathsf{T}$. The smallest number of rank-one terms R such that all values σ_r are non-zeros is called the rank of $\boldsymbol{M}$ and $\boldsymbol{M}$ is referred to as a low-rank matrix if R is smaller than $\min(n_1, n_2)$.

It can be checked out that such a decomposition exhibits now at most $R(n_1 + n_2 - 1)$ degrees of freedom (number of free parameters). Indeed, each rank-one component has n_1 parameters in $\boldsymbol{u}_r$, n_2 in $\boldsymbol{v}_r$, and one more is σ_r.

However vectors $\boldsymbol{u}_r$ and $\boldsymbol{v}_r$ are normalized, removing one degree of freedom in each, meaning a rank-one component has $n_1 + n_2 - 1$ degree of freedom.

The problem is that this decomposition is not unique as soon as R is strictly greater than one. *Uniqueness* is a key feature of any learning model as soon as the sought parameters must be physically interpreted. For instance, say one wants to interpret parameter matrices $\boldsymbol{U}$ and $\boldsymbol{V}$ as collections of emission and excitation spectra in fluorescence spectroscopy, respectively. The existence of several solutions prohibits this interpretation, as only one of the possible solutions may actually correspond to the desired spectra.

To see this lack of uniqueness, consider any orthonormal $R \times R$ matrix $\boldsymbol{Q}$. Then we also have $\boldsymbol{M} = \boldsymbol{U\Sigma Q}(\boldsymbol{VQ})^{\mathsf{T}}$. This shows that other decompositions of the form $\boldsymbol{M} = \boldsymbol{U}'\boldsymbol{\Sigma}'\boldsymbol{V}'^{\mathsf{T}}$ hold true, if we define $\boldsymbol{V}' = \boldsymbol{VQ}$, $\boldsymbol{\Sigma}'$ the diagonal matrix containing the norm of each column of $\boldsymbol{U\Sigma Q}$ and $\boldsymbol{U}' = \boldsymbol{U\Sigma Q\Sigma}'^{-1}$. Thus, for ensuring uniqueness, orthogonality between columns of matrices $\boldsymbol{U}$ and $\boldsymbol{V}$ is generally imposed, which yields the Singular Value Decomposition (SVD):

$$M_{ij} = \sum_{r=1}^{R} \sigma_r \, U_{ir} V_{jr}, \text{ such that } \boldsymbol{U}^{\mathsf{T}}\boldsymbol{U} = \boldsymbol{I}_{n_1} \text{ and } \boldsymbol{V}^{\mathsf{T}}\boldsymbol{V} = \boldsymbol{I}_{n_2}, \tag{6.4}$$

where $\boldsymbol{I}_{n_i}$ is the identity matrix of size $n_i \times n_i$. The SVD can also be written in a compact form: $\boldsymbol{M} = [\![\boldsymbol{\Sigma}; \boldsymbol{U}, \boldsymbol{V}]\!] := \boldsymbol{U\Sigma V}^{\mathsf{T}}$, with left orthogonal matrices $\boldsymbol{U}$ and $\boldsymbol{V}$, and where $\boldsymbol{\Sigma}$ is diagonal $R \times R$ with positive entries $\Sigma_{rr} = \sigma_r$. As a notation convention, in the rest of the chapter we preferably denote orthogonal matrices with letters $\boldsymbol{U}$ and $\boldsymbol{V}$, and general matrices with other letters.

Another convenient way to write the SVD is the following

$$\boldsymbol{M} = \sum_{r=1}^{R} \sigma_r \, \boldsymbol{D}_r, \tag{6.5}$$

where $\boldsymbol{D}_r$ are rank-one unit-norm matrices, which are orthogonal to each other. The SVD is *unique* if all singular values are distinct. Of course, rank-1 terms in the sum (6.4) or (6.5) can be permuted, because the sum is commutative. This induces a permutation among columns of matrices $\boldsymbol{U}$ and $\boldsymbol{V}$, which can be fixed by sorting singular values in decreasing order in matrix $\boldsymbol{\Sigma}$.

The number of degrees of freedom in the SVD is $R(n_1 + n_2 - R)$, which can be compared to the $R(n_1 + n_2 - 1)$ degrees of freedom we would have if orthogonality was not imposed. More precisely, there are $n_1R - R(R+1)/2$ degrees of freedom in $\boldsymbol{U}$, $n_2R - R(R+1)/2$ in $\boldsymbol{V}$, and R in $\boldsymbol{\Sigma}$. The sum of these three terms is $R(n_1 + n_2 - R)$. Indeed, it is sufficient to think of each column of matrices $\boldsymbol{U}$ and $\boldsymbol{V}$ as normalized and orthogonal to all previous columns, meaning there are for each matrix $\sum_{r=1}^{R} r$ additionally fixed degrees of freedom. This reduced number of degrees of freedom provides intuition as to why SVD is often unique, while unconstrained low-rank matrix factorization is not. Note that imposing non-negativity in matrix $\boldsymbol{\Sigma}$ permits to fix sign indeterminacies among columns of $\boldsymbol{U}$ and $\boldsymbol{V}$, but does not reduce the number of degrees of freedom.

6.2.2 Canonical Polyadic Decomposition, PARAFAC/CanDecomp

Let us now define the Canonical Polyadic Decomposition (CPD) as an extension of SVD for higher-order tensors. In what follows, we will see that for third-order tensors, it is possible to drop the orthogonality constraint of the SVD and still obtain a unique decomposition. This makes the CPD appealing in practice for source separation [8], or for addressing other identification problems [9–12].

Denote by $\boldsymbol{S}$ a three-way diagonal array with σ_r as diagonal entries. For a large enough R, a $n_1 \times n_2 \times n_3$ tensor $\boldsymbol{\mathcal{T}}$ with entries $\mathcal{T}_{ijk}$ can always be decomposed as:

$$\mathcal{T}_{ijk} = \sum_{r=1}^{R} \sigma_r \, A_{ir} B_{jr} C_{kr}, \tag{6.6}$$

where factor matrices $\boldsymbol{A}$, $\boldsymbol{B}$, and $\boldsymbol{C}$ have unit-norm columns, and where weights σ_r may be imposed to be real positive. Indeed, just like with matrices, one may write

$$\boldsymbol{\mathcal{T}} = \sum_{r_1, r_2, r_3}^{n_1, n_2, n_3} \mathcal{T}_{r_1, r_2, r_3} \mathcal{E}(r_1, r_2, r_3) \tag{6.7}$$

where $\mathcal{E}(r_1, r_2, r_3)_{ijk} = \delta_{i,r_1} \delta_{j,r_2} \delta_{k,r_3}$ is a tensor with zeros everywhere except a one in position (r_1, r_2, r_3), and $\delta_{i,r}$ is the Kronecker symbol. Clearly, any tensor $\boldsymbol{\mathcal{T}}$ admits a decomposition in separable terms (6.6) as soon as R is large enough. A first naive upper bound on R is therefore $n_1 n_2 n_3$. On the other hand, since the above decomposition always exists for large enough R, there always exists a *minimal value* of R, which is called the *rank* of tensor $\boldsymbol{\mathcal{T}}$. It will be denoted rank$\{\boldsymbol{\mathcal{T}}\}$, as for matrices. Moreover, decomposition (6.6) with minimal number of terms R is the exact *Canonical Polyadic Decomposition* (CPD) of tensor $\boldsymbol{\mathcal{T}}$. The exact CPD is therefore the decomposition of a tensor $\boldsymbol{\mathcal{T}}$ into a minimal sum of rank-one tensors. A graphical representation of the CPD is given in Figure 6.3. This can also be conveniently written as:

$$\boldsymbol{\mathcal{T}} = \sum_{r=1}^{R} \sigma_r \, \boldsymbol{\mathcal{D}}_r, \tag{6.8}$$

$$\mathcal{T} = \sigma_1 \boldsymbol{a}_1 \otimes \boldsymbol{b}_1 \otimes \boldsymbol{c}_1 + \cdots + \sigma_R \boldsymbol{a}_R \otimes \boldsymbol{b}_R \otimes \boldsymbol{c}_R$$

Figure 6.3 A graphical representation of the CPD. A tensor $\mathcal{T}$ is expressed as a minimal sum of R rank-one tensors, which can themselves be expressed as outer products $\sigma_r \boldsymbol{a}_r \otimes \boldsymbol{b}_r \otimes \boldsymbol{c}_r$.

with $\boldsymbol{D}_r = \boldsymbol{a}_r \otimes \boldsymbol{b}_r \otimes \boldsymbol{c}_r$, where $\boldsymbol{a}_r$ (resp. $\boldsymbol{b}_r$ and $\boldsymbol{c}_r$) denote the unit-norm columns of matrix factor $\boldsymbol{A}$ (resp. $\boldsymbol{B}$ and $\boldsymbol{C}$), and $\otimes$ denotes the outer-product defined as

$$[\boldsymbol{a} \otimes \boldsymbol{b} \otimes \boldsymbol{c}]_{ijk} = a_i b_j c_k. \tag{6.9}$$

The notation used is summarized in pages xxiii and xxiv. As for the matrix SVD, we can decide to sort values σ_r in decreasing order to fix the permutation ambiguity stemming from addition commutativity; note the similarity with (6.4).

A necessary condition for this decomposition to be unique, up to permutations and signs ambiguities, is that the number of degrees of freedom in the left-hand side of equation (6.6) be at least as large as that in the right-hand side. In general there are $n_1 n_2 n_3$ elements in tensor $\mathcal{T}$, and $R(n_1 + n_2 + n_3 + 1)$ parameters in the CPD with $3R$ normalization constraints. In other words, we must have:

$$n_1 n_2 n_3 \geq R(n_1 + n_2 + n_3 - 2). \tag{6.10}$$

As soon as $R < \frac{n_1 n_2 n_3}{n_1+n_2+n_3-2}$, this necessary condition holds, and it becomes possible for the CPD to be unique. We shall subsequently see that this condition is not sufficient to ensure CPD uniqueness, but that for small enough R, the orthogonality constraint is generally not required to obtain a unique decomposition. This is unlike the matrix case, where uniqueness cannot be attained without additional constraints except when $R = 1$. In addition, condition (6.10) may hold true even if R exceeds $\min\{n_1, n_2, n_3\}$, which is not possible under orthogonality constraints. The bound $R \leq n_1 n_2 n_3/(n_1 + n_2 + n_3 - 2)$, induced by the counting (6.10) of degrees of freedom, is studied in [13].

Strangely enough, condition (6.10) is not sufficient to guarantee uniqueness of the CPD. It has been shown that the CPD (6.8) is unique[1] provided

1 Again, if we refer to (6.6) instead of (6.8), uniqueness is to be understood up to permutation of terms in the sum; we made the same observation for matrices in Section 6.2.1 about Eq. (6.5).

the rank is not too large [14–16]. In particular, uniqueness is ensured if:

$$R \leq \frac{1}{2}(\text{krank}\{\boldsymbol{A}\} + \text{krank}\{\boldsymbol{B}\} + \text{krank}\{\boldsymbol{C}\} - 2). \tag{6.11}$$

In the *sufficient condition* above, krank$\{\boldsymbol{A}\}$ denotes Kruskal rank[2] of matrix $\boldsymbol{A}$. Uniqueness of the CPD (6.8) can be ensured under conditions weaker than (6.11); see the recent paper [17] and references therein.

Equation (6.6) can be seen in many ways, among which two are common for a practical use.

- First, it is a tensor factorization model that seeks a small number R of separable patterns to describe the data. This point of view is usually used for justifying the use of tensor decomposition techniques in machine learning as an exploratory technique.
- Second, it is a parameter identification technique where physically meaningful matrices $\boldsymbol{A}, \boldsymbol{B}$, and $\boldsymbol{C}$ are of interest for a further task. In other words, the CPD may be seen as a source separation technique. For instance, matrix $\boldsymbol{A}$ may stand for a spectral signature that enables the identification of the chemical compounds present in a mixture. This second approach contrasts with the first one since uniqueness here is a key feature for interpreting the results. It is mostly seen in bio-medical applications (metabolomics, neuroimaging) or in sensor arrays where the CP decomposition results from a set of physical equations.

The model is particularly useful when it coincides with the physical model of the data, because it provides meaningful solutions due to the uniqueness property.

At this stage, it is worth saying a word about terminology. Even if the CP decomposition has been introduced originally in 1927 by Hitchcock [18], it has been rediscovered in 1970 by Harshman [19] and by Carroll and Chang [20]. They gave it the names Parafac and Candecomp, respectively. To unify the terminology, Kiers [21] proposed the acronym CP, which can wisely stand for “Candecomp/Parafac,” as well as for “Canonical Polyadic.” In the sequel we shall refer to (6.6) and (6.8) as the exact *CP Decomposition* (CPD) when R is indeed minimal and reveals *tensor rank*. If we assume a multilinear model without column normalization, then an additional scaling indeterminacy appears:

$$\mathcal{T}_{ijk} = \sum_{r=1}^{R} A_{ir} B_{jr} C_{kr}, \quad \text{denoted by} \quad \mathcal{T} = [\![\boldsymbol{A}, \boldsymbol{B}, \boldsymbol{C}]\!]. \tag{6.12}$$

---

2 The Kruskal rank of a matrix is the largest number κ such that any subset of κ columns is full rank. Hence Kruskal rank cannot exceed rank. For almost all matrices, Kruskal rank is equal to rank.

Such a decomposition without unit-norm constraint is not unique and contains $2R$ free parameters (scaling factors). More precisely, without normalization constraints, the CPD model features $(n_1 + n_2 + n_3)R$ parameters since each rank-one component involves three vectors of sizes n_1, n_2, and n_3, and the norms of these vectors may be pulled apart without modifying the sum of rank-one terms, since

$$\forall \mu, \lambda, \nu \in \mathbb{R}, \quad \mu \boldsymbol{a} \otimes \lambda \boldsymbol{b} \otimes \nu \boldsymbol{c} = \mu\lambda\nu\,(\boldsymbol{a} \otimes \boldsymbol{b} \otimes \boldsymbol{c}). \tag{6.13}$$

Therefore, only $(n_1 + n_2 + n_3 - 2)R$ parameters can possibly be identified. More bibliographical pointers to the CPD may be found in [4, 22].

6.2.3 Manipulation of Tensors

In the following we introduce some commonly used manipulations of tensors, namely how to unfold them into matrices or vectors. These manipulations are useful in many different contexts, from the derivation of algorithms using linear algebra to programming. We also explain how these unfoldings link the outer product with the Kronecker product of matrices. The Kronecker product of two matrices $\boldsymbol{A} \in \mathbb{R}^{n_1 \times n_2}$ and $\boldsymbol{B} \in \mathbb{R}^{m_1 \times m_2}$ is denoted by $\boldsymbol{A} \boxtimes \boldsymbol{B} \in \mathbb{R}^{n_1 m_1 \times n_2 m_2}$ and is defined by:

$$\boldsymbol{A} \boxtimes \boldsymbol{B} := \left[\begin{array}{c|c|c|c} a_{11}\boldsymbol{B} & a_{12}\boldsymbol{B} & \dots & a_{1n_2}\boldsymbol{B} \\ \hline a_{21}\boldsymbol{B} & & & \\ \hline \vdots & & & \\ \hline a_{n_1 1}\boldsymbol{B} & & & a_{n_1 n_2}\boldsymbol{B} \end{array}\right]. \tag{6.14}$$

6.2.3.1 Vectorization

There are *a priori* a combinatorial number of ways to arbitrarily transform a tensor in $\mathbb{R}^{n_1 \times n_2 \times n_3}$ into a vector in $\mathbb{R}^{n_1 n_2 n_3}$. We refer to such an operator as a vectorization. However, it is quite clear that an arbitrary, pseudo-random vectorization will destroy any nice structure that the input tensors might have. In particular, if $\mathcal{T}$ follows a CPD, it is a reasonable ordeal to ask for its vectorized version to also satisfy a similar equation.

Taking this into consideration, there are still a few different ways to define a vectorization, and several co-exist in the literature. We propose to use the row-wise vectorization that exhibits a nice and simple link between the outer product and the Kronecker product as shown below. Let $\mathcal{T}$ be a $n_1 \times n_2 \times n_3$ tensor; we define the row-wise vectorization as follows:

$$\text{vec}(\mathcal{T})_{(i-1)n_3 n_2 + (j-1)n_3 + k} = T_{ijk}. \tag{6.15}$$

See Figure 6.4 for a graphical explanation of these formulas.

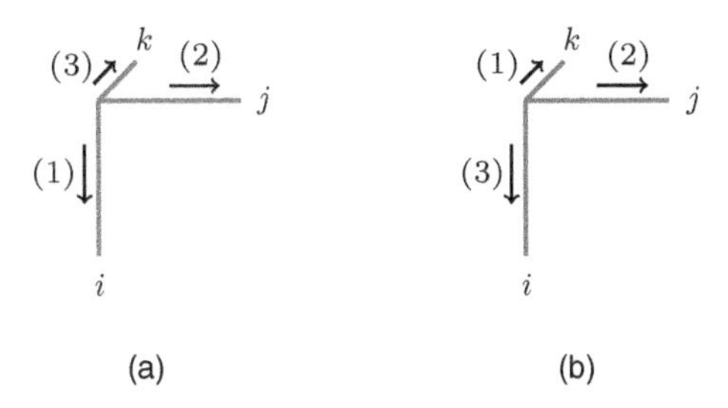

Figure 6.4 The suggested row-wise vectorization (b) reads the entries of the tensor along the last index in the lexicographic order. This contrasts with the column-wise vectorization (a) which is also often encountered. (a) Columnwise vectorization and (b) suggested vectorization.

Note that the above definition of the vectorization operator may not coincide with native implementation in some languages such as MATLAB, which is column-wise, but does coincide with the memory layout of other languages such as C. More details can be found in [23].

With the above definitions, we have in particular the property:

$$\text{vec}(\boldsymbol{a} \otimes \boldsymbol{b} \otimes \boldsymbol{c}) = \boldsymbol{a} \boxtimes \boldsymbol{b} \boxtimes \boldsymbol{c}, \tag{6.16}$$

which is not enjoyed by most other definitions, where terms need to be permuted. Because the vectorization operation is trivially linear, we have further that

$$\text{vec}\left(\sum_{r=1}^{R} \sigma_r \boldsymbol{a}_r \otimes \boldsymbol{b}_r \otimes \boldsymbol{c}_r\right) = \sum_{r=1}^{R} \sigma_r \boldsymbol{a}_r \boxtimes \boldsymbol{b}_r \boxtimes \boldsymbol{c}_r, \tag{6.17}$$

which nicely transposes the CPD to a Kronecker equation.

Furthermore, this Kronecker equation can itself be written in a more compact format, making use of the Khatri–Rao product. The Khatri–Rao product between two matrices $\boldsymbol{A} = [\boldsymbol{a}_1, \dots, \boldsymbol{a}_R]$ and $\boldsymbol{B} = [\boldsymbol{b}_1, \dots, \boldsymbol{b}_R]$ is nothing else than a column-wise Kronecker product:

$$\boldsymbol{A} \odot \boldsymbol{B} = \left[\boldsymbol{a}_1 \boxtimes \boldsymbol{b}_1 | \dots | \boldsymbol{a}_R \boxtimes \boldsymbol{b}_R\right] = \left[\begin{array}{c|c|c} A_{11}\boldsymbol{b}_1 & \dots & A_{1R}\boldsymbol{b}_R \\ \hline \vdots & \dots & \vdots \\ \hline A_{n_1 1}\boldsymbol{b}_1 & \dots & A_{n_1 R}\boldsymbol{b}_R \end{array}\right]. \tag{6.18}$$

Then it is easy to check that

$$\text{vec}\left(\sum_{r=1}^{R} \sigma_r \boldsymbol{a}_r \otimes \boldsymbol{b}_r \otimes \boldsymbol{c}_r\right) = (A \odot B \odot C)\boldsymbol{s}, \tag{6.19}$$

with $\boldsymbol{s}$ the vector of length R containing all values σ_r.

6.2.3.2 Matricization

Similar to vectorization, to be able to resort to known results borrowed from linear algebra, we shall sometimes need to store the elements of three-way arrays into two-way arrays (i.e. matrices). This can be done in various manners, but we shall retain three of them, namely those having as number of rows one dimension of the original tensor. The operation transforming a d-way array into a matrix is known as matrix unfolding, matrix flattening, or matricization [7, 23, 24]. If $\mathcal{T}$ is a tensor of dimensions $n_1 \times n_2 \times n_3$, we shall use the following matrix unfoldings $\boldsymbol{T}^{(p)}$ along mode p defined as

$$
\begin{aligned}
\boldsymbol{T}^{(1)} \text{ is } n_1 \times n_2 n_3 &: \mathcal{T}^{(1)}_{in} = \mathcal{T}_{ijk}, \text{ with } n = k + (j-1)n_2 \\
\boldsymbol{T}^{(2)} \text{ is } n_2 \times n_3 n_1 &: \mathcal{T}^{(2)}_{jp} = \mathcal{T}_{ijk}, \text{ with } p = k + (i-1)n_3 \\
\boldsymbol{T}^{(3)} \text{ is } n_3 \times n_1 n_2 &: \mathcal{T}^{(3)}_{kq} = \mathcal{T}_{ijk}, \text{ with } q = j + (i-1)n_1
\end{aligned}
\tag{6.20}
$$

These unfoldings are illustrated in Figure 6.5. Matricizations along mode p fold the pth dimension of the tensor on the rows of the matricized tensor. This maps the range of the tensor on the pth dimension to the column space of the matricized tensor.

Again, similar to the vectorization, tensors are matricized by selecting entries along the deepest index first. This way, matricizations and vectorizations do not permute the terms from tensor products to Kronecker products. Indeed, it holds that

$$
\begin{aligned}
[\boldsymbol{a} \otimes \boldsymbol{b} \otimes \mathbf{c}]^{(1)} &= \boldsymbol{a} \otimes \boldsymbol{b} \boxtimes \mathbf{c} \\
[\boldsymbol{a} \otimes \boldsymbol{b} \otimes \mathbf{c}]^{(2)} &= \boldsymbol{b} \otimes \boldsymbol{a} \boxtimes \mathbf{c} \\
[\boldsymbol{a} \otimes \boldsymbol{b} \otimes \mathbf{c}]^{(3)} &= \mathbf{c} \otimes \boldsymbol{a} \boxtimes \boldsymbol{b},
\end{aligned}
\tag{6.21}
$$

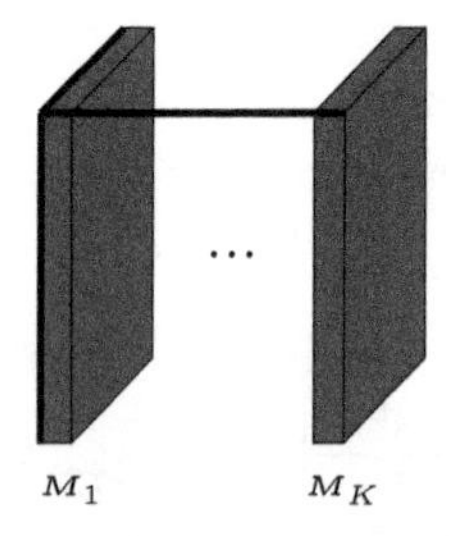

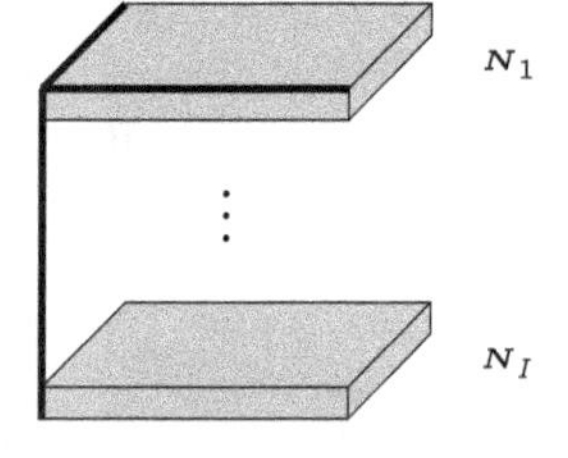

$$
\begin{aligned}
\boldsymbol{T}^{(1)} &= [\boldsymbol{M}_1 | \dots | \boldsymbol{M}_K] \\
\boldsymbol{T}^{(2)} &= [\boldsymbol{N}_1 | \dots | \boldsymbol{N}_I] \\
\boldsymbol{T}^{(3)} &= [\boldsymbol{N}_1^T | \dots | \boldsymbol{N}_I^T]
\end{aligned}
$$

Figure 6.5 Three unfoldings of tensor $\mathcal{T}$.

In terms of matrix unfoldings defined in (6.20), using (6.21) and the linearity of the matricization, the CPD can be written in three different ways:

$$\begin{aligned} T^{(1)} &= US^{(1)}(V \boxtimes W)^{\mathsf{T}} = U\text{diag}\{s\}(V \odot W)^{\mathsf{T}} \\ T^{(2)} &= VS^{(2)}(U \boxtimes W)^{\mathsf{T}} = V\text{diag}\{s\}(U \odot W)^{\mathsf{T}} \\ T^{(3)} &= WS^{(3)}(U \boxtimes V)^{\mathsf{T}} = W\text{diag}\{s\}(U \odot V)^{\mathsf{T}} \end{aligned} \tag{6.22}$$

As a side note, there exist several definitions of tensor-matrix bijective maps in the literature. What is important is to define the inverse map and related properties consistently.

6.2.3.3 Contractions and CPD

It is convenient to have at our disposal a compact notation to indicate summations on several indices. We shall assume the notation proposed in [7]:

$$\mathcal{T} = [\![\mathcal{G}; A, B, C]\!], \tag{6.23}$$

meaning just that $\mathcal{T}_{ijk} = \sum_{\ell mn} A_{i\ell} B_{jm} C_{kn}\ \mathcal{G}_{\ell mn}$. Note that another notation has been proposed in [25] and would be equally meaningful: $\mathcal{T} = (A, B, C) \cdot \mathcal{G}$. Some authors also write (6.23) as $\mathcal{T} = (A \otimes B \otimes C) \cdot \mathcal{G}$, which accounts for the fact that $\mathcal{T}$ is the image of $\mathcal{G}$ by the multilinear operator defined by $A \otimes B \otimes C$ [23]. In the remainder, only notation (6.23) will be used.

In particular, if we have that $\mathcal{T}_{ijk} = \sum_r A_{ir} B_{jr} C_{kr}$, then this could be denoted as $\mathcal{T} = [\![\mathcal{I}; A, B, C]\!]$, where $\mathcal{I}$ is a diagonal 3-way array with ones on its diagonal. In such a case, one can omit tensor $\mathcal{I}$ and just write:

$$\mathcal{T} = [\![A, B, C]\!]. \tag{6.24}$$

Note that the CPD (6.6) can thus be written as $[\![S; A, B, C]\!]$, where S here denotes a diagonal tensor whose entry (q, q, q) equals σ_q. Moreover, by setting $A' = [\sigma_1 a_1, \dots, \sigma_R a_R]$, it holds that

$$[\![A', B, C]\!] = [\![S; A, B, C]\!]. \tag{6.25}$$

In other words, the CPD may be written in several equivalent notations. In this chapter, we will use whichever format is more convenient in each situation.

6.2.4 The Chain Rule

There exists a useful algebra result making use of the compact contraction notation defined in Section 6.2.3.3. Although it can be proven almost trivially using more general results from tensor algebra, we shall formulate it and prove it in simple terms.

Property 6.1 (Chain Rule) *Given matrices* $\mathbf{A}$, $\mathbf{B}$, $\mathbf{C}$ *and* $\mathbf{A}'$, $\mathbf{B}'$, *and* $\mathbf{C}'$ *that are compatible for products pairwise, it holds that*

$$[\![[\![\mathcal{T}; \mathbf{A}, \mathbf{B}, \mathbf{C}]\!] ; \mathbf{A}', \mathbf{B}', \mathbf{C}']\!] = [\![\mathcal{T}; \mathbf{A}'\mathbf{A}, \mathbf{B}'\mathbf{B}, \mathbf{C}'\mathbf{C}]\!]. \tag{6.26}$$

We coin this property the chain rule. To prove this result it is sufficient to show that for any mode, here arbitrarily the first one,

$$[\![[\![\mathcal{T}; \mathbf{A}, \mathbf{B}, \mathbf{C}]\!] ; \mathbf{A}', \mathbf{I}, \mathbf{I}]\!] = [\![\mathcal{T}; \mathbf{A}'\mathbf{A}, \mathbf{B}, \mathbf{C}]\!], \tag{6.27}$$

and then apply this partial result sequentially. This partial result is obtained by observing that

$$\begin{aligned} [\![[\![\mathcal{T}; \mathbf{A}, \mathbf{B}, \mathbf{C}]\!] ; \mathbf{A}', \mathbf{I}, \mathbf{I}]\!]_{ijk} &= \textstyle\sum_{l'} A'_{il'} \sum_{lmn} A_{l'l} B_{jm} C_{kn} \mathcal{T}_{lmn} \\ &= \textstyle\sum_{lmn} \left(\sum_{l'} A'_{il'} A_{l'l} \right) B_{jm} C_{kn} \mathcal{T}_{lmn} \\ &= [\![\mathcal{T}; \mathbf{A}'\mathbf{A}, \mathbf{B}, \mathbf{C}]\!]_{ijk}, \end{aligned} \tag{6.28}$$

and since this proof clearly holds for modes 2 and 3 as well by symmetry, we have just proved Property 6.1.

The chain rule has an important corollary that we will use throughout the chapter:

Property 6.2 *Given invertible matrices* $\mathbf{A}, \mathbf{B}, \mathbf{C}$, *and two tensors* $\mathcal{T}$ *and* $\mathcal{G}$, *if the dimensions are compatible, then*

$$[\![\mathcal{T}; \mathbf{A}, \mathbf{B}, \mathbf{C}]\!] = \mathcal{G} \leftrightarrow \mathcal{T} = [\![\mathcal{G}; \mathbf{A}^{-1}, \mathbf{B}^{-1}, \mathbf{C}^{-1}]\!], \tag{6.29}$$

6.2.5 Multilinear Singular Value Decomposition

A second possibility to extend SVD to tensors is to keep orthogonality of factor matrices. In that case, a $n_1 \times n_2 \times n_3$ tensor $\mathcal{T}$ is decomposed as:

$$\mathcal{T} = [\![\mathcal{G}; \mathbf{U}, \mathbf{V}, \mathbf{W}]\!], \tag{6.30}$$

where the so-called core tensor $\mathcal{G}$ is of size $R_1 \times R_2 \times R_3$, smaller than $\mathcal{T}$, that is: $R_1 \leq n_1, R_2 \leq n_2, R_3 \leq n_3$, and factor matrices $\mathbf{U}, \mathbf{V}$ and $\mathbf{W}$ have orthogonal unit-norm columns. This is interesting only if at least one dimension is strictly smaller, or when the core is sparser, which means that a compression[3] has been performed. Again, it is then clear by just counting degrees of freedom that the diagonal form generally cannot be imposed[4] in $\mathcal{G}$.

---

3 For the moment, this compression is lossless. Lossy compression will be addressed in Section 6.5.3.1.

4 In fact, tensors that are orthogonally diagonalizable form a very small class, and their rank must be bounded by all their dimensions.

In terms of matrix unfoldings defined in (6.20), the multilinear SVD can be written in three different ways:

$$\begin{aligned} \boldsymbol{T}^{(1)} &= \boldsymbol{U}\boldsymbol{G}^{(1)}(\boldsymbol{V} \otimes \boldsymbol{W})^{\mathsf{T}} \\ \boldsymbol{T}^{(2)} &= \boldsymbol{V}\boldsymbol{G}^{(2)}(\boldsymbol{U} \otimes \boldsymbol{W})^{\mathsf{T}} \\ \boldsymbol{T}^{(3)} &= \boldsymbol{W}\boldsymbol{G}^{(3)}(\boldsymbol{U} \otimes \boldsymbol{V})^{\mathsf{T}} \end{aligned} \tag{6.31}$$

Indeed, it can be noticed that the multilinear SVD is nothing more than another decomposition of $\mathcal{T}$ into separable terms:

$$\mathcal{T} = \sum_{lmn}^{R_1,R_2,R_3} \mathcal{G}_{lmn} \boldsymbol{u}_l \otimes \boldsymbol{v}_m \otimes \boldsymbol{w}_n, \tag{6.32}$$

with $\boldsymbol{u}_l, \boldsymbol{v}_m, \boldsymbol{w}_n$ the columns of matrices $\boldsymbol{U}, \boldsymbol{V}, \boldsymbol{W}$, respectively. Therefore the unfoldings formulas are obtained from (6.21) by linearity. These three writings show that matrices $\boldsymbol{U}$, $\boldsymbol{V}$, and $\boldsymbol{W}$ are built with the left singular vectors of matrices $\boldsymbol{T}^{(1)}$, $\boldsymbol{T}^{(2)}$, and $\boldsymbol{T}^{(3)}$, respectively. They can hence be computed by matrix SVDs. Once they are known, the core tensor can in turn be computed as

$$\mathcal{G} = [\![\mathcal{T}; \boldsymbol{U}^{\mathsf{T}}, \boldsymbol{V}^{\mathsf{T}}, \boldsymbol{W}^{\mathsf{T}}]\!], \tag{6.33}$$

using (6.29) of Property 6.2. One defines the *multilinear rank* as the triplet of minimal values (R_1, R_2, R_3) such that (6.30) holds exactly. Then it can be shown that

$$\max\{R_1, R_2, R_3\} \leq \operatorname{rank}\{\mathcal{T}\} \leq \min\{R_1R_2, R_2R_3, R_3R_1\}. \tag{6.34}$$

Because the multilinear SVD is computed with the help of three SVDs, it enjoys the same uniqueness conditions. The multilinear SVD has been first suggested by Kroonenberg and Leeuw in 1980 [26], and further studied in 2000 by De Lathauwer *et al.* in [24] under the name of High-Order SVD (HOSVD). But the premises of multilinear SVD appeared earlier with the Tucker3 decomposition, which we address now.

6.2.6 Tucker

Tucker proposed much earlier [27], in 1966, a multilinear decomposition similar to (6.30) but without orthogonality constraints on factor matrices:

$$\mathcal{T} = [\![\mathcal{G}; \boldsymbol{A}, \boldsymbol{B}, \boldsymbol{C}]\!]. \tag{6.35}$$

The consequence of relaxing all constraints is that this decomposition – often referred to as Tucker3 – is not unique anymore, even if the size of the core is the same. The 3 in Tucker3 refers to the number of modes that are subspaced in the decomposition. The Tucker3 decomposition is also often called the Tucker format, by opposition to Tucker decomposition, because of the uniqueness issue [2]. Tucker3 decomposition formally encompasses both multilinear SVD and CPD, which appear as constrained versions. Other constraints such as non-negativity or sparsity could be thought of and would yield other decompositions, see Section 6.3.

The particular case when one mode is not reduced in dimension, which amounts to fixing the factor on that mode to the identity matrix, say $\boldsymbol{C} = \boldsymbol{I}$, is sometimes of interest and has received the name of Tucker2 (since now strict subspaces are defined in only two modes). It can be denoted $\mathcal{T} = [\![\mathcal{G}; \boldsymbol{A}, \boldsymbol{B}, \boldsymbol{I}]\!]$.

6.2.7 PARAFAC2

PARAFAC2 has been introduced in [28, 29]. It differs from the CPD (6.8) by the fact that a matrix factor, e.g. the first one, may not be the same for each matrix slice. Rather, these first-mode factors are related by an orthogonal transform:

$$\mathcal{T} = [\![\boldsymbol{A}(k), \boldsymbol{B}, \boldsymbol{C}]\!], \quad \boldsymbol{A}(k) = \boldsymbol{P}(k)\boldsymbol{H}, \quad \boldsymbol{P}(k)^T\boldsymbol{P}(k) = \boldsymbol{I}, \tag{6.36}$$

with a slight abuse of notation. In other words, if the data tensor is preprocessed as follows

$$\forall k \leq n_1, \; \mathcal{T}(k, :, :) \leftarrow \boldsymbol{P}(k)^T \mathcal{T}(k, :, :), \tag{6.37}$$

then it admits the CP decomposition $[\![\boldsymbol{H}, \boldsymbol{B}, \boldsymbol{C}]\!]$. Again, this means that PARAFAC2 is not a sum of separable terms, but becomes one after a linear transformation of each slice in the tensor. More details on the Parafac2 decomposition, from the coupled decomposition perspective, are given in a Block in Section 6.4.2.

6.2.8 Approximate Decomposition

Up to now, we have talked about *exact* decompositions. For instance, any tensor can be decomposed exactly as in (6.8). However, an exact representation is generally not suitable. Indeed, decomposition (6.8) is unique only if the tensor rank is not too large. On the other hand, (6.8) is exactly verified as long as the rank is large enough, but uniqueness may not be guaranteed. In fact, in the presence of noise, the minimal rank for (6.8) to hold may be larger than the Kruskal upper bound [30]. In addition, the underlying physical model is generally of interest only for a reasonably small value of R

that for instance may stand for the number of unknown sources in a source separation problem. For these reasons, a low-rank approximation is needed.

Instead of computing the exact CPD (6.6), a first idea is to minimize the objective

$$\Upsilon(\boldsymbol{S}, \boldsymbol{A}, \boldsymbol{B}, \boldsymbol{C}) = \|\mathcal{T} - \sum_{r=1}^{R} \sigma(r)\, \boldsymbol{a}(r) \otimes \boldsymbol{b}(r) \otimes \mathbf{c}(r)\|_F^2, \tag{6.38}$$

for a fixed[5] value of R, supposed to be smaller than the rank of $\mathcal{T}$. Also, define the Frobenius norm for tensors as $\|T\|_F^2 = \sum_{i,j,k=1}^{n_1,n_2,n_3} T_{ijk}^2$.

Unfortunately, the low-rank tensor approximation problem is generally ill-posed for tensors (if $R > 1$ and $d > 2$) [25, 31]. More precisely, a minimizer of (6.38) may not exist. This is in contrast with matrices, for which a low-rank approximation can be easily computed by truncating the SVD [32]. Some solutions for this are discussed in Section 6.3 relative to constrained decompositions. However, in practice, this fact is often overlooked. Section 6.5 details how to compute an approximate CPD by tentatively solving optimization problems similar to minimizing (6.38).

6.3 Constraints in Decompositions

Although tensor decomposition techniques have proven useful in a wide range of applications, they are seldom used as a black box model. Rather, problem-specific constraints are often applied on the factors of the decomposition. There are several main reasons to apply constraints to a tensor decomposition model [22]:

- Despite the identifiability properties of tensor decompositions, the computed parameters may not fulfill key properties of the sought factors, therefore hindering interpretability of the results. Imposing constraints, for instance non-negativity, may ensure interpretable results are obtained.
- When the parameters of a tensor decomposition model are not identifiable, constraints can restore identifiability, e.g. by reducing the size of the search space.

---

5 The problem of choosing the appropriate rank to obtain a meaningful approximation is application-dependent and typically very intricate. As detailed in Section 6.5.1.2, deflation strategies should not be employed in the general case.

- The underlying optimization problem of low-rank approximations can be shown to have a solution in the presence of constraints, while it may not in the general case as explained in Section 6.2.8.
- Estimation performance is increased in a noisy scenario.

Below, some of the most widely used constrained tensor decomposition models are introduced. Many constrained models are however not discussed here, since their derivation is either straightforward or of relatively lesser importance in source separation applications.

Among others, the following constraints have been explored in the literature: non-negativity [33–35], orthogonality [34, 36–40], smoothness [41–44], unimodality [33], simplex or sum-to-one [22, 45], dictionary and sparsity [46–49], coherence constraints to ensure existence of approximation [50]. Some of these constraints will be developed in subsequent paragraphs.

6.3.1 Non-negativity

When dealing with data acquired by measuring physical properties of natural processes, such as fluorescence or reflectance measurements, one of the most widely encountered a priori information available on the model parameters is non-negativity. Indeed, parameters related to many types of spectra or concentrations must be non-negative by definition, and cannot be easily interpreted if they are partially negative.

Of course, non-negativity constraints may be applied in any tensor decomposition model, but in this section we focus on the CPD, which is by far the most studied non-negative tensor decomposition model. At the end of this section, we discuss briefly the non-negative Tucker decomposition.

6.3.1.1 Non-negative CPD

Formally, a non-negative CPD is derived as follows:

$$\mathcal{T} = [\![\boldsymbol{A}, \boldsymbol{B}, \boldsymbol{C}]\!] \quad \text{and} \quad \boldsymbol{A} \succeq 0, \boldsymbol{B} \succeq 0, \boldsymbol{C} \succeq 0, \tag{6.39}$$

where the inequality signs $\succeq$ are to be understood entry-wise. Clearly, without noise nor modeling error, non-negative factors imply non-negative tensor data[6] $\mathcal{T}$, but a non-negative tensor could be written as a CPD model with negative entries in factor matrices. Therefore, it may occur that the rank of a tensor, which is the minimal number of columns in unconstrained factor matrices, is strictly smaller than the *non-negative rank*, which is the minimal number of columns in non-negative factor matrices and will be

6 In practice negative entries may appear in these tensors because of measurement error, which does not in principle prevent from fitting an approximate non-negative CPD.

denoted $\text{rank}^+\{T\}$. A simple example of discrepancy[7] between rank and non-negative rank is obtained by considering the following tensor (written slice-wise):

$$\mathcal{T} = \left[\begin{array}{cc|cc} 1 & 1 & 1 & 1 \\ 1 & 1 & 1 & 0 \end{array}\right], \tag{6.40}$$

which has rank 2, but non-negative rank 3. Indeed,

$$\mathcal{T} = \begin{bmatrix} 1 \\ 1 \end{bmatrix} \otimes \begin{bmatrix} 1 \\ 1 \end{bmatrix} \otimes \begin{bmatrix} 1 \\ 1 \end{bmatrix} + \begin{bmatrix} 0 \\ 1 \end{bmatrix} \otimes \begin{bmatrix} 0 \\ 1 \end{bmatrix} \otimes \begin{bmatrix} 0 \\ -1 \end{bmatrix}, \tag{6.41}$$

shows that $\text{rank}\{\mathcal{T}\} \leq 2$ (while it is clear that $\text{rank}\{\mathcal{T}\} > 1$). However it can be shown[8] that there exists no way to write $\mathcal{T}$ as the sum of two rank-one tensors with non-negative entries so that $\text{rank}\{\mathcal{T}\}_+ > 2$, and

$$\mathcal{T} = \begin{bmatrix} 1 \\ 1 \end{bmatrix} \otimes \begin{bmatrix} 1 \\ 1 \end{bmatrix} \otimes \begin{bmatrix} 1 \\ 0 \end{bmatrix} + \begin{bmatrix} 1 \\ 1 \end{bmatrix} \otimes \begin{bmatrix} 1 \\ 0 \end{bmatrix} \otimes \begin{bmatrix} 0 \\ 1 \end{bmatrix}$$
$$+ \begin{bmatrix} 1 \\ 0 \end{bmatrix} \otimes \begin{bmatrix} 1 \\ 1 \end{bmatrix} \otimes \begin{bmatrix} 0 \\ 1 \end{bmatrix}. \tag{6.42}$$

is a rank 3 non-negative CPD of $\mathcal{T}$.

Therefore, although the parameters of the PARAFAC model are identifiable, adding non-negativity constraints may change the solution to the decomposition problem entirely. It has been shown recently however that in a generic case, i.e. choosing a non-negative tensor at random, non-negative rank and the usual tensor rank match [51].

As stated earlier, non-negativity constraints are also important to make the approximation problem well posed. Indeed, non-negativity constraints prevent components cancelation referred to as "degeneracy" [52–55] by bounding the set of admissible parameters [34]. More precisely, under non-negativity constraints, the cost function (6.38) becomes coercive[9] and one can define a compact ball, possibly very large, within which the cost is upper-bounded. Because this cost is continuous in all parameters, it must reach its minimum value within that ball. In contrast, without non-negativity constraints, rank-one components can cancel out while growing to infinity so that such a compact ball may not exist.

---

7 Note that this discrepancy also exists for non-negative matrices [4]: non-negative rank can be strictly larger than rank. Non-negative matrices are used in Chapter 3

8 In a nutshell, any rank-one term in a non-negative decomposition of $\mathcal{T}$ must have a zero in the second entry of one of the vector component, e.g. $A_{2r} = 0$. However the second row, column and fiber of $\mathcal{T}$ are nonzero (they are all equal to $\begin{bmatrix} 1 \\ 0 \end{bmatrix}$). It can be observed that at least three rank-one terms are required to place these nonzero elements.

9 the cost grows to infinity in all directions of the parameter space.

In addition, non-negativity can guarantee uniqueness of the best low-rank approximation [35], and even the uniqueness of the CPD of the best low non-negative rank approximate [51]. For these reasons, non-negativity should be imposed each time it has a physical justification [34, 44, 56–60].

6.3.1.2 Non-negative Tucker Decomposition

Non-negativity constraints have also been extensively used along with the Tucker model introduced in Section 6.2.6:

$$\mathcal{T} = [\![\mathcal{G}; \boldsymbol{U}, \boldsymbol{V}, \boldsymbol{W}]\!] \quad \text{and} \quad \boldsymbol{U} \geq 0, \boldsymbol{V} \geq 0, \boldsymbol{W} \geq 0, \mathcal{G} \geq 0. \tag{6.43}$$

Similar to non-negative matrix factorization, by adding non-negativity constraints in the Tucker model, one hopes to obtain a uniquely defined tensor decomposition [61].

The field of applying constraints on Tucker models was pioneered by Smilde and Kiers in a series of papers [62–64]. They realized that it was mostly necessary to add several constraints such as non-negativity, forcing core elements to zero in order to obtain identified models.

But, in fact, little is known on that topic. It has been shown that in the restrictive case when the dimensions R_1, R_2, R_3 in the non-negative Tucker decomposition match the non-negative ranks of the factors $\boldsymbol{U}, \boldsymbol{V}, \boldsymbol{W}$ *and* the non-negative ranks of the unfoldings, the uniqueness of the non-negative Tucker model is equivalent to the uniqueness of three non-negative matrix factorizations of each unfolding of the tensor, which is a difficult condition to satisfy [65]. Therefore, the non-negative Tucker decomposition does not ship with powerful uniqueness properties. Nonetheless it can still prove useful as a nonlinear dimensionality reduction technique.

A control on the sparseness of factor matrices can be introduced, e.g. thanks to a ℓ_1 norm penalty, as for non-negative matrices [66]. The advantage is that it empirically leads to a unique solution, namely the sparsest [67, 68].

If an alternating algorithm is used, a proximal term can be inserted to guarantee local convergence [69], see Section 6.5.

6.3.2 Block Decompositions

In some applications, like fluorescence spectroscopy where components have the same concentration over all experiments, or like multipath propagation in antenna array processing, it may happen that one factor matrix in the CPD has collinear columns, which prevents CP uniqueness since one of Kruskal rank is then equal to 1 in (6.11). More precisely, let $\boldsymbol{D}$ a tensor be written as:

$$\mathcal{D} = \boldsymbol{a} \otimes \boldsymbol{b}_1 \otimes \boldsymbol{c}_1 + \boldsymbol{a} \otimes \boldsymbol{b}_2 \otimes \boldsymbol{c}_2, \tag{6.44}$$

then $\mathcal{D}$ can be equivalently written as a so-called Block Term Decomposition (or BTD in short) [70–72] by factorizing component $\boldsymbol{a}$:

$$\mathcal{D} = \boldsymbol{a} \otimes (\boldsymbol{b}_1 \otimes \boldsymbol{c}_1 + \boldsymbol{b}_2 \otimes \boldsymbol{c}_2) = \boldsymbol{a} \otimes \boldsymbol{B}\boldsymbol{C}^T. \tag{6.45}$$

A sum of terms similar to equation (6.45) leads to a decomposition of the form:

$$\mathcal{T} = \sum_{r=1}^{R} \boldsymbol{a}_r \otimes \boldsymbol{B}_r \boldsymbol{C}_r^T, \tag{6.46}$$

where matrices $\boldsymbol{B}_r$ and $\boldsymbol{C}_r$ are of respective sizes $n_2 \times L_r$ and $n_3 \times L_r$ for a $n_1 \times n_2 \times n_3$ tensor $\mathcal{T}$. Consequently, the number R of terms in such a BTD can be much smaller than tensor rank. Simultaneous to the discovery of block-term decomposition, model (6.46) was investigated under the name PARALIND [70], and the two names still coexist today.

Because of the relationship between equations (6.44) and (6.45), block-term decomposition is in fact a constrained CPD with colinear columns in one factor. It is to be noted that we refer here to a specific kind of block-term decomposition [9, 72, 73], but other more involved models were introduced in [71], which do not relate directly to CPD with collinear columns in factors.

An intrinsic property of block-term decompositions is that a rotation ambiguity is introduced in each block, since for any invertible matrix $\boldsymbol{P}$, $\boldsymbol{B}\boldsymbol{C}^T = \boldsymbol{B}\boldsymbol{P}\boldsymbol{P}^{-1}\boldsymbol{C}^T$. The uniqueness of the products $\boldsymbol{B}_r\boldsymbol{C}_r^T$ as well as the uniqueness of factor $\boldsymbol{A}$ have been studied in the literature [17, 74, 75].

In spirit, constraining the block-term decomposition model could be a solution to remove the rotational ambiguity of the blocks. To that end, sparsity-constrained block-term decomposition and coupled block-term decomposition have been studied [73, 76]. The coupling is however not as flexible as in [77]. Some authors have also developed a PARAFAC2 block-term decomposition [78].

Application-wise, the block-term decomposition has been used for biomedical image processing to detect epileptic seizures [79, 80], but also for hyperspectral unmixing, among others.

6.3.3 Structured Factors

6.3.3.1 Re-parameterization

As discussed above, a versatile way to impose constraints in a tensor decomposition model is to constrain the factors directly. For instance for non-negativity constraints, factors are required to have only non-negative entries. This can be efficiently imposed by merely parameterizing entries as squares [81]. It is also possible to impose more complicated constraints via parameterization.

First, a parametric model may be used to describe one or several factors. For instance, the factor $\boldsymbol{A}$ in a CPD may be a sinusoidal function so that $A_{ir} = \sin(2i\pi\rho + r\phi)$ and the new parameter set becomes ρ, ϕ [10]. Such parameterizations have been studied in the literature in the context of array processing, where factors are well represented by exponential maps [9, 11]. Damped exponentials also have been used to model factor matrices [12]. An exponential decay also appears in early literature of chemometrics [82].

Second, a basis of representation may be provided in a matrix format, and the constrained factors are then represented by coefficients in a new feature space. For instance, both non-negativity and smoothness may be imposed on factor matrix $\boldsymbol{A}$ by fixing a family of B-splines $\boldsymbol{D}$ of size $n_1 \times R_1$ for some small integer R_1 as described in [41], and impose that $\boldsymbol{A} = \boldsymbol{D}\boldsymbol{A}_c$. B-splines are piece-wise polynomial functions that are zero valued outside a given interval. Therefore, imposing $\boldsymbol{A}$ to live in the (non-negative) span of $\boldsymbol{D}$ necessarily implies that it is non-negative and smooth as a sum of polynomials. Such splines bases are widely used in psychometrics, where factors in the CPD are heavily constrained by user prior knowledge [83].

Interestingly, if such a matrix $\boldsymbol{D}$ is provided as *a priori* information, and that R_1 is smaller than R, then provided the columns of $\boldsymbol{D}$ are free, a compression similar to the Tucker compression discussed in Section 6.5.3.1 can be done using the QR decomposition of matrix $\boldsymbol{D}$ [41].

6.3.3.2 Dictionary Constraints

An important issue in source separation is the identification, or labeling, of the outputs. This can be done for instance by comparing the output factors with a library, also called a dictionary, of reference factors, like reference emission and excitation spectra of some well-known chemical compounds. However, if all chemical compounds are known in advance, it is also possible to exploit this dictionary inside the source separation algorithm to produce labeled outputs and improve identification accuracy. This dictionary may be very large and very redundant, since the source may be characterized by a family of correlated reference spectra.

Formally, given a dictionary $\boldsymbol{D}$, one wants to select columns of, say, factor $\boldsymbol{A}$ in the columns of $\boldsymbol{D}$, so that $\boldsymbol{A} = \boldsymbol{D}(:, \mathcal{K})$ for an index set $\mathcal{K}$ of size R. Such a combinatorial formulation was introduced in [49], but a former formulation using row-sparsity constraints can be found in [48]. Along with improving identification performances, dictionary constraints also make the low-rank approximation well-posed, similar to non-negativity constraints.

Two related problems remain open. First, when no library is provided a priori, how can a dictionary be learned from a set of tensors? Second, if a provided dictionary is not exactly adapted to the data at hand, what distance would best describe the discrepancy between the constrained factor matrix and the provided dictionary?

Note that important works have also been done about *tensor dictionary learning*, mostly focusing on learning a dictionary from a matrix data set that has a tensor structure. The problem of dictionary learning is, as of now, unrelated to what has been presented above and has yet found no application in chemometrics that we know of, but an interested reader can refer to [46, 47, 84].

6.4 Coupled Decompositions

One of the reasons why tensors have gained importance in signal processing is that the complexity and variety of sensors has skyrocketed in the recent years. On the other hand, most well-studied data analysis tools, for instance Principal Component Analysis, Factor Analysis, linear regression and sparse regression, are designed for two-way arrays. From a practical point of view, this means that only a relationship between two experimental parameters (e.g. time, wavelength) can be inferred. Nowadays multiple such parameters are involved in the measurement process, and one possible way to deal with this fact is to build matrices of data by stacking such parameters, thus overlooking the real intricate relationship between all sets of experimental parameters. PARAFAC, and other previously described models, does mine relationships between all experimental parameters through a collection of separable patterns.

An interesting way to understand the CP decomposition is to cast it as a simultaneous low-rank matrix decomposition of a collection of matrices with equality constraint on the mixing matrices as detailed in Section 6.4.1. This is also in line with how Richard Harshman developed the PARAFAC model based on the principle of parallel proportional profiles [85]. Thus the CPD is a reasonable tool for data fusion, i.e. the joint analysis of multiple data sets. However, it is clear that previously presented models may not be used to tackle any data fusion problem, since the separability assumption may be too strong to describe stacked heterogeneous data sets. Even if that was not the case, data sizes may be completely different due to varying sampling rates among sensors and experiments, and stacking would then not be feasible naively.

This section introduces a general framework for designing tensor decomposition models in a broad context of data fusion. Regression models between multiple blocks of variables will not be addressed here. Our main goal here is not to collect all existing tensor models that account for some specific types of fusion, but rather to give a taste of how to implement peculiar knowledge on the relationship between sources into a decomposition model.

Note that because of the wide range of problems that can be coined as data fusion, there exist numerous *a priori* unrelated models in chemometrics alone designed to address either multimodality or subject variability that we will not address in this chapter.

6.4.1 Exact Coupled Decomposition, A First Approach

Let us first illustrate these concepts on a simple data fusion model, namely the exact coupled decomposition. Suppose data from N sensors are collected in the form of N tensors $\mathcal{T}_n$ of order 3, with one shared experimental parameter. The exact coupled decomposition model supposes that each tensor shares at least one factor with all the others. If exactly one is shared, then for all n in $[1, N]$,

$$\mathcal{T}_n = [\![\boldsymbol{A}, \boldsymbol{B}_n, \boldsymbol{C}_n]\!] + \mathcal{E}_n, \tag{6.47}$$

where $\mathcal{E}_n$ is a noise tensor and $\boldsymbol{A}$ is the shared factor. Note that we have assumed, to simplify, that the shared experimental parameter's sampling rate is the same for each data set, which is not necessarily true in practice [77].

Although coupled tensor factorization stems from well-established concepts such as canonical-correlation analysis [86] and although data fusion with tensors was previously studied in chemometrics [87, 88], exact coupled decomposition was formally introduced much later [6, 77]. Exact coupled decomposition has been used in many application domains such as metabolomics [89] or recommender systems [90].

Exact coupled decomposition is not only a useful data mining tool, but is also the link between matrix factorization models and tensor decomposition. The PARAFAC model may indeed be cast as an exactly coupled matrix factorization model: let $\boldsymbol{M}_k$ be a collection of n_3 matrices of size $n_1 \times n_2$ such that $\boldsymbol{M}_k = \boldsymbol{A}\boldsymbol{D}_k\boldsymbol{B}^\text{T}$ where $\boldsymbol{A}$, $\boldsymbol{B}$ are $n_1 \times R$ and $n_2 \times R$ matrices, respectively, and $\boldsymbol{D}_k$ is a $R \times R$ diagonal matrix of weights. Then the tensor $\mathcal{T}$ obtained by stacking matrices $\boldsymbol{M}_k$ along a third mode, i.e.

$$\mathcal{T}(:, :, k) = \boldsymbol{A}\boldsymbol{D}_k\boldsymbol{B}^T, \tag{6.48}$$

follows a PARAFAC model

$$\mathcal{T} = [\![\boldsymbol{A}, \boldsymbol{B}, \boldsymbol{C}]\!], \tag{6.49}$$

where $\boldsymbol{C} = \left[\text{diag}(\boldsymbol{D}_1), \dots, \text{diag}(\boldsymbol{D}_K)\right]$. This can be easily checked by looking at the expanded formula

$$\mathcal{T}_{ijk} = \sum_{r=1}^{R} A_{ir} D_{k\,rr} B_{jr} = \sum_{r=1}^{R} A_{ir} B_{jr} C_{kr}. \tag{6.50}$$

In other words, computing the CPD of a three-way array is equivalent to finding the common factorization of a collection of matrices with individual component weights, or again to compute the exact coupled decomposition of matrices where all factors are coupled.

Writing an approximation problem for exact coupled decomposition is straightforward (but its solution may raise difficulties, as pointed out in Section 6.2.8). In the presence of i.i.d. Gaussian noise of variance σ_n on each entry of the tensor $\boldsymbol{\mathcal{T}}_n$, the data distribution $p(\boldsymbol{\mathcal{T}}_n|\boldsymbol{A}, \boldsymbol{B}_n, \boldsymbol{C}_n)$ is Gaussian, and therefore the maximum likelihood estimator of all factors is given by the following optimization problem:

$$\max_{\boldsymbol{A},\boldsymbol{B}_n,\boldsymbol{C}_n} \log p(\boldsymbol{\mathcal{T}}_n|\boldsymbol{A}, \boldsymbol{B}_n, \boldsymbol{C}_n) = \min_{\boldsymbol{A},\boldsymbol{B}_n,\boldsymbol{C}_n} \sum_{n=1}^{N} \frac{1}{\sigma_n^2} \|\boldsymbol{\mathcal{T}}_n - [\![\boldsymbol{A}, \boldsymbol{B}_n, \boldsymbol{C}_n]\!]\|_F^2. \tag{6.51}$$

Many different optimization algorithms may be used to compute the exact coupled model, which are derived from either alternating least squares or all-at-once descent algorithms introduced in Section 6.5.1.1. Important contributions to the identifiability of the parameters of the exact coupled tensor decomposition model have been made by Sørensen and De Lathauwer [74].

This exact coupling model is however nowhere near satisfying in most practical scenarios. In fact, realistic problems are more complicated and may feature:

- different tensor sizes on the shared mode,
- more complex variation slice-wise of the coupled factors,
- a stochastic coupling relationship.

To extend the ideas presented above and allow for customization of coupled multiway decomposition models, a more flexible framework than exact coupled decomposition is therefore needed, as elaborated now.

6.4.2 A General Framework for Data Fusion in Tensor Decompositions

As illustrated above, most data fusion methods for source separation make the assumption that a subset of parameters are linked. Describing how these parameters are linked is therefore the cornerstone of designing data fusion models. Understanding how to design tensor data fusion models is essential for linking various existing tensor decomposition models together. As an example, the block shown next shows how the PARAFAC2 model described in Section 6.2.7 can be written as a coupled matrix factorization model, shedding light on the link between PARAFAC and PARAFAC2.

Block 6.1 PARAFAC2 as a flexible tensor coupled decomposition

Suppose a collection of n_3 matrices $\boldsymbol{M}_k$ of sizes $n_1 \times n_2$ is to be jointly factorized using a coupled model, using a single coupled mode. Using above notations, this can be formalized as such:

$$\boldsymbol{M}_k = \boldsymbol{A}\boldsymbol{D}_k\boldsymbol{B}_k^{\mathsf{T}}, \tag{6.52}$$

which can be rearranged into

$$\boldsymbol{M} = \left[\boldsymbol{M}_1 \middle| \dots \middle| \boldsymbol{M}_{n_3}\right] = \boldsymbol{A}\left[\boldsymbol{D}_1\boldsymbol{B}_1 \middle| \dots \middle| \boldsymbol{D}_{n_3}\boldsymbol{B}_{n_3}\right] = \boldsymbol{A}\overline{\boldsymbol{B}}^{\mathsf{T}}, \tag{6.53}$$

by stacking "horizontally" matrices $\boldsymbol{M}_k$ (resp. matrices $\boldsymbol{D}_k\boldsymbol{B}_k$) into a large $n_1 \times n_2n_3$ matrix $\boldsymbol{M}$ (resp. a large $R \times JK$ matrix $\overline{\boldsymbol{B}}^{\mathsf{T}}$). Therefore, if matrices $\boldsymbol{B}_k$ share no relationship, the coupled model is simply equivalent to a large low-rank matrix factorization. On the other hand, equality between matrices $\boldsymbol{B}_k$ would yield the PARAFAC model as shown in equation (6.48). An intermediate constraint may therefore be sought, so that factors $\boldsymbol{B}_k$ are all related but not equal. The PARAFAC2 model suggests to fix the inner-products $\boldsymbol{B}_k^T\boldsymbol{B}_k$ across k. This yields the following flexible matrix coupling model:

$$\boldsymbol{M}_k = \boldsymbol{A}\boldsymbol{D}_k\boldsymbol{B}_k^{\mathsf{T}} \quad \text{and} \quad \boldsymbol{B}_k = \boldsymbol{P}_k\boldsymbol{E}, \tag{6.54}$$

where matrices $\boldsymbol{P}_k$ are $J \times R$ left-orthogonal matrices, i.e. $\boldsymbol{P}_k^{\mathsf{T}}\boldsymbol{P}_k = \boldsymbol{I}_R$, and $\boldsymbol{E}$ is a common $R \times R$ Grammian matrix (a Grammian matrix is a symmetric matrix containing all pairwise scalar products between several vectors). In other words, $\boldsymbol{M}_k\boldsymbol{P}_k = \boldsymbol{A}\boldsymbol{D}_k\boldsymbol{E}^{\mathsf{T}}$, which is nothing more than a CPD.

A convenient way to formalize more general relationships between parameter sets in multiway array decompositions is to resort to a Bayesian probabilistic formulation. In a Bayesian framework, decomposing multiple tensors $\{\mathcal{T}_n\}_{n \le N}$ means finding the parameters $\{\theta_n\}_{n \le N}$ so that the probability $p(\theta_1, \dots, \theta_N, \mathcal{T}_1, \dots, \mathcal{T}_N)$ is maximized over $\{\theta_n\}_{n \le N}$. As shown in [77], to rewrite this criterion in a useful form, the following hypothesis is required:

6.4.2.1 H1: Conditional Independence of the Data

The data arrays $\mathcal{T}_n$ are statistically independent *conditionally* to their decomposition parameters θ_n. This means that knowing the factors of a decomposition for $\mathcal{T}_n$, this tensor can be fully reconstructed, without using the other data sets.

Hypothesis **H1** is a technical assumption and can be assumed to be true in most practical data fusion problems. Moreover, to provide a Bayesian model for the coupled data sets, it is also necessary to know the joint densities of the coupled parameters $p(\theta_1, \dots, \theta_N)$, as well as the likelihood functions $p(\mathcal{T}_n|\theta_n)$, which contain the decomposition model for each data set. Under **H1** and given the likelihoods and the joint probability of the coupled parameters, deriving the Maximum A Posteriori (MAP) estimator boils down to maximizing the log-posterior, i.e.:

$$\theta_{MAP} = \underset{\theta_n}{\operatorname{argmax}} \sum_{n=1}^{N} \log\left(p(\mathcal{T}_n|\theta_n)\right) + \log\left(p(\theta_1, \dots, \theta_N)\right), \tag{6.55}$$

which can be used as the cost function in an optimization problem, see Section 6.5. In the vast majority of coupled decomposition problems, some blocks of parameters should be coupled, and some should not. For the latter, it is possible to marginalize their contribution to the joint distribution, so that only the joint distribution of the coupled parameters is used in (6.55).

Let us instantiate (6.55) with a simple demonstrative example. If each $\mathcal{T}_n$ follows a CPD model with Gaussian Noise of i.i.d. noise of variance σ_n^2, and if column-wise normalized factors $\boldsymbol{C}_n$ are coupled through the following model:

$$\boldsymbol{C}_n = \boldsymbol{C}^* + \boldsymbol{\Gamma}_n, \tag{6.56}$$

for a latent variable matrix $\boldsymbol{C}^*$, and a zero-mean Gaussian noise $\boldsymbol{\Gamma}_n$ whose entries have a variance $\sigma_{C_n}^2$, then Eq. (6.55) becomes

$$\underset{\boldsymbol{A}_n,\boldsymbol{B}_n,\boldsymbol{C}_n,\boldsymbol{C}^*}{\operatorname{argmin}} \sum_{n=1}^{N} \frac{1}{\sigma_n^2} \|\mathcal{T}_n - [\![\boldsymbol{A}_n, \boldsymbol{B}_n, \boldsymbol{C}_n]\!]\|_F^2 + \frac{1}{\sigma_{C_n}^2} \|\boldsymbol{C}_n - \boldsymbol{C}^*\|_F^2, \tag{6.57}$$

where $\boldsymbol{C}_n$ are normalized column-wise.

Note that a latent shared factor matrix $\boldsymbol{C}^*$ is added to the set of parameters. This was done by supposing the joint probability of the coupled parameters is known conditionally to a latent variable θ^*, for which a non-informative prior is used. Then using the Bayes law, Eq. (6.57) is obtained from (6.55). Also note that the normalization is important, since all factors $\boldsymbol{C}_n$ should relate to one matrix $\boldsymbol{C}^*$ with a given error measured by σ_{C_n}, and this coupling relationship is not invariant by scaling.

This flexible exact coupling example is simply meant for illustrating how a coupled decomposition model can be designed. In some practical cases, the probabilistic framework may be dropped and instead, and a deterministic coupling model can be used. All the examples of data fusion models described in Section 6.4.3 are indeed based on a deterministic description of the relationship between a subset of the decomposition variables. However, it should be borne in mind that, using a probabilistic framework, the modeling possibilities are much wider.

6.4.3 Examples of Coupled Decomposition Models

In what follows, we introduce coupled decomposition models that can be encountered in chemometrics. For most of them, solving the underlying optimization problem is non-trivial, but an interested reader can refer to the original publications for more details on this subject. To this list, one should add the PARAFAC2 model described in Section 6.2.7 and in Block 6.1 page 282.

6.4.3.1 Advanced Coupled Matrix Tensor Factorization

Advanced coupled matrix tensor factorization [91] was designed to adapt exactly coupled decomposition (6.47) for situations when only a portion of the components of the coupled factor $\boldsymbol{A}$ are shared across the data sets (thus the "advanced" adjective).

A naive way to design a so-called partially coupled decomposition model is to fix manually the subset of coupled columns of factors $\boldsymbol{A}_n$, such that

$$\boldsymbol{A}_n = [\boldsymbol{A} \mid \boldsymbol{A}_n^{nc}], \tag{6.58}$$

where matrix $\boldsymbol{A}$ contains the shared components, matrices $\boldsymbol{A}_n^{nc}$ the uncoupled ones, and the concatenation is horizontal. However, such a formulation suggests that the number of coupled components is known in advance, which may not be the case. Moreover, a specific ordering of the components is imposed which can bring some permutation problems in the decomposition algorithm.

Advanced coupled matrix tensor factorization proceeds differently. It makes use of a sparsity constraint on the norms of the components (recall formulation (6.25)), to impose this partial coupling, solving the following optimization problem[10]:

$$\begin{aligned}
&\underset{\boldsymbol{A},\boldsymbol{B}_n,\boldsymbol{C}_n,\boldsymbol{S}_n}{\operatorname{argmin}} \|\mathcal{T}_n - [\![\boldsymbol{S}_n;\boldsymbol{A},\boldsymbol{B}_n,\boldsymbol{C}_n]\!]\|_F^2 + \lambda\sum_{n=1}^{n_3}\|\mathrm{Diag}(\boldsymbol{S}_n)\|_1 \\
&\quad + \alpha\sum_{r=1}^{R}\left[(\|\boldsymbol{a}_r\|_2^2 - 1)^2 + \sum_{n=1}^{n_3}(\|\boldsymbol{b}_{nr}\|_2^2 - 1)^2 + (\|\boldsymbol{c}_{nr}\|_2^2 - 1)^2\right]
\end{aligned} \tag{6.59}$$

10 the original publication gave a slightly different optimization problem featuring a relaxation of the ℓ_1 norm.

where $\boldsymbol{S}_n$ are diagonal tensors containing the values σ_{nr} of the components intensities for each tensor $\mathcal{T}_n$ with $n \leq n_3$, while λ and α are hyperparameters tuned by the user. By forcing normalization of the factor matrices, the components amplitude stored in $\boldsymbol{S}_n$ truly reflect the importance of component r in data block n. If a component $\boldsymbol{a}_r$ in the shared matrix $\boldsymbol{A}$ should not be used in data block $\mathcal{T}_n$, then the score of that component in the decomposition of this data block σ_{nr} may be set to zero. This motivates the use of a sparsity-inducing penalization such as the ℓ_1 norm on the entries of the diagonal tensors $\boldsymbol{S}_n$.

Advanced coupled matrix tensor factorization has been used successfully in metabolomics and brain imaging [89, 92].

On the theoretical side, the uniqueness of partially exact coupled decompositions has been studied in the case of matrix-tensor coupled decompositions [93]. Partial coupling is shown to reduce rotational ambiguities in the matrix decomposition without completely negating it. Furthermore, the presence of constraints, in particular non-negativity constraints, may further improve the identification properties of the partially coupled matrix-tensor models.

6.4.3.2 Shift PARAFAC and Others

When dealing with several data sets acquired in similar experimental conditions, it is natural to assume that in the presence of time measurements, some delay is to be accounted for across the various data sets. But such delays may actually depend on the source index, if some variability occurs in the behavior of each source along the various measurements. Modeling and estimating this delay is partially what the PARAFAC2 model described above does, but in a fairly general manner.

In 2003, Harshman *et al.* [94] introduced a more specific modeling of component shifts in a collection of coupled low rank data matrices. This model, coined as Shift-PARAFAC, may be cast in the coupled decomposition framework as follows:

$$[\boldsymbol{C}_n]_{k,r} = [\boldsymbol{C}^*]_{k+\tau_n,r}, \tag{6.60}$$

supposing the coupled factor is $\boldsymbol{C}$ and the amount of shift is proportional to the sampling rate. It is also possible to design an arbitrary shift amount, i.e. not an increment of the indices, by resorting to interpolation between the latent factor and its shifted instances. The computation of the Shift PARAFAC model can be done rather efficiently by resorting to the Fourier transform of the data [95], since a shift in time domain becomes a product in the Fourier domain. The Shift-PARAFAC model has been used mainly to decompose functional magnetic resonance imaging (fMRI) data [95], so as to account for variations in the activation profiles of brain sources.

It is worth noting that a few other similar relationships have been explored in the literature. In particular, distortions due to time contraction or dilatation are discussed in the Warped Factor Analysis model [96, 97]. Also, both Shift-PARAFAC and Warped Factor Analysis differ from data alignment approaches like Ico-shift [98, 99] which preprocess the data to remove any delay among the related data slices. Indeed, if only the data slices are shifted, then implicitly all the components are supposed to have the same delays.

6.4.3.3 GSVD

The Generalized Singular Value Decomposition model has been proposed by Van Loan [100] as a generalization of the SVD to more than a single matrix. It was one of the first attempts at defining a joint diagonalization technique. Its main usage in source separation has been for genomics, where generalized singular value decomposition (GSVD) has been applied notably by Alter to discriminate cancerous DNA from sane DNA [101].

GSVD resembles an exactly (i.e. without noise) coupled decomposition of two matrices:

$$\begin{aligned} \boldsymbol{M}_1 &= \boldsymbol{U}_1\boldsymbol{\Sigma}_1\boldsymbol{V}^T \\ \boldsymbol{M}_2 &= \boldsymbol{U}_2\boldsymbol{\Sigma}_2\boldsymbol{V}^T, \end{aligned} \tag{6.61}$$

but with orthogonality constraints imposed on the non-coupled matrices $\boldsymbol{U}_n$. Note that without such constraints, computing the exact coupled decomposition of a collection of matrices amounts to a single matrix factorization of the stacked matrices $\boldsymbol{M}_n$, which does not admit a unique solution (up to scaling and permutation). This is in contrast with non-orthogonal Joint Approximate Diagonalization of matrices; see [102] and references therein.

On the other hand, the parameters of the GSVD are identifiable, and a closed-form solution is available to compute it when the data are not corrupted by noise. For these reasons, GSVD can be used as an exploratory model, in a similar spirit as Principal Component Analysis, rather than being cast as a physical modeling of the two data sets.

Notably, GSVD was also extended to deal with multiple data matrices [103], and with two coupled third-order tensors [104]. In both cases, the relationship with coupled models is not as straightforward.

6.5 Algorithms

This section aims at giving a short overview of simple and well-understood optimization algorithms that are known to work for computing tensor decomposition models. As a warning to already-informed readers however, research on optimization techniques for tensor decompositions is extremely prolific, and surveying all the available methods while discussing their pros and cons would require another book in itself. Therefore only our understanding of mainstream approaches is described below. On the other hand, to readers who simply want to make use of well-designed toolboxes for source separation problems can turn to the following programs:

- Tensor toolbox[11] (Matlab): this open-source toolbox features a particular care to processing and storaging large sparse tensors, and implements several basic tensor routines. It can therefore also be used as a backend onto which building one's own code.
- N-way toolbox[12] (Matlab): a simple yet comprehensive open-source toolbox implementing the Alternating Least Squares to compute both non-negative and unconstrained CP, as well As other related regression models and the PARAFAC2 decomposition.
- Tensorly[13] (Python): a tensor decomposition collaborative open-source toolbox that mimics the scikit-learn syntax. It supports the use of several backends such as numpy or pytorch, and is geared toward machine learning applications.
- Tensorlab[14] (Matlab): a toolbox, which can identify many of the well-known tensor models, including CPD, block-term decomposition, some joint decompositions, as well as imposing various constraints on the factors. It is based on a nonlinear least squares solver and develops its own syntax.
- PLS_Toolbox[15] (Matlab): A very comprehensive, open-source (and commercial) toolbox for general chemometric modeling. It includes tools for CPD, Tucker and PARAFAC2 including various constrained versions. It also includes older direct methods for CPD modeling based on generalized eigenvalue decomposition such as the generalized rank annihilation method.

Again, many other toolboxes exist. Several lists updated regularly are available online.[16] Some original codes can also be found on authors' home pages, but are generally not part of a toolbox; see for instance the TensorPackage.[17]

11 www.tensortoolbox.org/
12 www.models.life.ku.dk/nwaytoolbox
13 tensorly.org/
14 www.tensorlab.net/
15 www.eigenvector.com
16 www.tensorworld.org/toolboxes/, tensornetwork.org/software/
17 www.gipsa-lab.grenoble-inp.fr/~pierre.comon/TensorPackage/tensorPackage.html

6.5.1 Unconstrained Tensor Decomposition

Among all tensor decomposition models, the most studied by far in terms of optimization strategies is the CPD. In the unconstrained case, both iterative and direct algorithms have been designed in the literature, leading to a wide variety of possible algorithms to choose from. Iterative methods are however the most common choice for approximate decompositions and are therefore the main focus of this section.

6.5.1.1 Iterative Algorithms for Approximate CPD

Let $\mathcal{T} \in \mathbb{R}^{n_1 \times n_2 \times n_3}$, and consider the following cost function[18]:

$$\Upsilon(\boldsymbol{A}, \boldsymbol{B}, \boldsymbol{C}) = \|\mathcal{T} - [\![\boldsymbol{A}, \boldsymbol{B}, \boldsymbol{C}]\!]\|_F^2. \tag{6.62}$$

The minimization of objective (6.62) is the problem we end up with, if we want to find the maximum likelihood estimates of factor matrices $(\boldsymbol{A}, \boldsymbol{B}, \boldsymbol{C})$, when the data follow a rank-R PARAFAC model corrupted by an additive isotropic Gaussian noise. Equivalently, the solution to (6.62) is the best rank-R approximation of $\mathcal{T}$ when it exists.

Since no closed form solution for the minimum of (6.62) is known in the general case, iterative methods rely on the following strategy:

- Provide an initial guess for factors $\boldsymbol{A}$, $\boldsymbol{B}$, and $\boldsymbol{C}$.
- Fix a subset (possibly empty) of the parameters and update the others.
- Stop when convergence is reached.

The main difference between various iterative methods is therefore the choice of fixed parameters and the update strategy.

Gradient Computation Most iterative strategies make use of the gradient of (6.62) for the update rule. The gradient of (6.62) is as follows[19]:

$$\begin{aligned}
\frac{\partial \Upsilon}{2\partial \boldsymbol{A}} &= -\boldsymbol{T}_{(1)}(\boldsymbol{B} \odot \boldsymbol{C}) + \boldsymbol{A}\left(\boldsymbol{B}^T\boldsymbol{B} \boxdot \boldsymbol{C}^T\boldsymbol{C}\right) \\
\frac{\partial \Upsilon}{2\partial \boldsymbol{B}} &= -\boldsymbol{T}_{(2)}(\boldsymbol{A} \odot \boldsymbol{C}) + \boldsymbol{B}\left(\boldsymbol{A}^T\boldsymbol{A} \boxdot \boldsymbol{C}^T\boldsymbol{C}\right) \\
\frac{\partial \Upsilon}{2\partial \boldsymbol{C}} &= -\boldsymbol{T}_{(3)}(\boldsymbol{A} \odot \boldsymbol{B}) + \boldsymbol{C}\left(\boldsymbol{A}^T\boldsymbol{A} \boxdot \boldsymbol{B}^T\boldsymbol{B}\right),
\end{aligned} \tag{6.63}$$

To derive these gradients easily, one may resort to the matricized versions of the PARAFAC model. Indeed, since the Frobenius norm acts entry-wise, the cost (6.62) is equivalently rewritten as

$$\Upsilon(\boldsymbol{A}, \boldsymbol{B}, \boldsymbol{C}) = \|\boldsymbol{T}_{(1)} - \boldsymbol{A}(\boldsymbol{B} \odot \boldsymbol{C})^T\|_F^2. \tag{6.64}$$

---

18 This section is written for third-order tensors without loss of generality.

19 The gradient w.r.t. a matrix is understood entry-wise.

The gradient of this cost function with respect to $\boldsymbol{A}$ is given by

$$\left(-\boldsymbol{T}_{(1)} + \boldsymbol{A}(\boldsymbol{B} \odot \boldsymbol{C})^T\right)(\boldsymbol{B} \odot \boldsymbol{C}). \tag{6.65}$$

At this stage, with some formula manipulation, one may note that for any indices r and q in $[1, R]$,

$$\left[(\boldsymbol{B} \odot \boldsymbol{C})^T(\boldsymbol{B} \odot \boldsymbol{C})\right]_{rq} = (\boldsymbol{b}_r \boxtimes \boldsymbol{c}_r)^T(\boldsymbol{b}_q \boxtimes \boldsymbol{c}_q) = (\boldsymbol{b}_r^T\boldsymbol{b}_q)(\boldsymbol{c}_r^T\boldsymbol{c}_q). \tag{6.66}$$

Therefore the second term in the gradient simplifies into $\boldsymbol{B}^T\boldsymbol{B} \boxdot \boldsymbol{C}^T\boldsymbol{C}$, which has a much lower computational complexity for small R since it is computed by multiplying an $n_i \times R$ matrix with its transpose, plus a few $R \times R$ element-wise products.

Then the bottleneck in the gradient computation is the matrix product $\boldsymbol{T}_{(1)}(\boldsymbol{B} \odot \boldsymbol{C})$, sometimes called Matricized Tensor Times Khatri Rao Product (MTTKRP), which has a naive complexity of $\mathcal{O}(Rn_1n_2n_3)$ since it is computed as the matrix product of a $n_1 \times n_2n_3$ matrix with a $n_2n_3 \times R$ matrix. The fast implementation of this costly product, and in general of tensor contractions, is the topic of many recent researches in high performance computing [105–110].

Alternating least squares The workhorse algorithm for identifying the PARAFAC model is the Alternating Least Squares algorithm. It is very easy to implement, and although several other algorithms are more reliable, alternating least square (ALS) still performs reasonably well in some cases. In its most simple form, ALS also features no parameter tuning. However, it fails to deliver in difficult scenarios, such as in the presence of near-collinear dependency among the factor's columns or rank under- (or over-) estimation. Notably, ALS is a particular case of the nonlinear block Gauss–Seidel method for solving nonlinear systems. It can be adapted to tackle very large data sets [111] and coupled decompositions [77].

The core principle of ALS is to minimize (6.62) with respect to each factor while the others remain fixed. Since PARAFAC is a multilinear model, it is linear with respect to each such block of parameters, and therefore the optimal solution with respect to one block only is known in closed form. Skipping the technical conditions for the existence of such a closed form solution, one may simply set the gradients (6.63) to zero to obtain the sequential update rules for the ALS. For instance, the estimate $\widehat{\boldsymbol{A}}$ of matrix $\boldsymbol{A}$ is given by

$$\widehat{\boldsymbol{A}} = \left(\boldsymbol{T}_{(1)}(\boldsymbol{B} \odot \boldsymbol{C})\right)\left(\boldsymbol{B}^T\boldsymbol{B} \boxdot \boldsymbol{C}^T\boldsymbol{C}\right)^{-1}. \tag{6.67}$$

A pseudo-code for ALS is given in Algorithm 6.1. The stopping conditions can be a fixed number of iterations, the relative decrease of Υ across successive iterations reaching a threshold, or any arbitrary condition that fits the needs of a particular application. Also, the inverse in the factor updates does not need to be explicitly computed. Rather, the least squares update can be computed by solving the linear system obtained by setting the gradients in (6.63) to zero.

Algorithm 6.1 A squeleton of the Alternating Least Squares algorithm

Input: Data tensor $\mathcal{T}$, Initial values $\boldsymbol{A}^{(0)}, \boldsymbol{B}^{(0)}, \boldsymbol{C}^{(0)}$

Set $k = 0$

while stopping condition is not met **do**

$$\boldsymbol{A}^{(k+1)} = \left(\boldsymbol{T}_{(1)}\left(\boldsymbol{B}^{(k)} \odot \boldsymbol{C}^{(k)}\right)\right)\left(\boldsymbol{B}^{(k)^T}\boldsymbol{B}^{(k)} \boxdot \boldsymbol{C}^{(k)^T}\boldsymbol{C}^{(k)}\right)^{-1}$$

$$\boldsymbol{B}^{(k+1)} = \left(\boldsymbol{T}_{(2)}\left(\boldsymbol{A}^{(k+1)} \odot \boldsymbol{C}^{(k)}\right)\right)\left(\boldsymbol{A}^{(k+1)^T}\boldsymbol{A}^{(k+1)} \boxdot \boldsymbol{C}^{(k)^T}\boldsymbol{C}^{(k)}\right)^{-1}$$

$$\boldsymbol{C}^{(k+1)} = \left(\boldsymbol{T}_{(3)}\left(\boldsymbol{A}^{(k+1)} \odot \boldsymbol{B}^{(k+1)}\right)\right)\left(\boldsymbol{A}^{(k+1)^T}\boldsymbol{A}^{(k+1)} \boxdot \boldsymbol{B}^{(k+1)^T}\boldsymbol{B}^{(k+1)}\right)^{-1}$$

Increment k = k+1

end while

Output: Final estimates $\boldsymbol{A}^{(k)}, \boldsymbol{B}^{(k)}, \boldsymbol{C}^{(k)}$

Importantly, ALS iterates of the objective function always converge since the cost function decreases at each iteration and is bounded below. However, this does not guarantee that iterates $(\boldsymbol{A}^{(k)}, \boldsymbol{B}^{(k)}, \boldsymbol{C}^{(k)})$ converge; if they do, the obtained solution is not either guaranteed to be a local minimum. For this to be true in the framework of local convergence,[20] some reasonable technical conditions on the Hessian matrix of Υ should be met [112], which sadly can hardly be checked in practice. Global convergence to a local minimum is also subject to theoretical technical conditions [113]. Further, in some pathological cases, the convergence speed of ALS can be sub-linear [113], a fact observed in difficult decomposition problems [114].

An important tweak on ALS, that often improves its convergence speed drastically, is to extrapolate factor estimates using current and previous estimates. This extrapolation procedure is standard in optimization [115], and is often called "Line Search" in the tensor community. The practical speed up makes it a nice feature of a good ALS implementation [22, 54, 114, 116, 117]. Note that extrapolation may lead to increasing the cost function at some iterations, and the so-called restart strategy that discards steps increasing the cost is the key to obtain an efficient acceleration in some of the works mentioned above [116, 117].

20 Local convergence means convergence if the starting point of the algorithm is in the neighborhood some local minimum.

First and second order descent algorithms All-at-once gradient-based methods are also a great choice for computing an unconstrained PARAFAC model. First, for first-order methods (based solely on the gradient for finding a descent direction), the complexity per iteration is the same as the ALS. Second, all the knowledge on descent methods applied to non-convex problems may be put to profit, in particular through convergence results or stochastic approaches for handling very large data sets [118]. Third, missing data can be dealt with using a mask of weights on the data, which is not feasible using the ALS algorithm. Since large data sets with a lot of missing data are common in machine learning, all-at-once descent algorithms have notably been preponderant in this context [119, 120].

There is no particular practical difficulty to computing a PARAFAC decomposition using well-known descent algorithms such as gradient descent, non-linear conjugate gradient descent [121], or Gauss–Newton once gradients (6.63) have been computed. However, some parameters such as the step size need to be tuned, which makes the ALS a simpler choice for novice users. On the other hand, for high precision works or difficult scenarios, resorting to second-order methods promoted to solve nonlinear least squares problems, like the Levenberg Marquardt algorithm [114], can prove rewarding. Indeed, the Jacobian matrix has a particular structure that can be used to speed-up the – otherwise time demanding – Hessian computation. Moreover, second-order methods have a guaranteed local convergence at quadratic speed.

Normalization Normalizing the columns in the CPD at each iteration is not always mandatory in practice, but it has two advantages: (i) it avoids scaling indeterminacies, and (ii) it helps avoid very small/large values in the factors, thus improving numerical stability as well as providing interpretable results. In practice since the CPD is scale-invariant, in the ALS Algorithm 6.1, one may normalize the columns of all factors at each inner iteration after the $\boldsymbol{C}$ matrix update and pull the product of these norms in one of the factors.

6.5.1.2 Deflation and N-PLS

Up to now, we supposed that the number of terms R in the decomposition is known. In practice this is rarely the case, but finding the optimal R has proven to be a difficult problem, mainly because contrary to matrices, best fitting models with adjacent rank values may have no relationship. The procedure consisting of computing R successive rank-1 approximations with the goal of obtaining a rank-R approximation is often referred to as *deflation*. This idea works for matrices, by subtracting the best rank-1 approximation at each iteration. But the reader should pay attention to the fact that this does not work for tensors, since subtracting the best rank-1 approximation generally does not reduce its rank [122–124]. Also note that it does not work either for matrices in $\mathbb{R}^+$ because of lack of stability by subtraction. Therefore, a naive strategy is to test several ranks and pick the one that works the best, but other methods have been proposed for specific applications [125–127].

Another way to perform deflation is by means of the N-PLS. Let us first describe what Partial Least Squares (PLS) is, i.e. for $N = 2$.

Given a $n \times p$ matrix $\boldsymbol{X}$ and a $n \times 1$ data vector $\boldsymbol{y}$, the goal of PLS is to find the part $\hat{\boldsymbol{y}}$ of $\boldsymbol{y}$ that is related to $\boldsymbol{X}$ in the form $\hat{\boldsymbol{y}} = \boldsymbol{X}\ \boldsymbol{b}$. The classical solution to finding this regression vector is well known and given by $\boldsymbol{b} = (\boldsymbol{X}^\mathsf{T}\boldsymbol{X})^{-1}\boldsymbol{X}^\mathsf{T}\boldsymbol{y}$. However, it is not desired to compute this expression: first, it can be computationally prohibitive, and second, matrix $\boldsymbol{X}$ can be ill-conditioned (e.g. if columns are close to collinear).

PLS aims at computing an approximation of weight $\boldsymbol{b}$ and regression $\hat{\boldsymbol{y}}$ by delivering at each iteration k the loading that provides a score vector with the highest possible cross-covariance with $\boldsymbol{y}$ (or the residual part of $\boldsymbol{y}$). It is robust with respect to ill-conditioning. One nice implementation of PLS is the least squares problems solved with QR factorization (LSQR) bidagonalization algorithm [128], which minimizes the objective $||\boldsymbol{y} - \boldsymbol{X}\boldsymbol{b}||^2$ by the conjugate gradient algorithm.[21] In fact, PLS yields at each iteration k the projection of $\boldsymbol{b}$ onto the subspace spanned by the k dominant eigenvectors of matrix $(\boldsymbol{X}^\mathsf{T}\boldsymbol{X})$. Hence, it can be stopped before convergence and can output at any time an approximation of the best regression.

The multi-linear partial least squares regression, also known as N-way PLS regression, is an extension of the two-way PLS regression described above [129]. It is based on sequentially extracting rank-one tensors from a given tensor $\boldsymbol{\mathcal{T}}$ and a given vector $\boldsymbol{y}$. For instance in trilinear PLS, $\boldsymbol{\mathcal{T}}$ is $I \times J \times K$ and $\boldsymbol{y}$ is a vector of size I [130]. The first rank one tensor has the property that the mode one component vector has maximal covariance with vector $\boldsymbol{y}$, to be predicted. More precisely, (i) a $J \times K$ projected matrix is computed $\boldsymbol{B}_k = \boldsymbol{\mathcal{T}}_k \bullet_1 \boldsymbol{y}_k$, and two weight vectors $(\boldsymbol{v}_k, \boldsymbol{w}_k)$ are computed as the dominant singular vectors of $\boldsymbol{B}_k$; then (ii) a deflation is performed by subtraction $\boldsymbol{y}_{k+1} = \boldsymbol{y}_k - \boldsymbol{t}$ with $\boldsymbol{t} = \boldsymbol{\mathcal{T}}_k \bullet_2 \boldsymbol{v}_k \bullet_3 \boldsymbol{w}_k$, and $\boldsymbol{\mathcal{T}}_{k+1} = \boldsymbol{\mathcal{T}}_k - \boldsymbol{t} \otimes \boldsymbol{v}_k \otimes \boldsymbol{w}_k$.

The process is repeated as long as new components improve the predictions, which is usually determined through cross-validation or similar tests. Note that in the algorithm described in [129], the rank-one approximation of a tensor is computed via two rank-one approximations of matrices, which is possible by breaking the role symmetry of the three modes.

21 This has been proved in exact arithmetic, i.e. if there are no rounding errors.

6.5.1.3 Exact Decomposition Methods

Another line of research on decomposition algorithms is to find exact decomposition algorithms, such that the remaining error in Eq. (6.62) is zero. Of course one can resort to the iterative methods described above. But in the noiseless case, there is no need to introduce a statistical framework, and therefore algebraic methods, that are not theoretically robust to noise, can provide fast and reliable solutions. Because in source separation, such exact decomposition problems have a relatively smaller importance, we shall here only provide a few references that an interested reader can refer to, e.g. [131–135], or in the world of chemistry [136–139]. In particular, in the absence of noise, only two matrix slices need to be used to compute a matrix factor of the CPD [140, 141].

6.5.2 Constrained Tensor Decomposition

Although unconstrained tensor decomposition algorithms are today quite well understood, constrained tensor decomposition algorithms on the other hand have been an important research theme over the last decades. It is practically impossible to summarize all the works on this topic in one section. For instance, the sole case of non-negative matrix factorization, which is a second-order constrained tensor factorization problem, has been discussed in the entire Chapter 3. Therefore, in this section, we shall simply sketch the research directions that have been pursued.

6.5.2.1 Constrained Least Squares

A first approach to constrained tensor decomposition is to modify the ALS algorithm presented in Section 6.5.1.1 to impose the constraints on the decomposition factors. Since ALS relies on solving least squares problems alternatively, this approach boils down to solving constrained least squares problems.

For non-negativity constraints, various efficient algorithms are available. The first non-negative least squares algorithm appeared in [142, 143] and made use of an active set strategy. It has been adapted to compute the non-negative CPD in [144]. More recent techniques employ exact block coordinate descent, using the observation that non-negative least squares can be solved exactly by clipping to zero for one-dimensional data [145]. These approaches have been used by Bro and Jong and Phan and coworkers for computing non-negative CPD, with a computational load comparable to the unconstrained CPD, respectively [44, 144].

To tackle a wider set of constraints, such as sparsity, it was suggested in [120] to solve each constrained least squares problem successively for each factor using a splitting of variables and a primal dual algorithm, namely the Alternating Direction Method of Multipliers (ADMM). The advantage of ADMM in this context, on top of the variety of problems it can tackle, is parallelism.

6.5.2.2 Projected Gradient and All-at-Once Proximal Methods

Similar to the unconstrained case, constrained tensor decompositions can be tackled using variants of gradient descent for constrained problems. The goal here is not to give a full description of constrained convex optimization, but in a nutshell, constrained first-order methods are based on projection operators. Namely, after a gradient step has been performed, the parameters are updated by projecting them onto the set of constraints. Another solution is to add a penalization term in the objective function to promote the constraint. Projected gradient and penalized approaches can be studied and improved under the hood of proximal operators, see Chapter 2 for more details on these optimization techniques and Chapter 3 for their application in the context of non-negative matrix factorization (NMF).

Various first-order methods have been used to compute the CPD, including penalized gradient [146] or proximal gradient [147]. Second-order methods, relying on an estimation of the curvature of the cost function using second-order derivatives, have been extensively used in conjunction with projection or penalization to be used in the Tensorlab toolbox [148].

6.5.2.3 Parametric Approaches

In the particular case of non-negativity, instead of explicitly imposing the constraint on the factors, several authors have suggested to parameterize the variables, for instance as squares, so that the non-negativity constraints are implicitly imposed; see e.g. [81, 146]. Actually, parametric approaches are a convenient way to handle structured factors in tensor decompositions, both with respect to formalism and optimization [149]. Notably, this kind of approach is used in several packages including TensorPackage and Tensorlab. See page 287 for links. See also [60] for a unit-norm parameterization.

6.5.3 Handling Large Data Sets

All the algorithms we presented above make the implicit assumption that the data set can fit into the computer memory, so that any data point can be accessed easily. However, when dealing with very large tensors, this may not be the case. Historically, a compression method coined as the Tucker compression, introduced below, served as both an acceleration method and a storage technique. It may however not be computable in reasonable time.

To cope with very large data sets, several strategies have been explored in the literature, such as sketching or randomized sampling [111]. However, this rapidly evolving topic is out of the scope of this chapter.

6.5.3.1 Multilinear SVD Compression

Given a large tensor $\mathcal{T}$ of size $n_1 \times n_2 \times n_3$ following an unconstrained unknown PARAFAC model of small rank R, computing the CP decomposition may prove quite time consuming. On the other hand, since the tensor is explained by a relatively small number of parameters, in fact by $R(n_1 + n_2 + n_3 - 2)$ parameters, it should be possible to reduce the data set to a more essential one, that can be stored and manipulated instead of the whole tensor.

Finding tensor representations for efficient storage or fast computation of decomposition models is actually a very active field of research, with representations such as the hierarchical decomposition or the tensor train format, see [150] and references therein. However in the context of source separation, and in particular in chemometrics, the most widely used representation method for storage and fast decompositions is the so-called Tucker compression or multilinear SVD compression, which is described in Section 6.2.5. Thus in what follows, we only describe this usual compression method, keeping in mind that newer approaches vastly widen the following discussion.

The idea behind multilinear SVD compression is to use the information that the rank of the tensor approximating the data is small with respect to its dimensions. Then because multilinear ranks, i.e. the ranks of the unfoldings, are always smaller than or equal to the tensor rank, the approximate tensor must have small multilinear ranks as well. Therefore, using a truncated singular value decomposition of each unfolding as a way to compute approximate multilinear SVD[22], a basis for each mode is obtained which can be used to project the data tensor onto a feature space of lower dimensions.

Formally, if a CPD $\mathcal{T} = [\![\boldsymbol{A}, \boldsymbol{B}, \boldsymbol{C}]\!]$ is sought, first a multilinear SVD $\mathcal{T} = [\![\mathcal{G}; \boldsymbol{U}, \boldsymbol{V}, \boldsymbol{W}]\!]$ is computed where $\boldsymbol{U}, \boldsymbol{V}$, and $\boldsymbol{W}$ are left orthogonal matrices of respective sizes $n_1 \times R_1, n_2 \times R_2$, and $n_3 \times R_3$, and the compressed dimensions R_i are larger than or equal to R. Then, using Property 6.2,

$$\mathcal{G} = [\![\mathcal{T}; \boldsymbol{U}^T\boldsymbol{A}, \boldsymbol{V}^T\boldsymbol{B}, \boldsymbol{W}^T\boldsymbol{C}]\!] := [\![\boldsymbol{A}_c, \boldsymbol{B}_c, \boldsymbol{C}_c]\!], \tag{6.68}$$

22 As explained in [24], the solution obtained by SVDs would not be optimal in a noisy setting. Nevertheless, this truncation procedure is generally broadly sufficient as a preprocessing before computing the exact CPD.

which is nothing more than a CPD of the smaller $R_1 \times R_2 \times R_3$ tensor $\mathcal{G}$. Once that CPD is computed, the original CPD of the larger tensor $\mathcal{T}$ can be recovered by $\boldsymbol{A} = \boldsymbol{U}\boldsymbol{A}_c$ and similarly on other modes.

In practice, given a large tensor $\mathcal{T}$ for which multiple approximate PARAFAC models of various ranks are to be computed, it is sufficient to compute the multilinear SVD of $\mathcal{T}$ with reasonably small multilinear ranks, which outputs matrices $\boldsymbol{U}, \boldsymbol{V}$ and $\boldsymbol{W}$. Then after computing the compressed core tensor $\hat{\mathcal{G}}$ once, $\hat{\mathcal{G}}$ becomes the new data set, to be decomposed using any PARAFAC model with compressed factors $\boldsymbol{A}_c$, $\boldsymbol{B}_c$, and $\boldsymbol{C}_c$ of small sizes $R_i \times R$. Tensor $\mathcal{T}$ can also be stored with small loss using its multilinear SVD compression, while using a PARAFAC model often leads to a more lossy compression. In practice, it is always recommended to compress the tensor to be decomposed [114].

As a side note, very few works study efficient compression and acceleration techniques in the presence of constraints. In our opinion, this topic is a promising line of research. Early works have been proposed for non-negative CPD [58].

Structured decompositions Another strategy to accelerate tensor decomposition algorithms is to simply write tensor $\mathcal{T}$ as a structured tensor using any tensor decomposition model, for instance $\mathcal{T} = [\![\mathcal{G}; \boldsymbol{U}, \boldsymbol{V}, \boldsymbol{W}]\!]$. Using such structure leads to faster computations and lowers the memory requirements just like the Tucker compression. For instance, using again Multilinear SVD, gradient (6.63) with respect to $\boldsymbol{A}$ is written as

$$\frac{\partial \Upsilon}{2\partial \boldsymbol{A}} = -\boldsymbol{U}\boldsymbol{G}^{(1)}\left(\boldsymbol{V}^T\boldsymbol{B} \odot \boldsymbol{W}^T\boldsymbol{C}\right) + \boldsymbol{A}\left(\boldsymbol{B}^T\boldsymbol{B} \boxdot \boldsymbol{C}^T\boldsymbol{C}\right), \tag{6.69}$$

and the data-factors product, which is the bottleneck, has now a reduced complexity if $\boldsymbol{U}$ has fewer columns than rows.

The computation speed-up is similar but smaller than using Tucker compression; however, the structured approximation technique extends trivially to any constrained decomposition of $\mathcal{T}$ which makes it attractive in practice [149].

Other cost functions for fitting the CPD As a last remark, it often occurs that the discrepancy between the data tensor $\mathcal{T}$ and the CPD is not efficiently measured by the Frobenius norm. In fact, a wide variety of distances may be used to fit a CPD, which may be obtained by taking the log-likelihood of the data.

Despite the large choice of distance, there is a trick to easily obtain the gradient of a cost function written as

$$f \circ g(\mathcal{T}, \boldsymbol{A}, \boldsymbol{B}, \boldsymbol{C}) = f(\mathcal{T} - [\![\boldsymbol{A}, \boldsymbol{B}, \boldsymbol{C}]\!]). \tag{6.70}$$

Indeed, the following chain rule may be used [151]:

$$\begin{aligned}
\frac{\partial f \circ g}{\partial \boldsymbol{A}}(\mathcal{T}, \boldsymbol{A}, \boldsymbol{B}, \boldsymbol{C}) &= -\nabla_f(\mathcal{T} - [\![\boldsymbol{A}, \boldsymbol{B}, \boldsymbol{C}]\!])(\boldsymbol{B} \odot \boldsymbol{C}) \\
\frac{\partial f \circ g}{\partial \boldsymbol{B}}(\mathcal{T}, \boldsymbol{A}, \boldsymbol{B}, \boldsymbol{C}) &= -\nabla_f(\mathcal{T} - [\![\boldsymbol{A}, \boldsymbol{B}, \boldsymbol{C}]\!])(\boldsymbol{A} \odot \boldsymbol{C}) \\
\frac{\partial f \circ g}{\partial \boldsymbol{C}}(\mathcal{T}, \boldsymbol{A}, \boldsymbol{B}, \boldsymbol{C}) &= -\nabla_f(\mathcal{T} - [\![\boldsymbol{A}, \boldsymbol{B}, \boldsymbol{C}]\!])(\boldsymbol{A} \odot \boldsymbol{B}),
\end{aligned} \tag{6.71}$$

which is nothing more than the usual composition chain rule $(f \circ g)'(x) = f'(g(x))g'(x)$ written for vector valued functions. Note that for f set to the Frobenius norm, Eq. (6.71) recovers the gradients shown in (6.63) since $\nabla_{\|\,\|_F^2}(x) = 2x$.

More techniques can be found in the literature [151] which tackle the gradient computation for more general cost functions to fit the CPD.

6.6 Applications

6.6.1 Preprocessing

Before analyzing data sets, it is often necessary or beneficial to preprocess data. This goes for multi-way data as well. And essentially, the preprocessing is not much different from that of matrix data. The reader is therefore referred to the literature for classical preprocessing such as scatter correction of infrared spectral data [152], baselining of Raman data [153], removal of Raman and Rayleigh scattering effects before analyzing fluorescence [59], and normalization of, e.g. omics data [154]. One aspect though merits some special attention.

Centering and scaling are perhaps the most often used preprocessing methods both for matrix and tensor data. In matrix data analysis there are certain traditional approaches for centering and scaling, and those approaches actually help making sure that the preprocessing achieves what is expected. In tensor analysis, it is slightly more complicated mainly because there are few traditions. Richard Harshman has written an excellent description of the common pitfalls in centering and scaling [155, 156]. Centering often serves two separate and independent purposes: to remove offsets in data and to make sure that the components are centered, e.g. for subsequent regression problems. Not all types of centering will achieve these two goals. Imagine as an example, that a tensor follows a three-component CPD model plus an offset. Such data cannot be modeled by a three-component CPD model directly. Rather, a four-component model would be able to model the data. Upon centering, it is expected that the rank four data will now be rank three meaning that the offset information has been removed. Subtracting, e.g. the overall average of the data would not have that effect [156]. It can be shown that only centering **across** one mode will be able to remove offsets. Centering across one mode means that the average of each column/row/tube is subtracted from that column/row/tube. Any other centering will introduce artifacts in the data that must then also be modeled. Likewise for scaling. Tensor data has to be scaled **within** a mode. That means that each slab of a three-way array has to be scaled by the same scalar. As for centering, scaling differently than within a mode will increase the rank artificially.

6.6.2 Fluorescence

In fluorescence excitation emission spectroscopy, each sample is excited at K excitation wavelengths and the emission subsequently measured at J emission wavelengths. Hence, for I samples an $I \times J \times K$ tensor $\mathcal{T}$ is obtained. If the samples contain, say R, chemical compounds that fluoresce, then the rank of the tensor should be R under ideal conditions up to the noise of the measurements. That is, if the sample is fairly dilute and does not contain an excessive number of other chromophores that absorb significantly [22, 157]. In practice, such data may contain artifacts that need to be handled before a chemically meaningful CPD model can be fitted. If the absorbance of the sample is too high, there may be inner filter effects that distort the signal. There are several methods available for correcting for this either explicitly or implicitly [158–160].

In addition to inner filter effects, it is common that FEEMs will have significant variation caused by Raman and Rayleigh scattering [159]. The Raman scattering is often of moderate size and, for many applications, it can be removed by simply subtracting an FEEM of the solvent from each FEEM. The Rayleigh scattering (Figure 6.6) cannot be handled this way, so usually those areas are removed by replacing the measurements with missing values or interpolating [160, 161].

The sample shown in Figure 6.7 comes from a dataset of 27 samples all containing varying concentrations of the four fluorophores Hydroquinone, Tryptophan, Phenylalanine, and Dopa. Since there are four chemical compounds, it is expected that a four-component CPD model would provide an adequate model of the data.

Indeed in this example, a four-component CPD model has a so-called core consistency of 88% indicating a valid model [125]. However, several aspects seem suspicious. First of all, it seems that there may be problems with local minima. Refitting the model ten times, the fit varies between three distinct values: 99.8% variation explained, 96.6%, and 95.5%. Only the best fitting

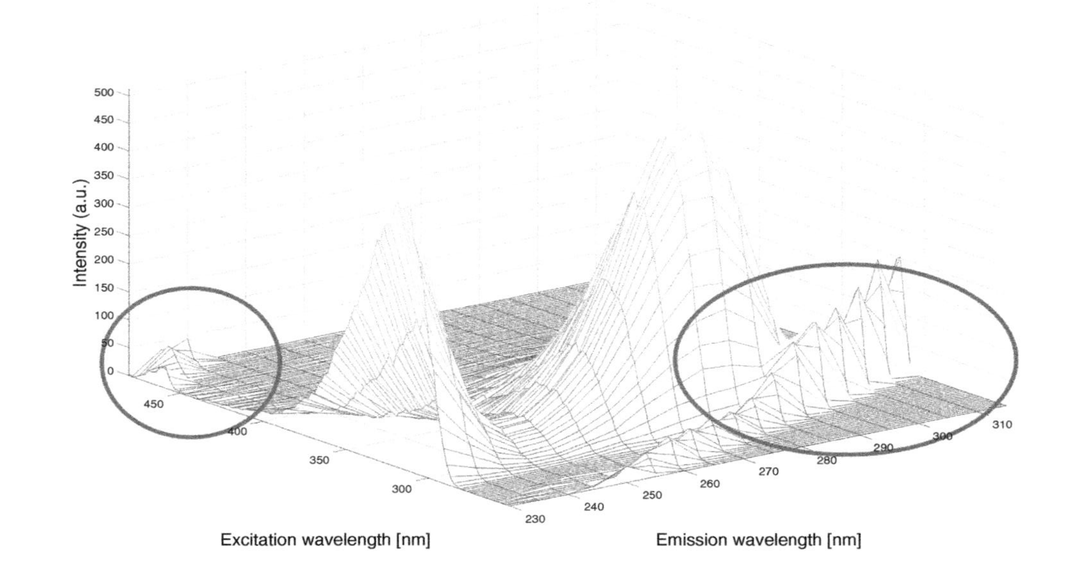

Figure 6.6 A fluorescence excitation emission matrix (FEEM). The two areas surrounded by ellipses reveal Rayleigh scattering.

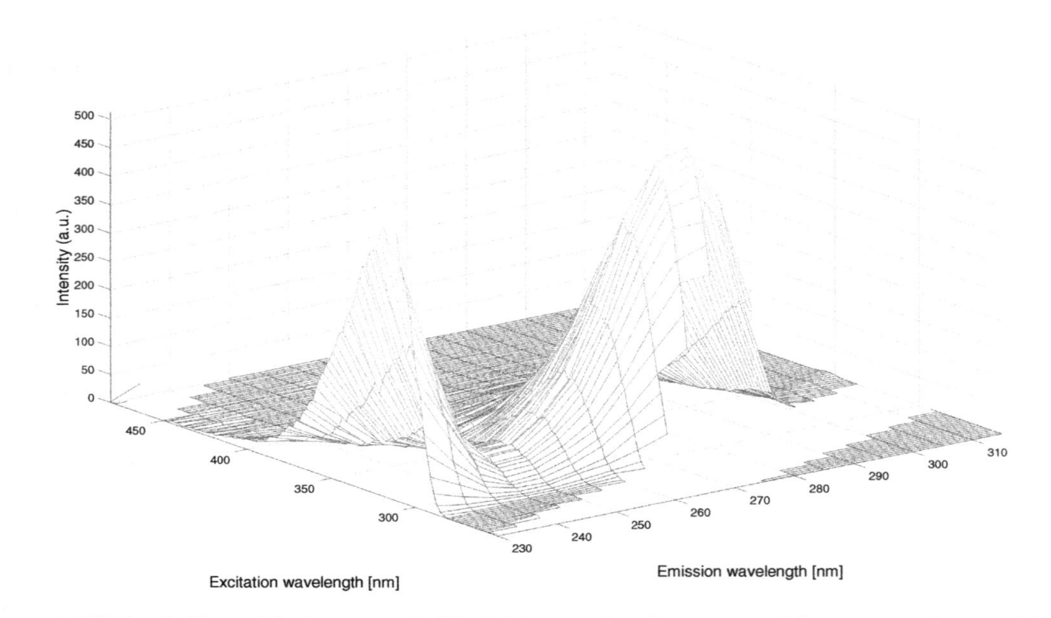

Figure 6.7 The same FEEM as in Figure 4.1 after removal of Rayleigh scattering. Some traces of Raman scattering are visible.

of those qualifies for being the actual CPD model, so the others have to be disregarded. Normally, local minima are not a huge problem for datasets that follow the CPD models well, but in this case where there is both a large amount of missing data and some outliers present, the algorithm apparently struggles. Investigating residuals and parameters, four outlying samples are identified and removed. The main reason for the outlying behavior is that the concentrations are quite high. Upon removing the four samples, a four-component model has a core consistency of 100%. Normally, it is advised to use models with the highest number of components with a sufficiently high core consistency [125].

It was investigated if the model was more stable and robust when using non-negativity constraints. Some of the estimated fluorescence spectra in the unconstrained model were slightly negative. Not enough to be a significant issue, but oftentimes, imposing *non-negativity* can also stabilize the model with respect to numerical problems. Indeed, a four-component model with non-negativity on all parameters did not show any local minima and had a perfect core consistency. Furthermore, the estimated emission and excitation spectra looked very similar to what would be expected from prior knowledge. The five-component model has a low core consistency, and some of the emission components come out identical which is not chemically meaningful. Hence, the four-component model seems a good candidate.

To verify the model, we perform a splithalf analysis where the data are split into two parts in the sample mode [162]. A four-component CPD model is fitted to the first 13 samples and independently to the last 14 samples. If the model is correctly specified the components should be the same in the two models. In Figure 6.8, the results of the two models are shown together with the overall model. As can be seen, the four estimated emission (top) and excitation (bottom) spectra are almost exactly identical even though they are estimated from different sample sets. This is a very convincing diagnostic for assessing the validity of the model.

As a final illustration of the ability to uncover the underlying chemistry, the scores are plotted in Figure 6.9. Each score is plotted against the known actual concentration of the corresponding chemical in each sample. As can be seen, the model is capable of recovering the concentrations up to a scaling; hence estimating the relative concentration of each compound.

6.6.3 Chromatography

Gas Chromatography with Mass Spectrometric detection (GC-MS) is a very common tool in analytical chemistry, e.g. for measuring hormones in food products, flavor compounds in wines, or proteins in blood. In simpler

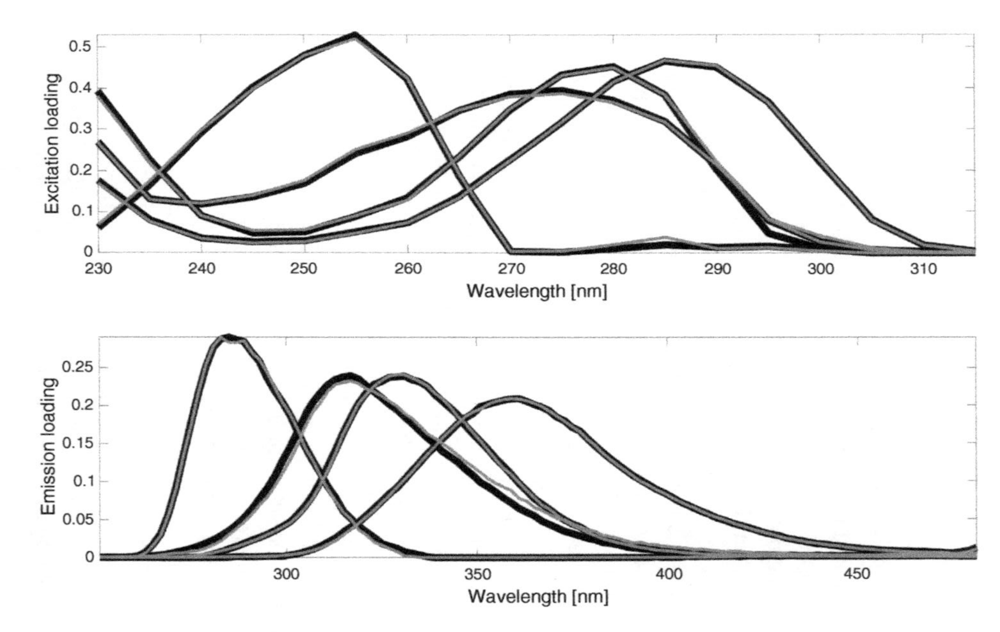

Figure 6.8 The results of split-half analysis (black: set 1; gray: set 2).

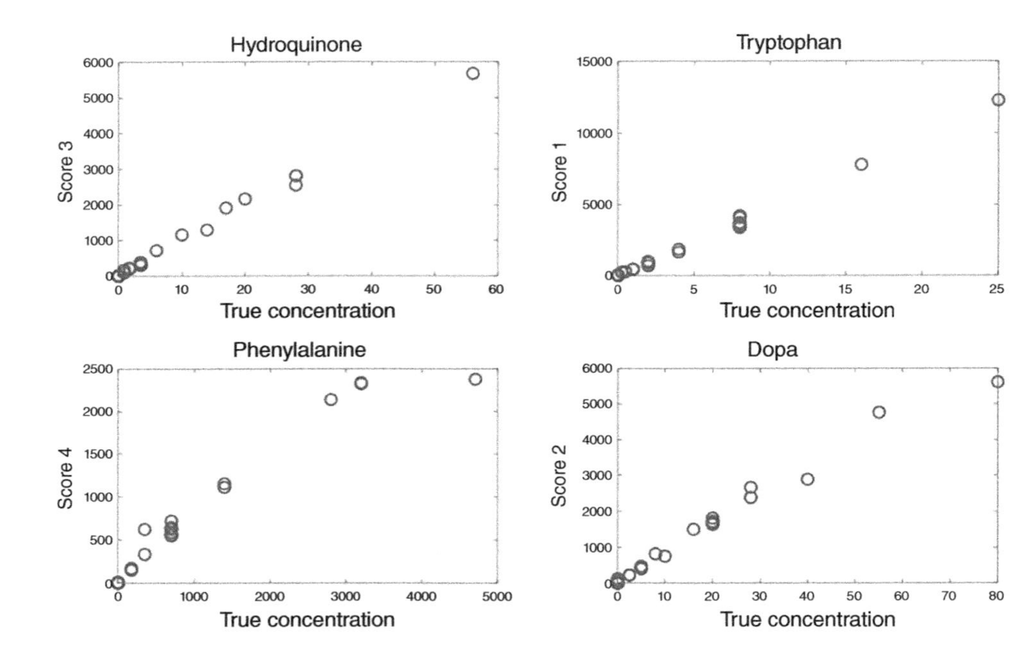

Figure 6.9 The four score vectors of a CPD model plotted against the corresponding actual concentrations.

cases, there is little need for much data analysis as the whole purpose of the chromatography is to ensure that different chemical constituents come out at different times. However as shown in Figure 6.10, sometimes the peaks of different chemicals are overlapping. Ideally, each chemical would be a baseline-resolved Gaussian curve, but when compounds overlap, the traditional approaches for handling the data often fail.

For ideal chromatographic data, fitting a CPD model would allow to resolve overlapping data. Each CPD component would consist of a component in the elution mode giving the elution profile and in the spectral mode giving the pure mass spectrum of each analyte. The sample mode would then give the relative concentration of each chemical in each sample/experiment [163, 164]. However, the CPD model requires that the elution profile of each chemical compound keeps the same shape across different samples. This is almost never the case in chromatography. Due to retention time shifts, the elution profile will change slightly from sample to sample. This is also evident in Figure 6.10, where the peak at approximately time 21.7 minutes varies. Further, there are a number of minor peaks around 21.8–22 minutes and it is difficult to discern exactly how many.

The PARAFAC2 model has been shown repeatedly to provide a good model for chromatographic data and fitting the model to interval indicated in Figure 6.10; it turns out that there are as many as seven components needed for describing the data. In Figure 6.11, the elution mode components are shown. There are 44 samples in the dataset; hence, there are 44 versions of each elution profile.

6.6.4 Other Applications

Tensor analysis has a long history in chemistry and there are many diverse fields of applications as also evidenced in older reviews [165]. The applications can be divided into typical groups. The first group consists of applications where hard modeling such as Beer's law is used to identify chemical information like pure spectra and concentrations. This can be used for untargeted approaches where many chemicals are being estimated at the same time [166] or in targeted approaches where one or a few compounds need to be quantified [167]. The models used are mostly CPD and PARAFAC2 but also sometimes alternatives such as restricted Tucker3 models [168] or methods based on rank annihilation [169–173]. Especially CPD is useful, e.g. for high-resolution nuclear magnetic resonance [174–178] as well as low-resolution magnetic resonance [179, 180]. Traditionally, CPD and variants have also been popular within electroencephalography [181–183]. For more exploratory purposes, it is common to use the Tucker3

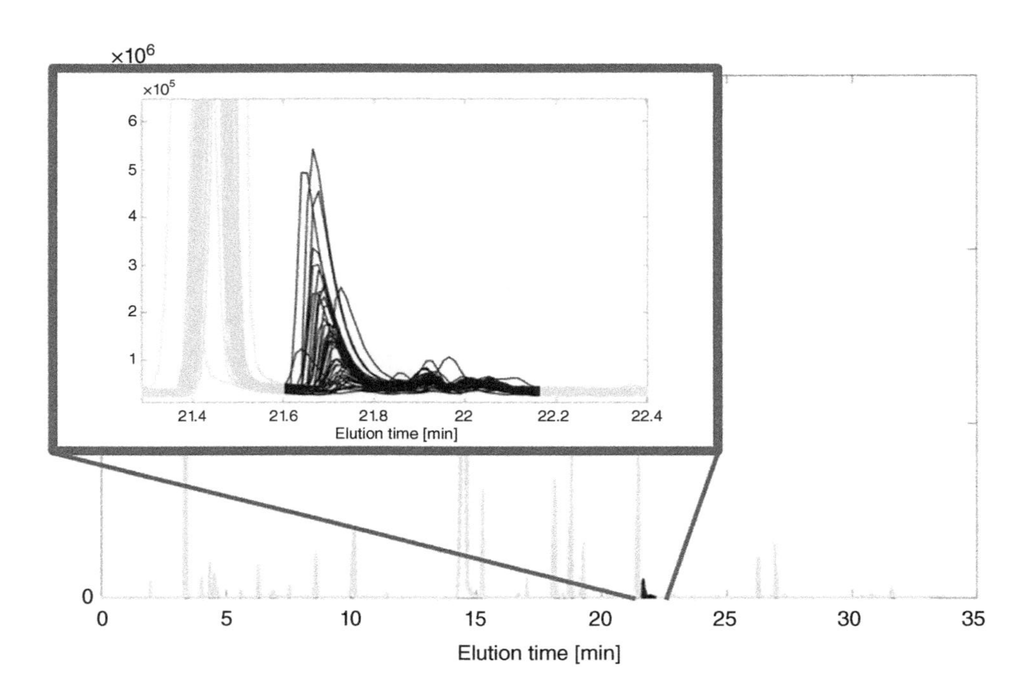

Figure 6.10 Example of a set of samples measured by GC-MS. The mass spectrum is summed at each time point so that the measurements of each sample become a vector called a TIC – Total Ion Current chromatogram. In the dark part, a time interval with overlapping peaks is shown.

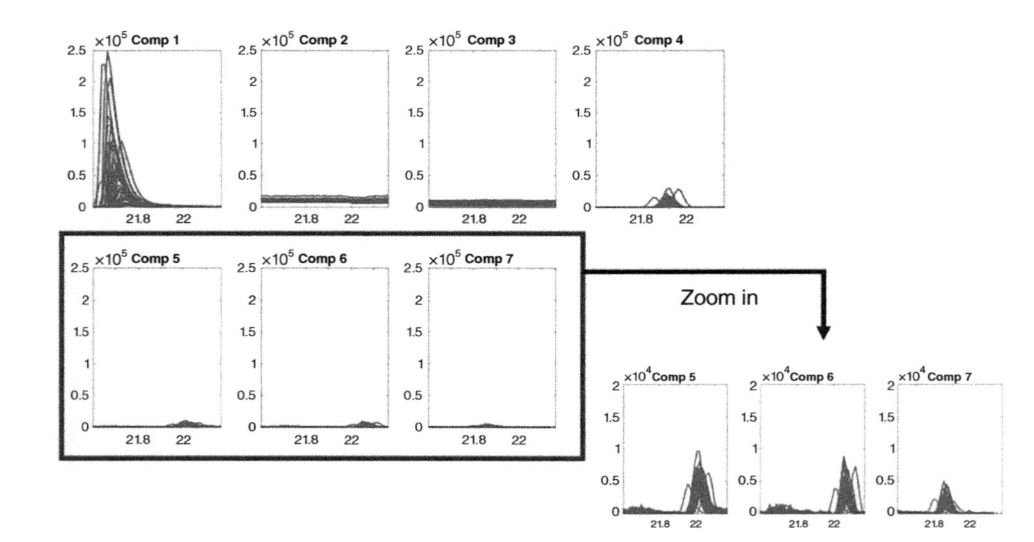

Figure 6.11 Estimated elution profiles from a seven-component PARAFAC2 model. Components 2 and 3 are describing baseline variation, whereas the remaining five components are describing different chemical compounds.

model often followed by some types of rotations of either the core or the factor matrices [36, 62]. Examples often come from environmental analysis [184–186], but variants of CPD are also used [187, 188]. Sensory profiling is a common approach for understanding human perception, e.g. in food analysis. The traditional sensory profiling data are a three-way structure consisting of a number of assessors assessing a number of items with respect to a number of attributes. The data can be analyzed with both CPD and Tucker3, but the Tucker2 model is often preferred because the extended core array allows meaningful interactions between components [189]. In batch process monitoring or multivariate statistical process monitoring in general, the aim is to understand and operate production processes. In early days, both Tucker and CPD, and even PARAFAC2 models were investigated [190, 191]. Nowadays though, the three-way data are often unfolded and analyzed as matrix data to better handle the complex dynamics that such data have. A third type of problem that occurs quite often in the chemical sciences is regression, which is also referred to as multivariate calibration. The classical problem is to replace a tedious and costly reference method with a prediction based on some more easily available data. The most popular algorithm for this is multi-way *partial least squares regression* [130], which has been used for a multitude of problems [176, 192–203]. An interesting alternative is the method SCREAM that combines the regression with the more flexible PARAFAC2 model [204].

References

1 Mac Lane, S. (2013) *Categories for the Working Mathematician*, vol. 5, Springer Science & Business Media.

2 Hackbusch, W. (2012) *Tensor Spaces and Numerical Tensor Calculus*, *Series in Computational Mathematics*, Springer-Verlag, Berlin, Heidelberg.

3 Landsberg, J.M. (2012) *Tensors: Geometry and Applications*, *Graduate Studies in Mathematics*, vol. 128, AMS Publications.

4 Comon, P. (2014) Tensors: a brief introduction. *IEEE Signal Processing Magazine*, 31 (3), 44–53. Special issue on BSS.

5 Bro, R. (2006) Review on multiway analysis in chemistry—2000–2005. *Critical Reviews in Analytical Chemistry*, 36, 279–293.

6 Acar, E. and Yener, B. (2009) Unsupervised multiway data analysis: a literature survey. *IEEE Transactions on Knowledge and Data Engineering*, 21 (1), 6–20.

7 Kolda, T.G. and Bader, B.W. (2009) Tensor decompositions and applications. *SIAM Review*, 51 (3), 455–500.

8 Comon, P. and Jutten, C. (eds) (2010) *Handbook of Blind Source Separation, Independent Component Analysis and Applications*, Academic Press, Oxford UK, Burlington USA. ISBN: 978-0-12-374726-6.

9 De Lathauwer, L. (2011) Blind separation of exponential polynomials and the decomposition of a tensor in rank-$(L_r, L_r, 1)$ terms. *SIAM Journal on Matrix Analysis and Applications*, 32 (4), 1451–1474.

10 Domanov, I. and De Lathauwer, L. (2016) Generic uniqueness of a structured matrix factorization and applications in blind source separation. *IEEE Journal of Selected Topics in Signal Processing*, 10 (4), 701–711.

11 Raimondi, F., Farias, R.C., Michel, O., and Comon, P. (2017) Wideband multiple diversity tensor array processing. *IEEE Transactions on Signal Processing*, 65, 5334–5346. Hal-01350549.

12 Sahnoun, S., Comon, P., and Usevich, K. (2017) Multidimensional ESPRIT for damped and undamped signals: algorithm, computations, and perturbation analysis. *IEEE Transactions on Signal Processing*, 65 (22), 5897–5910. Hal-01360438.

13 Chiantini, L., Ottaviani, G., and Vannieuwenhoven, N. (2014) An algorithm for generic and low-rank specific identifiability of complex tensors. *SIAM Journal on Matrix Analysis and Applications*, 35 (4), 1265–1287.

14 Kruskal, J.B. (1977) Three-way arrays: rank and uniqueness of trilinear decompositions, with application to arithmetic complexity and statistics. *Linear Algebra and its Applications*, 18, 95–138.

15 Sidiropoulos, N.D. and Bro, R. (2000) On the uniqueness of multilinear decomposition of N-way arrays. *Journal of Chemometrics*, 14, 229–239.

16 Stegeman, A. and Sidiropoulos, N. (2007) On Kruskal's uniqueness condition for the CP decomposition. *Linear Algebra and its Applications*, 420 (2–3), 540–552.

17 Domanov, I. and Lathauwer, L.D. (2017) Canonical polyadic decomposition of third-order tensors: relaxed uniqueness conditions and algebraic algorithm. *Linear Algebra and its Applications*, 513, 342–375.

18 Hitchcock, F.L. (1927) The expression of a tensor or a polyadic as a sum of products. *Journal of Mathematical Physics*, 6 (1), 165–189.

19 Harshman, R.A. (1970) Foundations of the PARAFAC procedure: models and conditions for an "explanatory" multimodal factor analysis. *UCLA Working Papers in Phonetics*, 16, 1–84.

20 Carroll, J.D. and Chang, J.J. (1970) Analysis of individual differences in multidimensional scaling via N-way generalization of "Eckart-Young" decomposition. *Psychometrika*, 35 (3), 283–319.

21 Kiers, H.A.L. (2000) Towards a standardized notation and terminology in multiway analysis. *Journal of Chemometrics*, 14 (3), 105–122.

22 Bro, R. (1997) PARAFAC, tutorial and applications. *Chemometrics and Intelligent Laboratory Systems*, 38, 149–171.

23 Cohen, J.E. (2015) About notations in multiway array processing. *arXiv:1511.01306*.

24 Lathauwer, L.D., Moor, B.D., and Vandewalle, J. (2000) A multilinear singular value decomposition. *SIAM Journal on Matrix Analysis and Applications*, 21 (4), 1253–1278.

25 Silva, V.D. and Lim, L.H. (2008) Tensor rank and the ill-posedness of the best low-rank approximation problem. *SIAM Journal on Matrix Analysis and Applications*, 30 (3), 1084–1127.

26 Kroonenberg, P.M. and Leeuw, J.D. (1980) Principal component analysis of three-mode data. *Psychometrika*, 45, 69–97.

27 Tucker, L.R. (1966) Some mathematical notes for three-mode factor analysis. *Psychometrika*, 31, 279–311.

28 Harshman, R.A. (1972) PARAFAC2: mathematical and technical notes. *UCLA working papers in phonetics* 22 (3044), 122215.

29 Kiers, H.A.L., ten Berge, J.M.F., and Bro, R. (1999) PARAFAC2: Part I. A direct fitting algorithm. *Journal of Chemometrics*, 13, 275–294.

30 Comon, P., Berge, J.M.F.T., DeLathauwer, L., and Castaing, J. (2009) Generic and typical ranks of multi-way arrays. *Linear Algebra and its Applications*, 430 (11–12), 2997–3007. Hal-00410058.

31 Lim, L.H. and Comon, P. (2014) Blind multilinear identification. *IEEE Transactions on Information Theory*, 60 (2), 1260–1280.

32 Golub, G.H. and Loan, C.F.V. (1989) *Matrix Computations*, The John Hopkins University Press.

33 Bro, R. and Sidiropoulos, N.D. (1998) Least squares algorithms under unimodality and non-negativity constraints. *Journal of Chemometrics*, 12, 223–247.

34 Lim, L.H. and Comon, P. (2009) Nonnegative approximations of nonnegative tensors. *Journal of Chemometrics*, 23, 432–441.

35 Qi, Y., Comon, P., and Lim, L.H. (2016) Uniqueness of nonnegative tensor approximations. *IEEE Transactions on Information Theory*, 62 (4), 2170–2183. *arXiv:1410.8129*.

36 Kiers, H.A.L. (1992) Tuckals core rotations and constrained Tuckals modelling. *Statistica Applicata*, 4 (4), 659–667.

37 Comon, P. (1994) Tensor diagonalization, a useful tool in signal processing, in *IFAC-SYSID, 10th IFAC Symposium on System Identification*, vol. 1 (eds M. Blanke and T. Soderstrom), Copenhagen, Denmark, pp. 77–82.

38 Martin, C.D.M. and van Loan, C. (2008) A Jacobi-type method for computing orthogonal tensor decompositions. *SIAM Journal on Matrix Analysis and Applications*, 30 (3), 1219–1232.

39 Krijnen, W.P., Dijkstra, T.K., and Stegeman, A. (2008) On the non-existence of optimal solutions and the occurrence of degeneracy in the Candecomp/Parafac model. *Psychometrika*, 73 (3), 431–439.

40 Uschmajew, A. (2010) Well-posedness of convex maximization problems on Stiefel manifolds and orthogonal tensor product approximations. *Numerische Mathematik*, 115, 309–331.

41 Timmerman, M.E. and Kiers, H.A.L. (2002) Three-way component analysis with smoothness constraints. *Computational Statistics and Data Analysis*, 440, 447–470.

42 Chen, Z., Chichocki, A., and Rutkowski, T.M. (2006) Constrained non-negative matrix factorization method for EEG alalysis in early detection of Alzheimer disease, in *ICASSP*, vol. V, Toulouse, pp. 893–896.

43 Berry, M.W., Browne, M., Langville, A.N., Pauca, V.P., and Plemmons, R.J. (2007) Algorithms and applications for approximate nonnegative matrix factorization. *Computational Statistics and Data Analysis*, 52 (1), 155–173.

44 Cichocki, A., Zdunek, R., Phan, A.H., and Amari, S.I. (2009) *Nonnegative Matrix and Tensor Factorization*, John Wiley & Sons.

45 Veganzones, M., Cohen, J.E., Farias, R.C., Chanussot, J., and Comon, P. (2016) Nonnegative tensor CP decomposition of hyperspectral data. *IEEE Transactions on Geoscience and Remote Sensing*, 54 (5), 2577–2588.

46 Roemer, F., Galdo, G.D., and Haardt, M. (2014) Tensor-based algorithms for learning multidimensional separable dictionaries, in *ICASSP*, pp. 3963–3967.

47 Boyer, R. and Haardt, M. (2016) Noisy compressive sampling based on block-sparse tensors: performance limits and beamforming techniques. *IEEE Transactions on Signal Processing*, 64 (23), 6075–6088.

48 Sahnoun, S., Djermoune, E.H., Brie, D., and Comon, P. (2017) A simultaneous sparse approximation method for multidimensional harmonic retrieval. *Signal Processing*, 137, 36–48.

49 Cohen, J.E. and Gillis, N. (2018) Dictionary-based tensor canonical polyadic decomposition. *IEEE Transactions on Signal Processing*, 66 (7), 1876–1889.

50 Sahnoun, S. and Comon, P. (2015) Joint source estimation and localization. *IEEE Transactions on Signal Processing*, 63 (10), 2485–2495.

51 Qi, Y., Comon, P., and Lim, L.H. (2016) Semialgebraic geometry of nonnegative tensor rank. *SIAM Journal on Matrix Analysis and Applications*, 37 (4), 1556–1580. Hal-01763832.

52 Mitchell, B.C. and Burdick, D.S. (1994) Slowly converging PARAFAC sequences: swamps and two-factor degeneracies. *Journal of Chemometrics*, 8, 155–168.

53 Paatero, P. (2000) Construction and analysis of degenerate PARAFAC models. *Journal of Chemometrics*, 14, 285–299.

54 Rajih, M., Comon, P., and Harshman, R. (2008) Enhanced line search: a novel method to accelerate PARAFAC. *SIAM Journal on Matrix Analysis and Applications*, 30 (3), 1148–1171.

55 Stegeman, A. (2006) Degeneracy in Candecomp/Parafac explained for $p \times p \times 2$ arrays of rank $p + 1$ or higher. *Psychometrika*, 71 (3), 483–501.

56 Cichocki, A., Zdunek, R., and Amari, S.I. (2008) Nonnegative matrix and tensor factorization. *IEEE Signal Processing Magazine*, 25 (1), 142–145.

57 Zhou, G., Cichocki, A., Zhao, Q., and Xie, S. (2014) Nonnegative matrix and tensor factorizations: an algorithmic perspective. *IEEE Signal Processing Magazine*, 31 (3), 54–65.

58 Cohen, J.E., Cabral-Farias, R., and Comon, P. (2015) Fast decomposition of large nonnegative tensors. *IEEE Signal Processing Letters*, 22 (7), 862–866.

59 Royer, J.P., Thirion-Moreau, N., Comon, P., Redon, R., and Mounier, S. (2015) A regularized nonnegative canonical polyadic decomposition algorithm with preprocessing for 3D fluorescence spectroscopy. *Journal of Chemometrics*, 29 (4), 253–265.

60 Cabral-Farias, R., Comon, P., and Redon, R. (2014) Data mining by nonnegative tensor approximation, in *IEEE MLSP*, Reims, France. Hal-01077801.

61 Phan, A.H. and Chichocki, A. (2008) Fast and efficient algorithms for nonnegative Tucker decomposition, in *International Symposium on Neural Networks*, vol. LNCS 5264, Springer-Verlag, Berlin, pp. 772–782.

62 Kiers, H.A.L. (1998) Recent developments in three-mode factor analysis: constrained three-mode factor analysis and core rotations, in

Data Science, Classification, and Related Methods, Studies in Classification, Data Analysis, and Knowledge Organization (eds C. Hayashi and K. Yajima), Springer, pp. 563–574.

63 ten Berge, J.M.F. and Smilde, A.K. (2002) Non-triviality and identification of a constrained Tucker3 analysis. *Journal of Chemometrics*, 16 (12), 609–612.

64 Tomasi, G. and Bro, R. (2009) Multilinear models: iterative methods, in *Comprehensive Chemometrics: Chemical and Biochemical Data Analysis*, vol. 2 (eds R.S. Brown and B. Walczak), Elsevier, Chapter 22, pp. 411–451.

65 Zhou, G., Cichocki, A., Zhao, Q., and Xie, S. (2015) Efficient nonnegative Tucker decompositions: algorithms and uniqueness. *IEEE Transactions on Image Processing*, 24 (12), 4990–5003.

66 Hoyer, P.O. (2004) Non-negative matrix factorization with sparseness constraints. *Journal of Machine Learning Research*, 5, 1457–1469.

67 Morup, M., Hansen, L.K., and Arnfred, S.M. (2008) Algorithms for sparse nonnegative Tucker decompositions. *Neural Computation*, 20 (8), 2112–2131.

68 Anandkumar, A., Hsu, D., and Kakade, M.J.S. (2015) When are overcomplete topic models identifiable? Uniqueness of tensor Tucker decompositions with structured sparsity. *Journal of Machine Learning Research*, 16, 2643–2694.

69 Xu, Y. (2015) Alternating proximal gradient method for sparse nonnegative Tucker decomposition. *Mathematical Programming Computation*, 7 (1), 39–70. *arXiv:1302.2559*.

70 Bro, R., Harshman, R.A., Sidiropoulos, N.D., and Lundy, M.E. (2009) Modeling multi-way data with linearly dependent loadings. *Journal of Chemometrics*, 23 (7–8), 324–340.

71 De Lathauwer, L. (2008) Decompositions of a higher-order tensor in block terms–Part II: definitions and uniqueness. *SIAM Journal on Matrix Analysis and Applications*, 30 (3), 1033–1066.

72 Guo, X., Miron, S., Brie, D., and Stegeman, A. (2012) Uni-mode and partial uniqueness conditions for Candecomp/Parafac of three-way arrays with linearly dependent loadings. *SIAM Journal on Matrix Analysis and Applications*, 33, 111–129.

73 Sorensen, M., Domanov, I., and De Lathauwer, L. (2015) Coupled canonical polyadic decompositions and (coupled) decompositions in multilinear rank-$(L_r, n, L_r, n, 1)$ terms - Part II: Algorithms. *SIAM Journal on Matrix Analysis and Applications*, 36 (3), 1015–1045.

74 Sorensen, M. and De Lathauwer, L. (2015) Coupled canonical polyadic decompositions and (coupled) decompositions in multilinear rank-($L_r, n, L_r, n, 1$) terms - Part I: Uniqueness. *SIAM Journal on Matrix Analysis and Applications*, 36 (2), 496–522.

75 Lahat, D. and Jutten, C. (2018) A new link between joint blind source separation using second order statistics and the canonical polyadic decomposition, in *14th International Conference on Latent Variable Analysis and Signal Separation (LVA-ICA)*, Springer, University of Surrey, Guildford, UK.

76 Caland, F., Miron, S., Brie, D., and Mustin, C. (2012) A blind sparse approach for estimating constraint matrices in Paralind data models, in *20th EUSIPCO*, Eurasip, Bucharest, pp. 839–843.

77 Farias, R.C., Cohen, J.E., and Comon, P. (2016) Exploring multimodal data fusion through joint decompositions with flexible couplings. *IEEE Transactions on Signal Processing*, 64 (18), 4830–4844.

78 Chatzichristos, C., Kofidis, E., and Theodoridis, S. (2017) PARAFAC2 and its block term decomposition analog for blind fMRI source unmixing, in *EUSIPCO*, Kos island, pp. 2081–2085.

79 Hunyadi, B., Camps, D., Sorber, L., Paesschen, W.V., Vos, M.D., Huffel, S.V., and Lathauwer, L.D. (2014) Block term decomposition for modelling epileptic seizures. *EURASIP Journal on Advances in Signal Processing*, 139, 1–19.

80 Aldana, Y.R., Hunyadi, B., Reyes, E.J.M., Rodriguez, V.R., and Van Huffel, S. (2018) Nonconvulsive epileptic seizure detection in scalp EEG using multiway data analysis. *IEEE Journal of Biomedical and Health Informatics*, 23 (2), 660–671.

81 Coloigner, J., Karfoul, A., Albera, L., and Comon, P. (2014) Line search and trust region strategies for canonical decomposition of semi-nonnegative semi-symmetric 3rd order tensors. *Linear Algebra and its Applications*, 450, 334–374.

82 Windig, W. and Antalek, B. (1997) Direct exponential curve resolution algorithm (DECRA): a novel application of the generalized rank annihilation method for a single spectral mixture data set with exponentially decaying contribution profiles. *Chemometrics and Intelligent Laboratory Systems*, 37 (2), 241–254.

83 Helwig, N.E. (2017) Estimating latent trends in multivariate longitudinal data via parafac2 with functional and structural constraints. *Biometrical Journal*, 59 (4), 783–803.

84 Dantas, C.F., Cohen, J.E., and Gribonval, R. (2019) Learning tensor-structured dictionaries with application to hyperspectral image denoising, in *2019 27th European Signal Processing Conference (EUSIPCO)*, IEEE, pp. 1–5.

85 Cattell, R.B. (1944) "Parallel proportional profiles" and other principles for determining the choice of factors by rotation. *Psychometrika*, 9 (4), 267–283.

86 Hotelling, H. (1936) Relations between two sets of variates. *Biometrika*, 28, 321–377.

87 Smilde, A.K. and Kiers, H.A.L. (1999) Multiway covariates regression models. *Journal of Chemometrics*, 13 (1), 31–48.

88 Smilde, A.K., Westerhuis, J.A., and de Jong, S. (2003) A framework for sequential multiblock component methods. *Journal of Chemometrics*, 17 (6), 323–337.

89 Acar, E., Bro, R., and Smilde, A.K. (2015) Data fusion in metabolomics using coupled matrix and tensor factorizations. *Proceedings of the IEEE*, 103 (9), 1602–1620.

90 Acar, E., Kolda, T.G., and Dunlavy, D.M. (2011) All-at-once optimization for coupled matrix and tensor factorizations. *CoRR*, **abs/1105.3422.**

91 Acar, E., Papalexakis, E.E., Gürdeniz, G., Rasmussen, M.A., Lawaetz, A.J., Nilsson, M., and Bro, R. (2014) Structure-revealing data fusion. *BMC Bioinformatics*, 15 (1), 239.

92 Acar, E., Levin-Schwartz, Y., Calhoun, V.D., and Adali, T. (2017) ACMTF for fusion of multi-modal neuroimaging data and identification of biomarkers, in *Signal Processing Conference (EUSIPCO), 2017 25th European*, IEEE, pp. 643–647.

93 De Lathauwer, L. and Kofidis, E. (2018) Coupled matrix-tensor factorizations-the case of partially shared factors, in *Proceedings of the Asilomar Conference on Signals, Systems and Computers*, accepted.

94 Harshman, R.A., Hong, S., and Lundy, M.E. (2003) Shifted factor analysis-Part I: Models and properties. *Journal of Chemometrics*, 17 (7), 363–378.

95 Mørup, M., Hansen, L.K., Arnfred, S.M., Lim, L.H., and Madsen, K.H. (2008) Shift-invariant multilinear decomposition of neuroimaging data. *NeuroImage*, 42 (4), 1439–1450.

96 Hong, S. (2005) Warped image factor analysis, in *1st IEEE International Workshop on Computational Advances in Multi-Sensor Adaptive Processing, 2005*, pp. 121–124.

97 Cohen, J.E., Cabral Farias, R., and Rivet, B. (2018) Curve registered coupled low rank factorization, in *14th LVA/ICA Conference, LNCS*, vol. 10891, Springer, University of Surrey, Guildford, UK.

98 Tomasi, G., Van Den Berg, F., and Andersson, C. (2004) Correlation optimized warping and dynamic time warping as preprocessing methods for chromatographic data. *Journal of Chemometrics*, 18 (5), 231–241.

99 Savorani, F., Tomasi, G., and Engelsen, S.B. (2010) *i*coshift: A versatile tool for the rapid alignment of 1D NMR spectra. *Journal of Magnetic Resonance*, 202 (2), 190–202.

100 Van Loan, C.F. (1976) Generalizing the singular value decomposition. *SIAM Journal on Numerical Analysis*, 13 (1), 76–83.

101 Alter, O., Brown, P.O., and Botstein, D. (2000) Singular value decomposition for genome-wide expression data processing and modeling. *Proceedings of the National Academy of Sciences of the United States of America*, 97 (18), 10101–10106.

102 Li, J., Usevich, K., and Comon, P. (2023) Convergence of gradient-based block coordinate descent algorithms for non-orthogonal joint approximate diagonalization of matrices. arxiv:2009.13377. *SIAM J. Matrix Anal. Appl.*, 44 (2), 592–621.

103 Ponnapalli, S.P., Saunders, M.A., Van Loan, C.F., and Alter, O. (2011) A higher-order generalized singular value decomposition for comparison of global mRNA expression from multiple organisms. *PLoS ONE*, 6 (12), e28072.

104 Sankaranarayanan, P., Schomay, T.E., Aiello, K.A., and Alter, O. (2015) Tensor GSVD of patient-and platform-matched tumor and normal DNA copy-number profiles uncovers chromosome arm-wide patterns of tumor-exclusive platform-consistent alterations encoding for cell transformation and predicting ovarian cancer survival. *PLoS ONE*, 10 (4), e0121396.

105 Smith, S. and Karypis, G. (2015) Tensor-matrix products with a compressed sparse tensor, in *Proceedings of the 5th Workshop on Irregular Applications: Architectures and Algorithms*, pp. 1–7.

106 Nelson, T., Rivera, A., Balaprakash, P., Hall, M., Hovland, P.D., Jessup, E., and Norris, B. (2015) Generating efficient tensor contractions for GPUs, in *2015 44th International Conference on Parallel Processing*, IEEE, pp. 969–978.

107 Shi, Y., Niranjan, U.N., Anandkumar, A., and Cecka, C. (2016) Tensor contractions with extended BLAS kernels on CPU and GPU, in *2016 IEEE 23rd International Conference on High Performance Computing (HiPC)*, IEEE, pp. 193–202.

108 Abdelfattah, A., Baboulin, M., Dobrev, V., Dongarra, J., Earl, C., Falcou, J., Haidar, A., Karlin, I., Kolev, T., Masliah, I., and Tomov, S.(2016) High-performance tensor contractions for GPUs. *Procedia Computer Science*, 80, 108–118.

109 Smith, D. and Gray, J. (2018) opteinsum-a python package for optimizing contraction order for einsum-like expressions. *Journal of Open Source Software*, 3 (26), 753.

110 Springer, P. and Bientinesi, P. (2018) Design of a high-performance GEMM-like tensor–tensor multiplication. *ACM Transactions on Mathematical Software (TOMS)*, 44 (3), 1–29.

111 Sidiropoulos, N., Papalexakis, E.E., and Faloutsos, C. (2014) Parallel randomly compressed cubes. *IEEE Signal Processing Magazine*, 31 (5), 57–70. Special issue on Big data.

112 Uschmajew, A. (2012) Local convergence of the alternating least squares algorithm for canonical tensor approximation. *SIAM Journal on Matrix Analysis and Applications*, 33 (2), 639–652.

113 Espig, M., Hackbusch, W., and Khachatryan, A. (2015) On the convergence of alternating least squares optimisation in tensor format representations. *arXiv preprint arXiv:1506.00062.*

114 Comon, P., Luciani, X., and De Almeida, A.L.F. (2009) Tensor decompositions, alternating least squares and other tales. *Journal of Chemometrics*, 23 (7–8), 393–405.

115 Nesterov, Y. (1983) A method of solving a convex programming problem with convergence rate O(1/k2). *Soviet Mathematics Doklady*, 27 (2), 372–376.

116 Mitchell, D., Ye, N., and De Sterck, H. (2018) Nesterov acceleration of alternating least squares for canonical tensor decomposition. *arXiv e-prints*, p. arXiv:1810.05846.

117 Ang, A.M.S., Cohen, J.E., and Gillis, N. (2020) Extrapolated alternating algorithms for approximate canonical polyadic decomposition, in *ICASSP 2020-2020 IEEE International Conference on Acoustics, Speech and Signal Processing (ICASSP)*, IEEE, pp. 3147–3151.

118 Schmidt, M., Le Roux, N., and Bach, F. (2017) Minimizing finite sums with the stochastic average gradient. *Mathematical Programming*, 162 (1–2), 83–112.

119 Acar, E., Dunlavy, D.M., Kolda, T.G., and Mørup, M. (2010) Scalable tensor factorizations with missing data, in *Proceedings of the 2010 SIAM International Conference on Data Mining*, SIAM, pp. 701–712.

120 Huang, K., Sidiropoulos, N.D., and Liavas, A.P. (2016) A flexible and efficient algorithmic framework for constrained matrix and tensor factorization. *IEEE Transactions on Signal Processing*, 64 (19), 5052–5065.

121 Paatero, P. (1999) The multilinear engine: a table-driven, least squares program for solving multilinear problems, including the *n*-way parallel factor analysis model. *Journal of Computational and Graphical Statistics*, 8 (4), 854–888.

122 Stegeman, A. and Comon, P. (2009) Subtracting a best rank-1 approximation does not necessarily decrease tensor rank, in *EUSIPCO'09*, Glasgow, Scotland. Hal-00435877.

123 Comon, P. (2009) Tensors vs matrices, usefulness and unexpected properties, in *15th IEEE Workshop on Statistical Signal Processing (SSP'09)*, Cardiff, UK, pp. 781–788. Keynote. hal-00417258.

124 Stegeman, A. and Comon, P. (2010) Subtracting a best rank-1 approximation does not necessarily decrease tensor rank. *Linear Algebra and its Applications*, 433 (7), 1276–1300. Hal-00512275.

125 Bro, R. and Kiers, H.A.L. (2003) A new efficient method for determining the number of components in PARAFAC models. *Journal of Chemometrics*, 17 (5), 274–286.

126 da Costa, J.P.C., Haardt, M., and Romer, F. (2008) Robust methods based on the HOSVD for estimating the model order in PARAFAC models, in *5th IEEE Sensor Array and Multichannel Signal Processing Workshop*, pp. 510–514.

127 Han, X., Albera, L., Kachenoura, A., Senhadji, L., and Shu, H. (2017) Low rank canonical polyadic decomposition of tensors based on group sparsity, in *25th European Signal Processing Conference, EUSIPCO 2017*, Kos, Greece, August 28 - September 2, 2017, pp. 668–672.

128 Wold, S., Ruhe, A., Wold, H., and Dunn, W.J. (1984) The collinearity problem in linear regression, the partial least squares (PLS) approach to generalized inverses. *SIAM Journal on Scientific and Statistical Computing*, 5 (3), 735–743.

129 Smilde, A., Bro, R., and Geladi, P. (2004) *Multi-Way Analysis*, Wiley, Chichester UK.

130 Bro, R. (1996) Multiway calibration, multilinear PLS. *Journal of Chemometrics*, 10, 47–61.

131 Luciani, X. and Albera, L. (2011) Semi-algebraic canonical decomposition of multi-way arrays and joint eigenvalue decomposition, in *2011 IEEE International Conference on Acoustics, Speech and Signal Processing (ICASSP)*, IEEE, pp. 4104–4107.

132 De Lathauwer, L., De Moor, B., and Vandewalle, J. (2004) Computation of the canonical decomposition by means of a simultaneous generalized Schur decomposition. *SIAM Journal on Matrix Analysis and Applications*, 26 (2), 295–327.

133 De Lathauwer, L. (2006) A link between the canonical decomposition in multilinear algebra and simultaneous matrix diagonalization. *SIAM Journal on Matrix Analysis and Applications*, 28 (3), 642–666.

134 Brachat, J., Comon, P., Mourrain, B., and Tsigaridas, E. (2010) Sym metric tensor decomposition. *Linear Algebra and its Applications*, 433 (11/12), 1851–1872.

135 Domanov, I. and Lathauwer, L.D. (2014) Canonical polyadic decomposition of third-order tensors: reduction to generalized eigenvalue decomposition. *SIAM Journal on Matrix Analysis and Applications*, 35 (2), 636–660.

136 Sanchez, E., Ramos, L.S., and Kowalski, B.R. (1987) Generalized rank annihilation method: I. Application to liquid chromatography-diode array ultraviolet detection data. *Journal of Chromatography A*, 385, 151–164.

137 Ramos, L.S., Sanchez, E., and Kowalski, B.R. (1987) Generalized rank annihilation method: II. Analysis of bimodal chromatographic data. *Journal of Chromatography A*, 385, 165–180.

138 Booksh, K.S., Lin, Z., Wang, Z., and Kowalski, B.R. (1994) Extension of trilinear decomposition method with an application to the flow probe sensor. *Analytical Chemistry*, 66, 2561–2569.

139 Smilde, A.K., Tauler, R., Saurina, J., and Bro, R. (1999) Calibration methods for complex second-order data. *Analytica Chimica Acta*, 398, 237–251.

140 Leurgans, S. and Ross, R.T. (1992) Multilinear models: applications in spectroscopy. *Statistical Sciences*, 17 (3), 289–319.

141 Leurgans, S., Ross, R.T., and Abel, R.B. (1993) A decomposition for three-way arrays. *SIAM Journal on Matrix Analysis and Applications*, 14 (4), 1064–1083.

142 Haskell, K.H. and Hanson, R.J. (1981) An algorithm for linear least squares problems with equality and nonnegativity constraints. *Mathematical Programming*, 21, 98–118.

143 Lawson, C.L. and Hanson, R.J. (1995) *Solving Least Squares Problems*, vol. 15, SIAM.

144 Bro, R. and Jong, S.D. (1997) A fast non-negativity-constrained least squares algorithm. *Journal of Chemometrics*, 11 (5), 393–401.

145 Gillis, N. and Glineur, F. (2012) Accelerated multiplicative updates and hierarchical ALS algorithms for nonnegative matrix factorization. *Neural Computation*, 24 (4), 1085–1105.

146 Paatero, P. (1997) A weighted non-negative least squares algorithm for three-way 'PARAFAC' factor analysis. *Chemometrics and Intelligent Laboratory Systems*, 38 (2), 223–242.

147 Fu, X., Gao, C., Wai, H.T., and Huang, K. (2019) Block-randomized stochastic proximal gradient for low-rank tensor factorization. *arXiv preprint arXiv:1901.05529*.

148 Vervliet, N., Debals, O., Sorber, L., Van Barel, M., and De Lathauwer, L. (2016) Tensorlab 3.0. Available online, https://www.tensorlab.net.

149 Vervliet, N., De Lathauwer, L., and Cocchi, M. (2018) Numerical optimization based algorithms for data fusion, in *Data Fusion Methodology and Applications* (ed M. Cocchi), Elsevier, pp. 693–697.

150 Grasedyck, L., Kressner, D., and Tobler, C. (2013) A literature survey of low-rank tensor approximation techniques. *GAMM-Mitteilungen*, 36 (1), 53–78.

151 Hong, D., Kolda, T.G., and Duersch, J.A. (2018) *Generalized Canonical Polyadic Tensor Decomposition. ArXiv:1808.07452.*

152 Martens, H. and Stark, E. (1991) Extended multiplicative signal correction and spectral interference subtraction: new preprocessing methods for near infrared spectroscopy. *Journal of Pharmaceutical and Biomedical Analysis*, 9 (8), 625–635.

153 Heraud, P., Wood, B.R., Beardall, J., and McNaughton, D. (2006) Effects of pre-processing of Raman spectra on *in vivo* classification of nutrient status of microalgal cells. *Journal of Chemometrics*, 20 (5), 193–197.

154 Goodacre, R., Broadhurst, D., Smilde, A.K., Kristal, B.S., Baker, J.D., Beger, R., Bessant, C., Connor, S., Capuani, G., Craig, A., Ebbels, T., Kell, D.B., Manetti, C., Newton, J., Paternostro, G., Somorjai, R., Sjöström, M., Trygg, J., and Wulfert, F. (2007) Proposed minimum reporting standards for data analysis in metabolomics. *Metabolomics*, 3 (3), 231–241.

155 Harshman, R.A. and Lundy, M.E. (1984) Data preprocessing and the extended PARAFAC model, in *Research Methods for Multimode Data Analysis* (eds J.H.H.G. Law, C.W. Snyder, and R.P. McDonald), Praeger, New York, pp. 216–284. HarsL84:rmmda centering scaling preprocess-ing.

156 Bro, R. and Smilde, A.K. (2003) Centering and scaling in component analysis. *Journal of Chemometrics*, 17 (1), 16–33.

157 Lee, C., Kim, K., and Ross, R.T. (1991) Trilinear analysis for the resolution of overlapping fluorescence spectra. *Korean Biochemical Journal (Korea Republic)*, 24, 374–379.

158 Kubista, M., Sjoback, R., Eriksson, S., and Albinsson, B. (1994) Experimental correction for the inner-filter effect in fluorescence spectra. *Analyst*, 119, 417–419.

159 Lakowicz, J.R. (1999) *Principles of Fluorescence Spectroscopy*, Kluwer Academic, New York.

160 Cohen, J.E., Comon, P., and Luciani, X. (2016) Correcting inner filter effects, a non multilinear tensor decomposition method. *Chemometrics and Intelligent Laboratory Systems*, 150, 29–40.

161 Elcoroaristizabal, S., Bro, R., Garcia, J.A., and Alonso, L. (2015) Parafac models of fluorescence data with scattering: a comparative study. *Chemometrics and Intelligent Laboratory Systems*, 142, 124–130.

162 Harshman, R.A. and DeSarbo, W.S. (1984) An application of PARAFAC to a small sample problem, demonstrating preprocessing, orthogonality constraints, and split-half diagnostic techniques. *Research Methods for Multimode Data Analysis p*, pp. 602–642.

163 Amigo, J.M., Skov, T., Coello, J., Maspoch, S., and Bro, R. (2008) Solving GC-MS problems with PARAFAC2. *TRAC Trends in Analytical Chemistry*, 27 (8), 714–725.

164 Amigo, J.M., Skov, T., and Bro, R. (2010) Chromathography: solving chromatographic issues with mathematical models and intuitive graphics. *Chemical Reviews*, 110, 4582–4605.

165 Bro, R., Workman, J.J., Mobley, P.R., and Kowalski, B.R. (1997) Review of chemometrics applied to spectroscopy: 1985–1995, Part III: Multi-way analysis. *Applied Spectroscopy Reviews*, 32, 237–261.

166 Khakimov, B., Amigo, J.M., Bak, S., and Engelsen, S.B. (2012) Plant metabolomics: resolution and quantification of elusive peaks in liquid chromatography–mass spectrometry profiles of complex plant extracts using multi-way decomposition methods. *Journal of Chromatography A*, 1266, 84–94.

167 Garcia, I., Sarabia, L., Ortiz, M.C., and Aldama, J.M. (2004) Three-way models and detection capability of a gas chromatography–mass spectrometry method for the determination of clenbuterol in several biological matrices: the 2002/657/EC European Decision. *Analytica Chimica Acta*, 515 (1), 55–63.

168 Smilde, A.K., Wang, Y., and Kowalski, B.R. (1994) Theory of medium-rank second-order calibration with restricted Tucker models. *Journal of Chemometrics*, 8, 21–36.

169 Hayashi, C. and Hayashi, F. (1982) A new algorithm to solve Parafac-model. *Behaviormetrika*, 11, 49–60.

170 Wilson, B.E., Sanchez, E., and Kowalski, B.R. (1989) An improved algorithm for the generalized rank annihilation method. *Journal of Chemometrics*, 3, 493–498.

171 Gerritsen, M.J.P., Tanis, H., Vandeginste, B.G.M., and Kateman, G. (1992) Generalized rank annihilation factor analysis, iterative target transformation factor analysis, and residual bilinearization for the quantitative analysis of data from liquid chromatography with photodiode array detection. *Analytical Chemistry*, 64, 2042–2056.

172 Faber, N.M., Buydens, L.M.C., and Kateman, G. (1994) Generalized rank annihilation method. I: Derivation of eigenvalue problems. *Journal of Chemometrics*, 8, 147–154.

173 Faber, N.M., Buydens, L.M.C., and Kateman, G. (1994) Generalized rank annihilation method. II: Bias and variance in the estimated eigenvalues. *Journal of Chemometrics*, 8, 181–203.

174 Orekhov, V.Y., Ibraghimov, I.V., and Billeter, M. (2001) MUNIN: a new approach to multi-dimensional NMR spectra interpretation. *Journal of Biomolecular NMR*, 20 (1), 49–60.

175 Gutmanas, A., Jarvoll, P., Orekhov, V.Y., and Billeter, M. (2002) Three-way decomposition of a complete 3D N-15-NOESY-HSQC. *Journal of Biomolecular NMR*, 24 (3), 191–201.

176 Dyrby, M., Petersen, M., Whittaker, A.D., Lambert, L., Nørgaard, L., Bro, R., and Engelsen, S.B. (2005) Analysis of lipoproteins using 2D diffusion-edited NMR spectroscopy and multi-way chemometrics. *Analytica Chimica Acta*, 531, 209–216.

177 Jansen, J.J., Bro, R., Hoefsloot, H.C.J., van den Berg, F., Westerhuis, J.A., and Smilde, A.K. (2008) PARAFASCA: ASCA combined with PARAFAC for the analysis of metabolic fingerprinting data. *Journal of Chemometrics*, 22, 114–121.

178 Bro, R., Viereck, N., Toft, M., Toft, H., Hansen, P.I., and Engelsen, S.B. (2010) Mathematical chromatography solves the cocktail party effect in mixtures using 2D spectra and PARAFAC. *TRAC Trends in Analytical Chemistry*, 29 (4), 281–284.

179 Engelsen, S.B. and Bro, R. (2003) Powerslicing. *Journal of Magnetic Resonance*, 163 (1), 192–197.

180 Engelsen, S.B., Pedersen, H.T., and Bro, R. (2006) Direct exponential curve resolution by slicing, in *Modern Magnetic Resonance* (ed. G.A. Webb), Springer-Verlag, Berlin, Heidelberg, New York, pp. 1823–1830.

181 Mørup, M., Hansen, L.K., Herrmann, C.S., Parnas, J., and Arnfred, S.M. (2006) Parallel factor analysis as an exploratory tool for wavelet transformed event-related EEG. *NeuroImage*, 29 (3), 938–947.

182 Acar, E., Aykut-Bingol, C., Bingol, H., Bro, R., and Yener, B. (2007) Multiway analysis of epilepsy tensors. *Bioinformatics*, 23 (13), 10–18.

183 Becker, H., Albera, L., Comon, P., Haardt, M., Birot, G., Wendling, F., Gavaret, M., Bénar, C.G., and Merlet, I. (2014) EEG extended source localization: tensor-based vs conventional methods. *NeuroImage*, 96, 143–157.

184 Basford, K.E., Kroonenberg, P.M., Cooper, M., and Hammer, G.L. (1996) Three-mode analytical methods for crop improvement programs, in *Plant Adaptation and Crop Improvement*, International Rice Research Institute, Chapter 14, pp. 291–305.

185 Stanimirova, I., Walczak, B., Massart, D.L., Simeonov, V., Saby, C.A., and Crescenzo, E.D. (2004) Statis, a three-way method for data analysis. Application to environmental data. *Chemometrics and Intelligent Laboratory Systems*, 73 (2), 219–233.

186 Pere-Trepat, E., Ginebreda, A., and Tauler, R. (2007) Comparison of different multiway methods for the analysis of geographical metal distributions in fish, sediments and river waters in catalonia. *Chemometrics and Intelligent Laboratory Systems*, 88 (1), 69–83.

187 Paatero, P. (1996) A weighted nonnegative least squares algorithm for three-way 'PARAFAC' factor analysis, in *2nd International Chemometrics InterNet Conference (InCINC'96)*.

188 Hopke, P., Xie, Y., and Paatero, P. (1999) Mixed multiway analysis of airborne particle composition data. *Journal of Chemometrics*, 13, 343–352.

189 Nœs, T. and Kowalski, B.R. (1989) Predicting sensory profiles from external instrumental measurements. *Food Quality and Preference*, 1 (4–5), 135–147.

190 Louwerse, D.J. and Smilde, A.K. (2000) Multivariate statistical process control of batch processes based on three-way models. *Chemical Engineering Science*, 55 (7), 1225–1235.

191 Wise, B.M., Gallagher, N.B., and Martin, E.B. (2001) Application of PARAFAC2 to fault detection and diagnosis in semiconductor etch. *Journal of Chemometrics*, 15 (4), 285–298.

192 Wittrup, C. (2000) Comparison of chemometric methods for classification of fungal extracts based on rapid fluorescence spectroscopy. *Journal of Chemometrics*, 14 (5–6), 765–776.

193 Zampronio, C.G., Gurden, S.P., Moraes, L.A.B., Eberlin, M.N., Smilde, A.K., and Poppi, R.J. (2002) Direct sampling tandem mass spectrometry (MS/MS) and multiway calibration for isomer quantitation. *Analyst*, 127 (8), 1054–1060.

194 Nilsson, J., de Jong, S., and Smilde, A.K. (1997) Multiway calibration in 3D QSAR. *Journal of Chemometrics*, 11, 511–524.

195 Heimdal, H., Bro, R., Larsen, L.M., and Poll, L. (1997) Prediction of polyphenol oxidase activity in model solutions containing various combinations of chlorogenic acid, (-)-epicatechin, O_2, CO_2, temperature and pH by multiway analysis. *Journal of Agricultural and Food Chemistry*, 45 (7), 2399–2406.

196 Coello, J., Maspoch, S., and Villegas, N. (2000) Simultaneous kinetic-spectrophotometric determination of levodopa and benserazide by bi- and three-way partial least squares calibration. *Talanta*, 53 (3), 627–637.

197 Hasegawa, K., Arakawa, M., and Funatsu, K. (1999) 3D-QSAR study of insecticidal neonicotinoid compounds based on 3-way partial least squares model. *Chemometrics and Intelligent Laboratory Systems*, 47, 33–40.

198 de la Pena, A.M., Mansilla, A.E., Gomez, D.G., Olivieri, A.C., and Goicoechea, H.C. (2003) Interference-free analysis using three-way fluorescence data and the parallel factor model. Determination of fluoroquinolone antibiotics in human serum. *Analytical Chemistry*, 75 (11), 2640–2646.

199 Tang, K.L. and Li, T.H. (2003) Comparison of different partial least-squares methods in quantitative structure-activity relationships. *Analytica Chimica Acta*, 476 (1), 85–92.

200 Ni, Y.N., Huang, C.F., and Kokot, S. (2004) Application of multivariate calibration and artificial neural networks to simultaneous kinetic-spectrophotometric determination of carbamate pesticides. *Chemometrics and Intelligent Laboratory Systems*, 71 (2), 177–193.

201 Bergant, K. and Kajfez-Bogataj, L. (2005) N-PLS regression as empirical downscaling tool in climate change studies. *Theoretical and Applied Climatology*, 81 (1–2), 11–23.

202 Durante, C., Cocchi, M., Grandi, M., Marchetti, A., and Bro, R. (2006) Application of N-PLS to gas chromatographic and sensory data of traditional balsamic vinegars of modena. *Chemometrics and Intelligent Laboratory Systems*, 83, 54–65.

203 Chow, E., Ebrahimi, D., Gooding, J.J., and Hibbert, D.B. (2006) Application of N-PLS calibration to the simultaneous determination of Cu^2+, Cd^2+ and Pb^2+ using peptide modified electrochemical sensors. *Analyst*, 131 (9), 1051–1057.

204 Marini, F. and Bro, R. (2013) SCREAM: a novel method for multi-way regression problems with shifts and shape changes in one mode. *Chemometrics and Intelligent Laboratory Systems*, 129, 64–75.

Index

Source Separation in Physical-Chemical Sensing, First Edition.
Edited by Christian Jutten, Leonardo Tomazeli Duarte, and Saïd Moussaoui.

m

n

o

p

r